Transforming Energy Systems

To Anna, Donal, Jacob and Lucy

Transforming Energy Systems

Economics, Policies and Change

Steven Fries

Senior Associate Fellow, Institute for New Economic Thinking, University of Oxford, Oxford, UK and Nonresident Senior Fellow, Peterson Institute for International Economics, Washington, USA

Edward **Elgar**
PUBLISHING

Cheltenham, UK • Northampton, MA, USA

Published by
Edward Elgar Publishing Limited
The Lypiatts
15 Lansdown Road
Cheltenham
Glos GL50 2JA
UK

Edward Elgar Publishing, Inc.
William Pratt House
9 Dewey Court
Northampton
Massachusetts 01060
USA

Paperback edition 2022

A catalogue record for this book
is available from the British Library

Library of Congress Control Number: 2021945324

This book is available electronically in the **Elgar**online
Economics subject collection
http://dx.doi.org/10.4337/9781800370371

ISBN 978 1 80037 036 4 (cased)
ISBN 978 1 80037 037 1 (eBook)
ISBN 978 1 0353 0907 8 (paperback)

Printed and bound by CPI Group (UK) Ltd, Croydon, CR0 4YY

Contents

Figures

Tables

Preface

Steven Fries was economic counsellor at Shell International Ltd. and group chief economist for Shell. The views expressed in the book are those of the author alone. They do not necessarily reflect those of acknowledged individuals, Shell International Ltd. or any other company of the Shell group of companies. All references in this book to external sources—primarily refereed journal articles, academic books and working papers and publications of governments, government agencies and intergovernmental and international organizations—and their analyses and conclusions do not necessary reflect those of the author.

The economics of transforming energy systems are analogous to orchestrating journeys and passengers—long-distance journeys sustained by changing technologies to provide low-carbon energy for transport, and vehicles to convert it into motion. These new technologies already include solar photovoltaic panels and battery electric automobiles. But a richer analogy for transforming energy systems is perhaps catching connecting trains to complete journeys, with low-carbon technologies providing connecting services to reach intended destinations for energy systems and the societies they serve. The key challenge is ensuring that sufficient connecting services—such as low-carbon electric power, rail network electrification and high-speed rail services—are available for societies sufficiently soon to both limit climate change and reach their intended societal destinations. In energy-related sectors, as with others, markets do much of the orchestrating of economic activities—both investments and consumer choices. But in energy system transformations, markets alone are insufficient to the task, especially in the early stages of change, because of multiple market imperfections, missing key markets, network effects and distributional impacts of change.

This book examines the experience with ongoing energy system transformations to learn lessons about effective roles of governments and policies as well as businesses, customers and markets in this crucially important journey. It aims to help ensure that the connecting low-carbon services that societies require are available in time for boarding and the sustainable destinations they offer attract most firms and households to them. To be avoided are both energy system transformations with too many passengers missing connecting services and reaching unintended destinations—with scrambles for a handful of electric taxis to get where they want to go—and climate change impacts to which societies and ecosystems would struggle to adapt.

In addition to the urgency of governments, businesses and households acting effectively and decisively to limit climate change, my motivation for writing this book arose from my experiences as an economist working in government (UK Department of Energy and Climate Change), industry (Shell) and an international financial institution (European Bank for Reconstruction and Development). The book aims to distil what has been learned from ongoing energy system transformations so that the immense challenges faced by societies in cutting net emissions of carbon dioxide and other greenhouse gases from energy systems in the decades ahead might be addressed more decisively. This learning from experience is informed by the rapidly accumulating body of academic research on climate change and energy system transformations. The perspective offered on this analysis and evidence is that of an applied economist, informed and shaped by experiences in transforming energy systems. The intended audience is policy makers in government and decision makers in industry and business. It is also hoped that the book may be of interest to academic and general audiences concerned with one of the most pressing and challenging issues of the day.

The book also reflects a lesson learned from the post-communist transformations in Eastern Europe and Central Asia. In that context, narrowly honed economic policy prescriptions, such as rapid liberalization of market prices and mass privatization, fell far short of the goals that some of its policy architects had intended. Amidst imperfect and incomplete markets, potentially time-inconsistent government policies, absent or inappropriately designed market-supporting institutions, and arrays of interests for and against reforms, such 'shock therapies' left some political and economic system transformations stuck and incomplete. More comprehensive reforms were sustained, including some with significant international support from the European Union, international financial institutions and others. Such a complex, systemic transformation was not guided alone by the invisible hand of market forces and private property rights. Rather, its sustained advance required comprehensive, coherent and credible government reform strategies supported by international engagement, coordination and cooperation (EBRD, 1999, pp. 4–12).

Similar issues and risks arise in transforming energy systems. A narrowly honed economic policy prescription of emissions pricing to internalize the environmental externality and government support for research and development to account for knowledge spillovers risks falling short, with transformations becoming stuck. This book thus examines evidence on both orthodox and heterodox domestic policies to transform domestic energy systems, as well as potential bases for international coordination of climate actions. The aim is to identify those energy-reform strategies that could accelerate and sustain energy system transformations and consolidate them with low-carbon technologies, renewable resources and alternative choices through mutually reinforcing actions of governments, businesses and households. The economic analysis and evidence that informs these reform strategies considers the multiple market imperfections and missing markets that beset energy systems and their distributional impacts, and balances risks of both market failures and government policy failures.

This book was written with the support of two people to whom I would like to express particular thanks: Jeremy Bentham and Sir David Hendry. They provided early encouragement, engaged in valuable discussions and provided helpful comments as the book took shape. It benefits too from comments and suggestions of Robert Ritz, Dirk Smit and Charlotte Taylor, who generously read the manuscript, and seminar participants at the Peterson Institute for International Economics. But one's understanding of such complex issues as climate change and energy system transformations is inevitably shaped by many ongoing interactions and discussions. For them, I would like to thank those with whom I have worked closely over many years on these issues in government and business, including Paul Bailey, Tijs Beek, Peter Betts, Neil Bush, Jason Eis, Vivien Geard, Martin Haigh, Peter Heijmans, David Hone, Malika Ishwaran, Cho Khong, Paro Konar-Thakkar, Eric Ling, Duncan Millard, Andrew Ray, Hugo Robson, Darci Sinclair, Oleksiy Tatarenko, Sam Thomas, Wim Thomas, Simon Virley, Alec Waterhouse, Mark Weintraub and Geraldine Wessing. I would also like to thank Rebecca Surender for her patient support and impatient encouragement while writing this book.

In addition, the book benefits from an understanding of energy systems and their societal contexts embodied in a series of Shell energy scenarios (www .shell.com/scenarios), although it is not based on them. They include the *New Lens Scenarios—A Shift in Perspective for a World in Transition* (2013), *Better Life with a Healthy Planet—Pathways to Net Zero Emissions* (2016) and *Sky—Meeting the Goals of the Paris Agreement* (2018). All of them envisaged societies transforming their energy systems and achieving net zero emissions from energy, but in differing enabling contexts—social, political, economic and technological—and over varying time horizons and climate change outcomes. The most recent Shell scenarios—*The Energy Transformation Scenarios* (2021)—explore potential impacts of the global novel coronavirus pandemic for societal efforts to achieve net zero emissions. All of these scenarios examine critical uncertainties that lie ahead and use stretching assumptions and projections to understand their implications. They are not predictions of the future, but rather tools to explore uncertainties and learn from future possibilities. This book aims to learn lessons from ongoing energy system transformations by examining accumulating evidence about them.

Introduction to *Transforming Energy Systems*

The Paris Agreement put limiting climate change front and centre for societies, where it remained even amidst the novel coronavirus pandemic. This agreement is an urgent call to action, and transforming energy systems to low-carbon alternatives is one of the most crucial actions in halting climate change. These alternatives are those energy resources, technologies and choices that create an alternative to traditional use of fossil fuels to eliminate their net emissions of carbon dioxide and other greenhouse gases. But changing from the current energy system, based largely on fossil-fuel use that disrupts the Earth's carbon cycle and climate, to low-carbon alternatives is perhaps the most profound challenge that societies face. Current technologies that use fossil fuels make dwellings, commercial buildings and factories so much more productive; transport of people and goods easier; and useful materials more abundant and affordable than a century ago. Fossil fuels do much of the work in modern societies, weaving deeply through the fabric of economic activities and household lives. The challenge is to maintain these benefits while pivoting to alternative low-carbon energy resources, technologies and choices.

The foundational investments in transforming energy are those that create low-carbon alternatives to incumbent technologies. Much as an 'industrial enlightenment' of scientific knowledge and its engineering application enlivened the first and second industrial revolutions and created modern energy systems, the awakening of low-carbon technologies today emerges from investments in knowledge and new capabilities. These foundational investments create quasi-public goods—the knowledge and new capabilities necessary to limit climate change while maintaining modern activities. To the extent they are successful, these upfront public and private investments enable all businesses and households to invest in low-carbon alternatives to current technologies to support their economic and household activities. In advanced industrialized countries, as well as China and Russia, transforming energy would require almost all businesses and households to invest in low-carbon alternatives. It also means energy businesses would need to invest in supplying low-carbon electricity and fuels when and where customers demand. In most developing counties, transforming energy would mean pursuing development

increasingly through renewable resources and cost-effective low-carbon technologies and building capabilities to adopt them.

This book's perspective on energy system transformations is through the
lens of the policy and investment decisions of governments, businesses and
households that halting climate change would require, including investments
in knowledge and new capabilities. Because of the timescales of transforma-
tional change and the immense scale of investment it would require, there is
no quick fix for energy, but it can be transformed and decisively so to limit
climate change. Encouragingly, awakened variable renewables generation
technologies have already begun to make a difference, like low-carbon
electricity from solar photovoltaic (PV) panels and wind turbines as well as
lithium-ion battery packs and battery electric vehicles (Le Quéré et al., 2019).
There are important strides too in improving energy and material efficiencies
of existing technologies and recycling materials in products at the end of their
lives. These changes reduce emissions now as well as facilitate changes to
low-carbon alternatives. It is crucially important to learn from these successes
in transforming energy systems, as well as their shortcomings.

But if the societal goals of the 2015 Paris Agreement of the United Nations
Framework Convention on Climate Change are to be achieved—to limit global
warming to well below 2°C, with a stretch goal of 1.5°C, and achieve net zero
emissions of carbon dioxide and other greenhouse gas emissions from human
activities in the second half of this century—energy transformations would
need to accelerate. Their scope too would need to widen to encompass most
countries and all energy-related sectors as well as agriculture, forestry and
other land-use activities. There is much to do to achieve this long-run goal, yet
little consensus in economics on the way forward. There is broad agreement
in economics on using emissions pricing to internalize the environmental
externalities and government policy support for research and development
(R&D) to compensate for well-evidenced knowledge spillovers from this
activity. Their rationales are readily grasped and widely understood, but not
yet adequately implemented, especially for emissions pricing. Moreover, there
is little agreement in economics on use of other policy instruments, such as
market-creating and industry-supporting (industrial) policies, even though
they have been widely used in practice to advance energy system transformations. They are misdirected and inefficient policies to some, key parts of the
policy mix to others.

THE ECONOMICS OF TRANSFORMING ENERGY SYSTEMS

A narrowly honed, orthodox framing of the economics problem rests on
the social cost of carbon dioxide emissions and governments imposing

a Pigouvian tax on emissions levied in line with this cost (Pigou, 1932, p. 142). The social cost is the present value of expected future social and economic losses associated with greenhouse gas emissions as best assessed. A root cause of climate change is that societies' energy choices in using fossil fuels have long ignored their consequences for the Earth's carbon cycle and climate. Governments' pricing of emissions in this way, or through equivalent measures, would internalize the environmental externality by requiring businesses and households to consider these environmental costs in their decisions. These decisions include what technological innovations to pursue and what goods and services to provide and purchase in markets. This policy paradigm is typically broadened to include another market failure that could hold back low-carbon technologies. This failure arises from well-evidenced knowledge spillovers from innovations and an inability of businesses to capture sufficient economic benefits from successful innovations.

The standard policy prescription in economics for limiting climate change is thus a Pigouvian tax on carbon dioxide emissions and government support for R&D activities to reward sufficiently investments in innovation. With these two fixes for two key market failures, competitive businesses would in principle develop and supply and their customers would choose the lower- and low-carbon technologies that are required to limit climate change at least cost. In this economics framing, the dynamic process of technological innovation is directed and motivated by temporary market power and rents arising from newly differentiated products or created cost advantages in production, including any arising from emissions pricing. Such Schumpeterian processes— recurring bouts of creative destruction—create and manage the market rents that provide returns to upfront investments in innovation and market creation for new technologies that survive the test of competitive market selection (Schumpeter, 1942, p. 83). This test can be sufficient to establish the social value of innovations and market creation for low-carbon technologies, but this is not necessarily so.

This crisp policy prescription is sufficient in principle to achieve the Paris goals only if businesses and households have well-founded expectations of facing goal-consistent emissions pricing and there are no other market imperfections impeding the advance of low-carbon alternatives. In policy practice, achieving the levels and scope of emissions pricing deemed an adequate reflection of their social costs or government climate goals has been challenging. The reasons for this policy shortcoming lie more in the realm of incomplete markets, adverse distributional impacts, arbitrary political shifts and potential time inconsistencies in policy implementation than with lack of understanding of its readily grasped implications for economic efficiency. In addition to this policy shortcoming, there are three reasons why Schumpeterian market dynamics are unlikely to provide a sufficient impetus to investments in innovation and creation of markets for low-carbon technologies, even with government R&D support. One is that knowledge spillovers arise not only from R&D activities but also from deploying and using low-carbon technologies—learning from experience. These knowledge spillovers can arise in manufacturing sectors that produce low-carbon technologies as well as within transforming energy systems. They can also arise among customers of low-carbon technologies; for example, through demonstration effects from early technology adopters.

A second is that some low-carbon technologies benefit from substantial scale economies, especially those that can be mass produced with potential for wide deployment across sectors and countries. Some such scale economies are internal to businesses, such as those that arise from larger-scale manufacturing plants. For them, firms could reasonably be expected to earn returns on upfront investments in market creation to achieve them, assuming they face few capital market imperfections. But some scale economies are external to businesses, as suggested by geographical clustering of manufacturing innovation-intensive technologies, including low-carbon ones. Perhaps more significantly, there are also network effects and external scale economies in transforming energy systems, with investments by early adopters of low-carbon technologies contributing to lower costs for those that follow. These effects, for example, can arise from shared infrastructures supporting low-carbon technologies. Cost complementarities among low-carbon technologies that work together to lower overall energy system costs can also create profit spillovers among businesses and customers, which are not necessarily fully internalized in individual decisions. While business models and commercial relationships among firms could be expected to evolve and internalize at least some of these spillovers over time, they are neither immediate nor perfect market mechanisms.

A third reason why Schumpeterian market dynamics are likely to fall short in transforming energy is that at its centre are government-designed and government-regulated markets for electric power. Ubiquitous solar irradiance and wind resources are available at sufficient scale to substitute substantially for fossil-fuel resources, and they are most readily converted into electric power, increasingly cost effectively. But wholesale power markets are inherently imperfect, with their government design balancing efficient scarcity pricing to remunerate fixed investments against potential abuses of market power when markets are tight. The assumption that these government-designed and government-regulated power markets would alone reward large upfront investments in developing and deploying renewable generation and complementary flexibility technologies glosses over missing-market and time-inconsistency problems. For example, governments have an incentive to encourage such sunk investments, but not necessarily to reward them through adequate scarcity and emissions pricing once made. Such time-inconsistency problems are a long-standing feature of power markets—they are not confined to emissions pricing. Key policy interventions in transforming electricity systems address this issue through commitment devices, such as long-term investment contracts for renewables generation that benefit from government support.

Narrowly cast structural reforms that ignore potentially relevant market imperfections and the institutional, political and societal contexts in which they take place could reasonably be expected to disappoint. This policy lesson has been hard-learned from other structural reform experiences—be they in financial sectors or economy-wide in the post-communist transitions of Eastern Europe. Well-designed policies to transform energy systems need to consider carefully the inherently imperfect contexts in which they are implemented. To this end, this book examines now-substantial evidence on energy system transformations that have been underway for several decades, especially in the electric power and automobile sectors. It focuses on not only emissions pricing but also non-price policy interventions such as market-creating and industry-supporting (industrial) policies for low-carbon technologies, especially during their early deployment in initial markets. Adapting energy market institutions—government-designed energy markets, their regulations and infrastructures—to low-carbon alternatives is also necessary and given close consideration. As are regulations and standards for energy and material efficiencies, together with evidence on market imperfections and behavioural 'anomalies' that would warrant such policies.

ENERGY-REFORM INTERESTS AND STRATEGIES

In economics, the optimal goal for the climate is its stabilization by balancing the marginal social benefits of avoiding expected losses from further climate change and the marginal social costs of sooner and deeper emissions cuts. The social cost of carbon dioxide emissions is the emissions price that balances these marginal social benefits and costs. A narrow market-based reform strategy that in principle could support this outcome is straightforward. Fix the two market failures to enable markets to address the fundamental knowledge and environmental problems, then allow competitive market dynamics to determine the optimal extent of climate change. In this perspective, the eventual climate outcome would be determined by the pace and extent of innovation, supported by government R&D incentives, and investment responses to current and expected emissions prices that reflect the social cost of carbon dioxide emissions. However, this social cost is inherently and highly uncertain, subject to value judgements about potential climate impacts on vulnerable groups and across generations, and can be expected to change over time with advances in climate science and low-carbon technologies. Moreover, its imposition through emissions pricing is vulnerable to adverse distributional impacts, shifts in political sentiments and potential time inconsistencies in government policy making, especially in early stages of change. With few opportunities for private investors to manage long-run emission-price risks—for example, because markets for future emissions prices are missing—investment horizons could shorten and backload responses.

Missing from the parsimonious policy mix are thus overcoming missing markets for emission prices and strengthening government commitment devices, as well as addressing other market imperfections that impede development and early deployment of low-carbon technologies. Climate-change outcomes from incomplete policy frameworks would be expected to fall short of what could be reasonably considered optimal. That said, caution in economics about broadening policy prescriptions arises from inconclusive theoretical insights when there are multiple market failures, evidence deemed to be weak on their significance and concern about potential government policy failures in addressing them.

Climate science provides an alternative framing of the societal goal, which is avoiding climate changes to which societies and ecosystems would struggle to adapt. The Paris Agreement interpreted this goal as keeping the likely rise in average surface temperatures this century to well below 2°C, and making efforts to limit this change to no more than 1.5°C. The emissions corollary of this goal to limit global warming is to achieve net zero emissions of carbon dioxide and other greenhouse gases from human activities in the second half of this century. Limiting the likely temperature change to 1.5°C would require achieving net zero emissions by around mid-century, and a number of countries have already set such goals by 2060, 2050 or indeed sooner. Their rationales are rooted more firmly in climate science than still-contentious economics of social benefits and costs. Although, to the extent that it could be reasonably assessed, the optimality of these emissions goals is subject to the same inherent uncertainties and challenges as the social cost of carbon dioxide emissions and its primary implementation through emissions pricing.

The challenge is to create energy-reform strategies that manage effectively these inherent uncertainties, market imperfections and distributional impacts, with the climate-science-framing offering some advantages. The long-run energy system implications of such climate stabilization and quantitative emissions goals are clear. They are time-bound system transformations to net zero emissions of carbon dioxide from traditional use of fossil fuels and industrial processes to halt their contribution to climate change. They create a clear focal point for businesses and their long-run expectations and investment decisions—they need to invest in advancing and deploying low-carbon alternatives. In contrast, an emissions price and its expected pathway over time can direct the focus of governments and businesses more towards short-run emissions cuts than long-run system transformations. This reflects in part high uncertainty around long-run emissions prices.

To the extent they are credible, these quantitative climate goals are comparable to large, expected fossil-fuel price shocks, without their adverse short-run impacts on economic activity and income distributions. Much as previous oil price shocks and energy security concerns spurred innovations in renewable technologies, alternative-drivetrain vehicles and energy efficiencies, net zero emissions goals can strengthen incentives for private investments in innovation and market creation for low-carbon alternatives. But on their own these policy goals may only send a relatively weak long-run price signal for want of credibility from implementing policies.

A heterodox domestic energy-reform strategy can help build such credibility—arguably more so than a narrowly honed orthodox economic policy prescription. The latter approach assumes that the Schumpeterian engine fires on all cylinders in all aspects of energy systems. But the approach could disappoint if this engine sputters; for example, in the electricity sector which would likely form the new core of low-carbon energy systems. If emissions pricing and R&D supports alone had formed the policy mix so far, there would have been the risk of slower advances in new renewable generation and other low-carbon technologies. This outcome, in turn, would have weakened the credibility of reforms, further dulling incentives to develop and deploy alternative technologies. An alternative heterodox policy mix, which has been pursued by most countries initiating low-carbon technology disruptions, includes market-creating and industry-supporting (industrial) policies for low-carbon technologies. These countries tend to be those that specialize in innovation and manufacturing low-carbon technologies, and these policies largely aim to subsidize capital and/or operating costs of relatively expensive investments by early adopters of low-carbon alternatives. These heterodox policies hold the potential to be more self-reinforcing of climate goals, including through sequencing of such industrial policies and emissions pricing.

Industrial policies can overcome at early stages of system transformations missing markets for future emission prices, serve as government commitment devices and build support for sustained energy reforms. By directly creating low-carbon technology demands, especially in countries that specialize in innovations and manufacturing low-carbon technologies, such as China, Germany, Japan, South Korea and United States, these policies direct and support upfront investments by firms in knowledge, innovation and market creation for these technologies. Through home-market effects on production of domestic demands, they foster growing activities and economic interests that benefit from low-carbon technologies and help build policy credibility. These emerging economic interests can help to sustain reforms and the advance of low-carbon alternatives. To the extent they are effective and successful, these policies expand economic interests in low-carbon alternatives that can promote stronger reforms over time, including any subsequent ratcheting up of emissions pricing. These industrial policies can also address knowledge spillovers beyond those from R&D and compensate for cost spillovers from external scale economies and network effects. However, these policies also carry policy failure risks, such as creating market inefficiencies and directing innovations and market creation away from those that would have been pursued in notionally perfect markets. The risks of such policy failures must be balanced against the risks of reform strategies being too narrowly cast and becoming stuck at low levels of climate ambition.

Comprehensive and coherent reform strategies are necessary not only to create self-reinforcing interests in energy transformations, but also to manage policy costs and distributional impacts. Scope for reform strategies to manage policy costs arises from cost spillovers not internalized by firm strategies, market prices or other policies. For example, industrial policy supports for alternative-drivetrain vehicles advanced initial development and deployment of lithium-ion battery packs in the transport sector, for which oil products have high unit energy costs. As battery pack costs decline, this technology is increasingly spilling over to the electric power sector, for which coal and natural gas have low unit energy costs. This targeted and sequenced approach saves real resource costs in advancing low-carbon alternatives. The burden on industrial policies is also eased in those sectors with significant potential for product differentiation and willingness to pay for low-carbon technologies by early adopters, as in automobile markets.

In a heterodox energy-reform strategy with a goal of a time-bound commitment to net zero emissions, emissions pricing complements an early focus on market creation and industry support in three important ways. Firstly, cost-effective, target-consistent emissions pricing extends the long-run shadow price of the binding emissions constraint, brings it forward in time and promotes shifts in relative prices economy-wide to reflect the future binding constraint. This is especially important in sectors that are expected to be relatively hard to decarbonize; it is less so but still important in sectors that are easy to decarbonize after allowing for dynamic impacts of industrial policies. Secondly, emissions pricing creates a market-based exit from industrial policies, helping to manage their policy costs and strengthen their efficiency. Thirdly, target-consistent emissions pricing differentiated across sectors—between relatively easy- and hard-to-decarbonize sectors—can help manage adverse distributional impacts of emissions pricing, especially on households. More targeted interventions to support adversely affected communities from structural changes, such as coal-mining communities and low-income households living in thermally inefficient dwellings, is also fundamental to managing effectively adverse distributional impacts of change. An early policy focus on energy and material efficiencies as warranted by market imperfections also create economic benefits that can help offset costs of other policies.

The institutional contexts for transforming energy systems, in addition, shape the effectiveness of reform strategies. For example, adapting government-designed energy markets and regulations to changing technologies is necessary for efficient investments in variable renewables like solar PV and wind turbines and complementary flexible-power technologies. Such adaptations are key to minimize overall, system-wide electricity costs. Moreover, economy-wide institutions to provide transparency of overall climate goals, reform strategies and their implementation promote greater public understanding of reforms and their goals and impacts. Such transparency promotes self-discipline in policy making and helps governments develop reputations for effective, goal-oriented reforms. Independent government advisory councils on climate are economy-wide examples of such institutions. Energy market monitoring agencies serve this role in government-designed and regulated electric power and natural gas markets.

But important as comprehensive, coherent and credible energy-reform strategies are for countries transforming their energy systems, each country acting in isolation would disappoint. Climate change is a global environmental externality, and all countries must act to achieve net zero emissions if the climate is to be stabilized. Moreover, country capabilities and interests in low-carbon alternatives necessary to eliminate net emissions from energy vary widely. Some countries have comparative advantages and specializations in innovations and manufacturing. Examples of such production capabilities include mass manufacturing of potential general-purpose technologies for low-carbon energy systems, such as solar PV panels and lithium-ion battery packs, as well as more complex technologies such as wind turbines. These countries have stronger interests in the development and early deployment of low-carbon technologies than others, and they are more likely to initiate industrial-policy-driven market disruptions of incumbent technologies. For them, innovations and low-carbon technology production are potential new sources of economic growth and international comparative advantage. At the same time, low-carbon technologies enable countries to derive value from heretofore largely untapped renewable resources such as solar irradiance and wind to produce modern energy. Many developing countries are abundantly endowed with these resources. Yet other countries have strong societal interests in limiting climate change with few reinforcing or opposing domestic economic interests. Many countries have abundant coal resources and use coal intensively for electric power generation and industrial production, with interests in maintaining the energy status quo. Most countries import crude oil and oil products; few are major oil exporters.

These widely ranging domestic interests in the energy status quo, alternative low-carbon technologies and choices, renewable resource endowments and environmental protection shape international actions on climate change. The Nationally Determined Contributions (NDCs) of countries that signed the Paris Agreement—their climate action goals and plans—are representations of these varied interests at country level. The agreement's mechanism for peer review of NDCs create opportunities for countries to coordinate their climate actions and engage in more formal cooperation to advance their implementation. Such coordination can also arise outside the formal mechanisms of the Paris Agreement. In fact, there has been significant tacit coordination of—and some formal cooperation on—market-creating and industry-supporting policies for solar PV panels, wind turbines, lithium-ion battery packs and battery electric vehicles.

THE LINE OF ARGUMENT

Part I of the book sets out the societal, environmental and technological contexts for transforming modern energy systems.

Chapter 1 begins with a brief history of energy capture—the human capability to harness vast primary energy resources and use them well—and the emergence of modern societies through two industrial revolutions enabled by accumulating scientific and engineering knowledge. They transformed societies and lifted living standards to unprecedented heights. Almost all firms and households in advanced industrialized countries and many in developing ones now depend on modern energy supplies derived primarily from fossil resources for their economic and household activities. The two industrial revolutions and more recent innovations also provide valuable insights into market-based dynamics of energy-related technological change. For example, innovations that proved commercially successful emerged initially in countries with capacity for innovation and economic specializations in primary industries and manufacturing. The pace of initial market penetration is shaped by incumbent technology characteristics and those of alternatives. They then diffused widely across countries over time, with some acceleration.

Chapter 2 characterizes modern energy systems at the scale of individual firms and households, highlighting the interdependencies between their choices and those of state-owned and private firms that supply energy. It shows that energy is demanded for the useful services and materials that it enables. These benefits from energy flow through the energy-using equipment and appliances in buildings, transport vehicles and industrial processes for making useful materials like steel, cement, plastics and chemicals. The demand for energy is thus derived from the demand for other things. To transform energy systems, while maintaining living standards that benefit from current energy use and creating opportunities for sustainable growth, it is necessary to change technologies in much of the energy-using and energy-producing capital stock, including those for producing electric power and fuels. The chapter also explains why current technologies that use fossil fuels in traditional ways disrupt the Earth's carbon cycle and change its climate.

Energy systems must achieve net zero emissions of carbon dioxide if they are to stop putting pressure on the climate, and Chapter 3 begins with the climate science behind this requirement. Alternative energy end-use and supply technologies that could contribute to this outcome are then surveyed. The core of such systems is increasingly clear with recent advances in solar PV, wind turbines and batteries. At their core would be electricity systems with zero or negative emissions and substantial generation shares from renewable resources, along with widespread electrification of economic activities. This core is now seen as relatively 'easy to decarbonize'. But other activities, especially in heavy industry and commercial transport, appear 'hard to decarbonize', needing low-carbon fuels such as hydrogen or carbon dioxide-removal technologies. While two decades ago most energy emissions then appeared hard to decarbonize, advancing low-carbon alternatives supported by innovation and industrial policies are expanding opportunities for change.

Part II recognizes that advances in low-carbon alternatives are endogenous developments in markets with multiple imperfections, have substantial risks and distributional impacts (positive and negative) and need comprehensive government policies to address and manage them (Stiglitz, 2019; Stern, 2021; and Stern and Stiglitz, 2021). It also recognizes the need to adapt market-supporting institutions for energy systems, including government-designed energy markets and regulations.

Chapter 4 begins by observing that the directions for technology advances in sectors now seen as easy to decarbonize took their lead from a series of oil price shocks and energy security concerns over several decades. It then examines evidence on market imperfections confronting low-carbon alternatives. Some are similar to those for new technologies in general, such as R&D knowledge spillovers. Their more distinguishing features are external scale economies and network effects from deploying them in transforming energy systems. Such positive cost spillovers from early deployment of low-carbon alternatives are a key feature of ongoing transformations. Departing from the orthodoxy of using government supports for R&D and emissions pricing alone to advance low-carbon alternatives, heterodox industrial policies were used to effectively address these and other market imperfections affecting their advance. This outcome is consistent with more general evidence on the effectiveness of industrial polices, especially when attuned to market competition.

While the need for pricing carbon dioxide emissions is clear, its calibration is not. Two concepts for emissions pricing are cost-effectiveness consistent with a quantitative emissions target and the social cost of carbon dioxide emissions. Chapter 5 examines their empirical foundations. They include the long-run marginal cost of abatement consistent with the target, recognizing that it subsumes some inherent uncertainties in the social-cost approach. Substantial inherent uncertainties in social costs and benefits drive a vast range of estimated social costs of carbon emissions, which is also examined. It is argued that a quantitative emissions target and cost-effective emissions pricing provide a more stable and informative focal point for government policies and long-run investment choices of businesses and households. Nevertheless, actual emissions pricing falls short of the levels and scope deemed adequate by either framing, though countries with more ambitious pricing benefit from low-carbon legacies such as hydroelectric and nuclear power.

Adapting energy markets, institutions and infrastructures to low-carbon alternatives is necessary to advance them, especially tailoring electricity markets' designs to variable renewables like solar PV and wind generation and complementary flexibility technologies for system balancing. Chapter 6 examines how electricity market designs balance controlling market power of generators in tight markets while ensuring electricity prices reflect the true scarcity value of electricity over time to remunerate system investments. Such scarcity pricing would also in principle need to remunerate upfront investments in advancing low-carbon power technologies in the absence of market-creating policies for them. Electricity market reforms such as capacity remuneration mechanisms aim to strike the market-design balance and ensure adequate investment in low-carbon generation capacity and flexibility. Natural gas infrastructures also need to adapt to low-carbon alternatives. A key challenge is coordination of investments in low-carbon alternatives in existing buildings, a task aided by upfront investments in building thermal efficiency.

Chapter 7 explores gaps between observed energy and material efficiencies and their feasible potentials. Gaps between actual and technically feasible energy efficiencies are relatively wide for buildings, household appliances and vehicles and narrow for heavy industries and commercial transport, although not all technically feasible efficiency gains are cost-effective. Evidence on causes of potential economic inefficiencies focuses on household investment decisions and energy choices regarding dwellings, appliances and automobiles. There is significant evidence that informational asymmetries, split incentives and behavioural anomalies contribute to economic inefficiencies. Evidence on energy efficiency policies finds that those that are well designed improve efficiency and overall welfare, including by directing innovations towards this somewhat neglected product characteristic. Similar issues arise for material efficiency, although available evidence is less extensive. Looking forward, efficiency policies should reflect expected shifts in relative prices arising from government climate goals and their implementing policies, especially those for building thermal and material efficiencies.

Part III explores the political economy of transforming energy systems, including domestic economic interests, societal and political values and institutions, and develops comprehensive domestic energy-reform strategies and approaches to their international coordination.

The political economy context for transforming energy systems is shaped by domestic economic interests, societal and political values, and institutions. While recognizing interests in the energy status quo, Chapter 8 focuses on interests in low-carbon alternatives and how they shape investments in knowledge and new capabilities necessary for low-carbon energy systems and implementation of policies that support them. Country capabilities and specializations in innovations and manufacturing are central to these foundational investments. Countries and firms that specialize in these activities are those that have largely initiated the technological disruptions to current energy systems, primarily by implementing market-creating and industry-supporting (industrial) policies. In addition, low-carbon technologies produce value from heretofore untapped renewable wind and solar resources, and many developing countries are relatively well endowed with them. They are potential sources of long-run economic growth and new comparative advantages for developing countries in a world with net zero emissions from energy.

Chapter 9, which concludes the book, distils from the preceding analysis and evidence key elements of energy-reform strategies. They are comprehensive in addressing market imperfections, risks and distributional impacts to advance and guide low-carbon alternatives to net zero emissions goals. They include industrial policies and emissions pricing consistent with these goals, appropriately sequenced and differentiated among easy- and hard-to-abate sectors. They also include adapting government-designed energy markets, regulations and infrastructures to low-carbon alternatives. Moreover, countries' comparative advantages and specializations in advancing low-carbon alternatives mean that acting in isolation would be costly and clearly fail to address the global environmental externality. The chapter concludes by exploring ways that countries, building on their domestic energy-reform strategies, can coordinate internationally climate actions and accelerate them towards achieving the Paris Agreement goals. While these approaches vary with technology characteristics and market structures, they point to the importance of sector-focused initiatives in accelerating change alongside countries' efforts.

PART I

Modernity, the climate and net zero emissions

1. Energy capture and modernity

Our ability to capture vast amounts of energy and use it well is—in the sweep of human history—relatively recent (Cook, 1971). This capability has transformed living standards and productivity at work beyond all conceivable imagination just 250 years ago. Before then, and from at least the emergence of human language and culture about 70,000 years ago, human material existence changed comparatively little. The agricultural revolution changed the way humans obtained food and organized societies, but this revolution barely raised living standards over more than ten millennia, at least in a sustained way. In contrast, the impacts of two industrial revolutions—the first fuelled by coal and the second by oil and electricity—over two centuries were revolutionary. They transformed societies and lifted living standards to unprecedented heights. While there were important societal and political underpinnings to the industrial revolutions, they were enabled from an energy and technology perspective by accumulating scientific knowledge and its engineering application—an industrial enlightenment (Mokyr, 2002, p. 35). With these innovations, buildings became better built and more comfortable, well-lit and productive places to live and work. Road and sea transport became faster, safer and more affordable, and rail and air transport possible. Productivity-enhancing equipment and appliances spread through factories, offices and households. In modern societies, fossil fuels do much of the work.

This transformation to modern societies, however, is having profound impacts on the Earth—its oceans, surface and atmosphere—as human capabilities flourish and populations grow. Economic development is largely a response to pressing physical and material needs, which have been understandably in sharp relief. Wider impacts on the environment typically have been neglected until course corrections were compelled, such as the gradual deforestation of Britain for agricultural land, building materials and fuel over centuries, and eventual recourse to coal (R.C. Allen, 2009, pp. 84–90). But it is the current use of fossil fuels that is associated with poor local air quality, especially in urban areas, and with increasing the risks of climate change due to rising concentrations of carbon dioxide and other greenhouse gases in the Earth's atmosphere (IEA, 2016; IPCC, 2018).

The contexts for transforming current energy systems are complex, however. Some relate to the social, political and economic contexts for change, of which the interplay between two industrial revolutions and societal transformations provides powerful historical illustrations. The industrial enlightenment and technological innovations it spawned underpinned the first sustained rise in living standards, a development pathway initiated in a few countries and soon followed in many others. Over time, successive waves of technological innovation accelerated economic growth, industrialization and urbanization. These economic and societal transformations were also associated with profound political changes, including the emergence of modern nation states. In modern societies, economies and energy systems are now deeply interdependent. Almost all firms and households in advanced industrialized countries and many in developing ones depend on modern energy supplies for their economic and household activities.[1] Thus, to transform current energy systems, most firms and households must change to low-carbon energy and technologies. But such changes raise challenging economic, social and political issues. They include vast scales of change in terms of the number and value of investment decisions, 'lock-in' of some incumbent technologies in investment decisions of individual firms and households, and distributional impacts within and across societies.

Other critical aspects to transforming energy arise from its country, sectoral and temporal dimensions. Again, the two industrial revolutions along with more recent energy-related innovations provide useful examples. Technological innovations that eventually proved commercially successful tended to emerge initially in those countries with capacity for innovation and economic specializations in primary industries and, over time, manufacturing. The pace of sectoral change in these initial markets tended to be slow, as the technologies and products were refined and adapted to specific uses. However, in subsequent markets, the pace of sectoral change accelerated, benefiting from experiences gained in initial markets and earlier innovations. Government interventions also accelerated some innovations and commercialization of energy-related technologies in the twentieth century. Similar country, sectoral and temporal variations are seen in current changes to low-carbon technologies, including the characteristics of incumbent technologies and their low-carbon alternatives. But before describing the historical backdrop to transforming current energy systems and assessing the temporal and spatial dimensions of change, it is useful to explain what energy is.

ENERGY

Energy is everywhere. The Sun showers the Earth's surface in two hours with more electromagnetic energy than all the primary energy humans capture in a year (Tsao et al., 2006). Solar irradiance and the Earth's gravity drive the water cycle—evaporation, cloud formation, precipitation and flowing water in streams and rivers. Differences in the Earth's surface temperatures, particularly between large areas of land and water, cause winds and ocean waves as its gravity is sufficiently forceful to retain an atmosphere. The Moon's gravity and its orbit around the Earth cause tides to ebb and flow. There is also the Earth's internal heat, which comes from radioactive decay of uranium and other isotopes in its crust and mantle as well as heat from its formation and molten metal inner core (Davies and Davies, 2010). However, the Sun is the dominant source of energy at the Earth's surface (more than 99 per cent).

Most life on Earth obtains its energy directly or indirectly from the Sun, although only a small fraction of solar irradiance is absorbed in this way and converted annually into chemical energy stores such as biomass (0.1 per cent; Barber, 2009). Plants on land and phytoplankton at sea use sunlight to convert elements—mostly nitrogen, phosphorous and potassium from soils and carbon, hydrogen and oxygen from air and water—into new biomass. Crop plants and some algae are relatively efficient at this photosynthesis process, converting 1–3 per cent of the sunlight that reaches these plants or organisms into energy stores (Blankenship et al., 2011). Animals digest plants or other animals to obtain energy in chemical form, mostly carbohydrates, fats and proteins. Fossil fuels are energy stores from fossilized remains of ancient plants and phytoplankton exposed to heat and pressure within the Earth's crust over millions of years.

Energy is also everything. Mass and energy are related, with a small amount of mass equivalent to a vast amount of energy. The energy of matter with mass is proportionate to its mass times the speed of light squared (Einstein, 1905; Rainville et al., 2005). Nuclear reactions are examples of this relationship working in one direction. Within the Sun, the fusion of hydrogen nuclei into lighter helium nuclei releases a vast amount of electromagnetic energy. On a very much smaller scale, fission within a nuclear power plant reactor splits apart the nuclei of uranium isotopes, reducing their mass slightly and yielding a substantial amount of energy.

While energy is everywhere and everything, it is hard to pin down precisely what it is. Energy takes many forms—electromagnetic (ultraviolet radiation, visible light and infrared radiation), chemical (food, feed and fuels), thermal (heat), kinetic (motion), electrical, nuclear and gravitational. There are many natural and man-made processes for converting one form of energy to another. For example, humans and other animals convert food and feed into life and motion. Burning biomass, like wood, converts fuel into heat. The internal combustion engine converts heat from burning liquid fossil fuels into motion. A nuclear power plant converts fissile material into electricity through, first, its conversion into heat and then motion. A working definition of energy is the property of transforming matter that has capacity to perform work, such as causing heat, light, motion and interaction of molecules, including life on Earth.[2]

One unit of measure for energy, a joule (J), is defined by an amount of work, specifically the amount of force required to accelerate a kilogram of mass by one second per metre over the length of a full metre.[3] Power is the flow of energy over time and is measured in joules per second or Watts (W). Another useful measure of energy is a Calorie (large calorie or kcal), which is the amount of heat required to raise the temperature of one kilogram of water (at sea level) by one degree Celsius (C), which is a measure of heat intensity (temperature). To give a sense of scale, moderately active adult females and males requires 2400 kcal and 2800 kcal per day, respectively, which is equivalent to 10 million J (or 10 megajoules (MJ)) and 11 MJ per day (FAO, 2004, pp. 35–52). The world average per capita daily dietary energy supply from food is about 2900 calories or 12 MJ per day.[4] This daily amount totals to just over four gigajoules (GJ) per year. In comparison, current average annual per capita final energy use in advanced industrialized and developing countries, excluding food, is about 110 GJ and 35 GJ, respectively.[5]

ENERGY AND SOCIETAL TRANSFORMATIONS

Brief narratives of the agricultural revolution and two industrial revolutions highlight the contexts for historic energy-related innovations and their impacts on economic activity and living standards, organization of societies and development of political order. They draw out both positive and negative aspects of change, and point to factors assessed to have contributed to both material and societal transformations at key historical junctures. The back story to the underpinning technological changes, which span 12,000 years, is much human trial and error with countless more and mostly untold failures than successes, which is likely why the timespan to reach modernity is so long.

The Agriculture Revolution and Settlement

While humans foraged for food and gathered wood and other biomass for fuel for most of their history, about 12,000 years ago—as the last glaciation ended and the climate warmed naturally—humans began to engage in agriculture and live in settlements.[6] Archaeological evidence has unearthed remains of early human settlements in an arc of rolling hills curving around the Tigris and Euphrates rivers in the Middle East as well as the Indus River and Huang He (Yellow River) valleys in Asia. While the transition from hunting and gathering to agriculture and settlement is not well understood, fixed dwellings were likely enabled by domestication of wild grasses and other plants with seeds rich in carbohydrates and some proteins—barley, lentils and wheat in the Middle East and millet and rice in Asia. Domestication of animals for food—

initially sheep and subsequently cattle and pigs in the Middle East—added regular sources of proteins and fats to diets.

Settled communities, however, co-existed alongside hunting and gathering tribes for millennia. This observation suggests that benefits (controllable food supplies) and costs (risks of disease and predation) in making the transition to settlements were finely balanced. They only tilted towards settlement where conditions were favourable and because of higher human fertility rates in settlements (Scott, 2017, pp. 87–92 and 113–5). Over time, human agriculture and settlements spread across the world as agricultural capabilities developed and adapted to local environments. For example, settlements in what is today Mexico were associated with domestication of corn and potatoes in Bolivia and Peru.

Subsequent energy-related technological changes took the form of domestication of animals for work—water buffalo in Asia around 6000 years ago; asses, oxen and camels in the Middle East 5000 years ago; and horses in Asia and the Middle East around 4000 years ago. Development of tools such as wooden seed drills, leather harnesses for draught animals and wooden ploughs to till soils enabled an intensification of agriculture to feed both domesticated animals and larger populations. Human capabilities and technologies continued to expand gradually. Progress took many forms—irrigation canals and fertilizer from organic wastes, crop rotations, wooden wheeled carts, pottery and bricks fired in kilns and charcoal and smelting of metal ores. The first major renewable energy technologies beyond biomass and draught animals appeared in the Asia and the Middle East around 2000 to 1500 years ago: ships capable of sailing into the wind, and windmills. By 1000 years ago, these renewable technologies along with water wheels were widely used in Asia, Europe and the Middle East.

Human material progress over millennia, though, was largely matched by expanding populations and average living standards barely rose in a sustained way over time. For example, by 1700 CE (all dates in Common Era), before industrialization began in earnest, Holland and England, then among the richest nations in the world, had estimated real per capita income levels of about $1,920 and $1,660 (all real incomes are in 2011 International $), equivalent to around $5 per day.[7] Their annual growth rates over the preceding five centuries averaged about 0.1 per cent, albeit interspersed with many growth episodes and reversals (Fouquet and Broadberry, 2015). By 1700, estimated per capita income in India reached $1,200, Germany $910 and Japan $840. To give a sense of scale, the monetary equivalent of a subsistence income level, including the value of informal activities such as subsistence farming, is about $700, equivalent to $1.90 per day (Bolt et al., 2018).

In 1700, average per capita energy use in Holland and England, which were still largely agricultural societies, was in the range 25 GJ to 30 GJ per year

(Malanima, 2006; Warde, 2007, pp. 131–8). This energy was largely garnered through food and feed, draught animals, wood for fuel and other rudimentary renewables. They included watermills and windmills for processing food, making textiles and crushing metal ores; charcoal for smelting metal ores; and sailing ships for transport. England also used some coal for heating dwellings.

From the agricultural revolution to the first industrial revolution, sustained gains in living standards were glacial and subject to Malthusian pressures (Galor, 2005). However, intensification of agriculture, growing yet vulnerable settlements and development of trade contributed to the development of much more complex societies. They included not only settlements but also early state formation with some central authority that raised resources and organized capacity for human coercion and warfare (Carneiro, 1970; Fukuyama, 2011, pp. 458–83, Scott, 2017, pp. 128–39). These political developments involved taxation of agricultural production, conquest of others' settlements and slavery, with theft and human exploitation garnering greater prosperity for some at the expense of many others during the Malthusian epoch.

The First Industrial Revolution: Coal, Steam and Modern Nation States

The epochal transition to modern fossil fuels from traditional biomass and rudimentary renewables began as forests became depleted from intensive use of wood for fuel and building material and land clearance for agriculture.[8] Initially, coal was used simply as a substitute for wood as a fuel for heat. This substitution first took place at scale in England as forests became more distant from cities, with a growing population, mounting pressures on land use from agriculture and deforestation. Also, coal seams were visible at the surface and close to the sea, and thus low-cost to mine and transport by ship. The transition to coal gathered pace in the eighteenth century with the invention and refinement in England of the steam engine, an innovation that emerged amidst a growing culture of scientific investigation and its mechanical applications as well as the increasing diffusion of knowledge through society (Mokyr, 2002, pp. 56–76; Jacob, 2014, pp. 20–31). The engine's capacity to do work offered a substitute to draught animals and workers as their costs and wages began to rise and the costs of energy services from coal and steam technologies declined (R.C. Allen, 2009, pp. 138–44; Jacob, 2014, pp. 66–82). This was the first machine to convert heat in the form of steam from burning fuel into useful motion. It worked by expanding steam from a boiler into a cylinder to drive a reciprocating piston, and its first use was to pump water from coal mines to boost production. Coal and coke were also increasingly used in smelting metal ores as a substitute for wood and charcoal as forests became depleted and wood relatively expensive.

As their designs became more efficient through experience-based learning, steam engines became more widely and intensively used, and by the middle of the nineteenth century coal had displaced wood as the main fuel. These technologies enabled and expanded mechanized production (powered billows for blast furnaces, displaced waterwheels to boost food and textile mill capacities) and transport (steam ships and rail locomotives). This steady refinement and spread of coal and steam technology through the English economy was facilitated by supportive institutions that emerged from a fraying of the English monarch's authority. The spreading and balancing of political powers in England, including development of the rule of law with some independence from the crown, laid important societal, political and institutional foundations for innovation and entrepreneurship as well as private investment and finance (Acemoglu and Robinson, 2012, pp. 182–212). Markets for commodities also expanded with growing international trade and European colonization of the Americas and parts of Africa and Asia. While it is perhaps by a sequence of happenstances that industrialization through entrepreneurship and private investment began in England, the demonstration effect was soon seen in other countries.

The direction and pace of technological innovation was increasingly shaped by entrepreneurial initiatives, expanding market opportunities, international trade and the diffusion of knowledge within and across countries. Coal and steam technologies soon spread across Europe, where coal resources were accessible and technologies could be adapted to local circumstances—initially Belgium and northern France, and subsequently the Ruhr Valley in Germany (Fernihough and O'Rourke, 2014). The knowledge and technology spread well beyond Europe to the United States and Japan by human travel, while Switzerland industrialized on the back of its significant innovation capabilities, renewable energy resources and imported coal. Coal and steam technologies reached into more economic sectors too, scaling up production of food, leather, textiles and metals (mostly iron) in factories and transport by steam ship and locomotive. In Britain, which was then at the technological frontier, growth in average real per capita incomes rose at an annual average rate of 0.5 per cent from 1700 to 1870, a significant acceleration from the pre-industrial rate.

By the end of this period—the first industrial revolution—the real average per capita income in Britain reached \$3,850.[9] Those in the United States and Belgium reached \$3,740 and \$3,370 in 1870, respectively, as these countries closed the technological gap with Britain. In France and Germany, they were around \$2,400 in 1870. The composition of economic activity changed as well. By 1870 in Britain, industrial production and services (including transport) each accounted for about 35 per cent of total output. Belgium and France also attained industry shares of total output of at least 30 per cent, with Germany just below (Fremdling and Solar, 2010). However, the Netherlands, which had gained early prosperity from shipbuilding, trading and productive agriculture, was only a latecomer to the industrial revolution, owing to diminished capacities for innovation and political upheavals from the French Revolution (Mokyr, 2000).

Industrialization in the technologically leading countries and their sustained growth in per capita output were fuelled by a surge in energy use. Average annual per capita energy use in Great Britain rose to about 120 GJ in 1870, up from 30 GJ in England in 1700, a four-fold increase (Warde, 2007, pp. 131–8). At the same time, per capita incomes rose by almost 2.5 times. Taken together, they point to a very high additional energy use to fuel economic growth—a factor of 1.7—at the technological frontier.[10] This underscores both the societal imperatives of expanding production and consumption possibilities beyond those achieved in the Malthusian epoch, and the inefficiencies of the new technologies in converting energy into useful work and output. In fact, the productivity of energy in terms of real gross domestic product (GDP) per GJ fell in Great Britain to $32/GJ in 1870 from $55/GJ in England in 1700, when it was still a largely agricultural society with some coal for heating. There was, however, much technological potential to improve energy productivity.

Coal and steam technologies enabled not only growing industrialization but also increasing urbanization which helped to create a social basis for modern nation states to emerge. Britain and Belgium, along with the Netherlands, had urban population shares in 1870 of 35–40 per cent, while the Western European average reached about 15 per cent. Industrialization and urbanization worked against the feudal political and social order that had prevailed in Western Europe for centuries and facilitated the formation of new social groups—workers, students, professionals and managers (Fukuyama, 2014, pp. 40–51). In anonymous cities, people gained more fluid identities that were no longer necessarily bound to village or family. New identities took new forms, including nationalism, and modern nation states sought to forge national identities in several ways. They included building railroad networks in the nineteenth century that served to link cities, towns and peoples (Fukuyama, 2014, pp. 165–84). These technological, social, political and institutional changes helped lay the foundations for the next wave of private innovation and investment and the second industrial revolution. They also abetted Western imperialism and national rivalries, and the industrialization of human conflict (Morris, 2010, pp. 490–526).

The Second Industrial Revolution: Oil, Electricity and Consumer Societies

The initial impetus for tapping crude oil was to find an alternative to whale oil for lighting, which was becoming increasingly scarce due to over-hunting, much as forest depletion prompted the initial switch to coal from wood in England. Kerosene, which is refined from crude oil by distillation, was a good alternative for oil lamps. While the existence of crude oil was long known from natural oil seeps and ponds, the first wells to tap into underground reservoirs beneath them were drilled in the mid-1800s in Azerbaijan and Pennsylvania. But the development that really spurred on the oil industry was the invention and refinement of internal combustion engines in the 1880s and 1890s in Germany and France. In contrast to steam engines, which rely on external combustion and boilers to produce steam, internal combustion engines burn liquid fossil fuels which are injected together with air into cylinders and burned to drive reciprocating pistons within them. They were more compact and efficient machines than steam engines, and fuels from distilled crude oil were more energy-dense than coal.

As with steam engines, initial applications of internal combustion engines were for stationary purposes, but as their design became more efficient they began to be used for road transport to power automobiles and trucks by the beginning of the twentieth century. While Germany and France saw their early designs take shape, it was in the United States where affordable, mass-produced automobiles gained broad societal use through assembly-line production, modern business organization, consumer finance and marketing pioneered by automobile manufacturers. Automobiles and gasoline rapidly replaced the horse and biomass in the early decades of the twentieth century as the main mode of road travel in the United States, facilitated by significant government investment in paved roads and fuelled by a growing oil industry. While automobiles and gasoline soon offered faster transport at lower cost per mile than horses and biomass, they also became part of a rapidly growing American consumer culture (Gordon, 2016, pp. 62–171). From virtually nil in 1900, US automobile ownership reached 187 per 1000 of population by 1930, a level not reached in Western Europe and Japan until the 1960s or later (Cain, 2006; Dargay et al., 2007).

The development of motorized road transport occurred in the context of a much broader wave of rapid technological progress, as the interactions among accumulating scientific knowledge, its engineering applications and experienced-based learning intensified. Modern energy and machines trans-formed many economic activities and sectors at the same time. They included breakthroughs in iron- and steel-making in Britain and Germany that raised quality and lowered costs, allowing wider applications in buildings and bridges and rapid expansion of railroads and shipping. The increased precision of machine tools and standardized measures enabled metal parts to be made to standard shapes and sizes and to close tolerances necessary for complex machines and mass production. This innovation, for example, transformed the manufacturing of transport vehicles—not only automobiles and trucks, but also trains, ships and in time airplanes. Paper-making machines made possible the low-cost production of paper and in turn mass communication through printed media. The invention of the telephone in America made instantaneous long-distance voice communication possible over wires, while that of the radio in Britain dispensed with wires for voice transmission.

But perhaps the most important new technology developed in the latter part of the nineteenth century was electricity. While a systematic understanding of electricity had accumulated through the efforts of many scientists since the late eighteenth century, a key breakthrough was the discovery in Britain that magnetism could produce a steady flow of electricity, and this opened the way for producing electric power from a rotating magnetic field. The steam turbine, also developed in Great Britain in the late 1800s, became the primary device for rotating a magnetic field around a stationary set of conductors (wires wrapped around an iron core) to produce a continuous flow of electricity. These turbines produce rotating motion from steam created at high pressure from a boiler and directed through the turbine blades to rotate them. The main initial use of electric power was lighting for commercial buildings and factories, which substantially improved both lighting quality and fire safety. Electromagnetic motors also expanded substantially the use of electricity to power machines in factories, replacing steam engines and belt drives. The commercialization of electricity generation, distribution, metering and electric lighting and motors was initiated in London and New York, integrating these technologies into electricity systems.

Electricity systems transformed not only commercial building and factories but also households. The connecting of large numbers of buildings and end users to energy networks began in the United States and Western Europe in the early 1900s. Coal and wood had been the main fuels for heating and hot water in houses for centuries, while an opened window provided cooling and ventilation if needed. By the middle of the twentieth century, the transformation of American households was extensive. More than half the population lived in urban areas by 1940 and virtually all urban households had access to electricity, three-quarters to natural gas for heating and hot water, and more than half had central heating (Gordon, 2016, pp. 115–26). Electricity, refrigeration and freezing transformed the production, transportation, retailing and household storage of food. The proliferation of electrical appliances transformed how household activities were undertaken and became the cultural norm of an American home. Western Europe and Japan lagged behind the United States in these developments, owing in part to the dislocations of war, but nevertheless followed similar albeit delayed pathways.

Development of energy networks in the United States and Western Europe to serve households, commercial buildings and factories also marked a change in how energy and energy-using technologies were developed and provided.

The early inventors and entrepreneurs in electricity integrated the entire value chain from electricity generation, transmission and distribution to customers by creating vertically integrated utilities. These early efforts were focused on major cities—Berlin, Chicago, London and New York—where per capita incomes, concentrations of economic activity and density of potential electricity demand were high. They were also shaped by the interplay between entrepreneurial and political interests in the sanctioning of public utilities. In New York and Chicago, the interests of technology tended to dominate the politics; in London it was the reverse (Hughes, 1983, pp. 461–5). In Berlin before the First World War, there was effective coordination of political and technological power, facilitating growth in both supply and demand necessary for achieving the significant economies of scale and lowering costs for customers. Similar considerations led major electrical equipment manufacturers to supply and standardize electrical appliances and promote the social norm of a modern household and industrial society in advertising to develop demand for both electrical equipment and appliances (Hughes, 1989; Nye, 1990, pp. 259–77).

As these energy, technology, societal and institutional developments transformed many areas of economic activity, the United States and Western Europe saw sharp accelerations in output growth and rising living standards. By 1940, US real per capita income almost tripled to $10,450 from its 1870 level, achieving an average annual growth rate of 1.5 per cent.[11] This rate and extent of sustained improvement in living standards was then without historical precedent. Real per capita incomes also rose in Western Europe, but at a slightly slower average rate of 1.3 per cent, reaching $9,260 in Britain in 1940, $7,570 in Germany and $6,060 in France (both in 1939). Per capita incomes in the United States and Western Europe doubled in less than 70 years. The two previous doublings of per capita incomes in Western Europe took about 150 years and 2000 years.

Average per capita energy use in the United States reached about 220 GJ in 1940, up from about 110 GJ in 1870 (excluding food).[12] While growth in output and living standards accelerated, the energy intensity of economic growth fell at the technological frontier to 0.6 from 1.7 in the first industrial revolution. This reflects the fact that the second wave of technologies for converting energy into useful outputs were much more efficient than those in the first. As the United States took over from Britain at the technological frontier, especially in scaling up and commercializing new energy technologies, GDP per unit of energy in the United States rose to $48/GJ in 1940 from $32/GJ in 1870. These energy and technology advances also scaled up military conflicts between nation states and fuelled energy insecurities in some for want of access to primary energy resources.

Globalization: Logistics, Aviation and Digitalization

Globalization—the growing worldwide interconnections among peoples and nations—is perhaps the defining political, economic and social development of the late twentieth century and early twenty-first century (Steger, 2017, pp. 11–17). Energy plays an important physical role in connecting places and peoples, but its indirect impact on globalization has been perhaps as great.

This recent wave of globalization emerged gradually from the ashes of the Second World War. The political response to this catastrophic intensification of human conflict among industrialized nations was the creation of a new liberal world order among nation states supported by several multilateral institutions. This approach to international relations aims to use markets more, and political coercion and military force less, to mediate relationships among nation states, with the multilateral institutions overseeing rules for these markets and providing a forum for cooperation among governments (the United Nations). The key markets in this world order are for international investment and finance, foreign exchange (money) and trade in goods. Their supporting institutions are the World Bank, International Monetary Fund and General Agreement on Tariffs and Trade (subsequently the World Trade Organization). The political ascendance of liberalism in the 1980s in the West and the so-called Washington consensus applied to these markets a heavy dose of market liberalization, which accelerated and deepened globalization to an extent last seen when the first era of globalization peaked in 1914 at the start of the First World War (Rodrik, 2011, pp. 24–44). The political collapse of the Soviet Union, significant economic liberalization in China and market-oriented reforms in India also greatly expanded the scope for these markets and global interconnections. While these political developments supported rapid expansion in trade and capital flows, and catch-up growth in some developing countries, mostly in Asia, they also occasioned a significant widening of income inequality in many advanced industrialized countries (Alvaredo et al., 2018, pp. 40–77).

Energy and energy-using technologies also played an important supporting role in the current wave of globalization. Key innovations lowered costs of maritime shipping significantly and air transport dramatically as the demand for these services rose with rising real incomes around the world and more open markets (Hummels, 2007). In shipping, the innovations related not to ship propulsion but rather to shipping fleet management and containerization—in other words, to the logistics of shipping. The rapid growth in international trade from the 1980s—at a pace much faster than overall economic growth—increased density of trade along major shipping routes and contributed to the emergence of hub ports in Asia, Europe and North America. At the same time, containerization of freight cargoes combined to boost ship utilization and cut freight-handling costs. In aviation, the costs of air passenger fares (per kilometre travelled) fell precipitously with the introduction and refinement of the jet engine, especially during the 1960s and 1970s. A key breakthrough was the development of highly efficient turbofan engines, which contributed to a significant decline in the relative price of air passenger transport and a ten-fold increase in its volume since 1970.[13]

The other technologies that have enabled vast interconnections among peoples and places are digital technologies for handing information, computing and communicating digital content. The first computers were developed in the United States and United Kingdom for military purposes during the Second World War. The general-purpose mainframe computers developed from them found their initial commercial use in sectors that managed vast amounts of information—banking, insurance and airlines. But it was the development of personal computers in the United States in the 1980s, along with new software that made their use widely accessible and productive, that brought computers out of the back office, onto office desks and into households. At the same time, the US military developed technology for linking computers through wires—intranets—for transmitting electronic data and protocols for linking together networks. These technologies enabled basic digital communication such as email over telephone lines. The vast amount of information communicated through the internet today is made possible by development at the European Organization for Nuclear Research of the World Wide Web, a set of protocols for creating and sharing information across the internet. These innovations, together with fibre optics and advanced mobile telephone networks, transformed how information and media are accessed. They also significantly expand the potential for greater efficiencies in provision of services from energy and materials and alternatives to them.

SPATIAL AND TEMPORAL DIMENSIONS OF ENERGY TRANSFORMATIONS

Two centuries of developing and expanding modern energy systems in advanced industrialized countries, including recent innovations and developments, provide insights into the spatial and temporal dimensions of these transformations. The new energy resources and technologies of the first industrial revolution emerged initially in England and Northwest Europe, and those of the second in Western Europe and the United States. From these initial markets, where the technologies were conceived, first tested and refined through commercial use, they spread more widely. Over time, they diffused across sectors and countries as they industrialized, a process which continues in developing countries, and with new waves of innovation like information and communication technologies. This process of energy system transformations and industrialization spanned one-half to a full century. For example,

many of the breakthrough technologies of the first and second industrial revolutions took around half a century to reach market 'maturity' (Perez, 2002,
pp. 49–56). Large-scale, capital-intensive infrastructures typically take more
than half a century to establish and reach most cities and towns—most firms
and households (Grübler et al., 1999). However, at the level of individual
technologies, there is much variation in timescales for their development, commercialization and diffusion, including among initial and subsequent markets.

The risks of climate change give rise to a timebound imperative to transform modern energy systems to low-carbon alternatives, and timescales for
this change are tight—much tighter than the historical energy revolutions.
Looking back, many factors shaped these past changes, including characteristics of the technologies themselves and environments in which they were
deployed. Energy-related technologies are very heterogenous, ranging from
energy supply and complex industrial processes to relatively simple energy
end-use technologies. These various technologies require very different types
of innovation and commercialization efforts. They also deploy into widely
varying market contexts, ranging from initial to subsequent markets and from
competitive to regulated ones. Such differences among energy-related technologies and contexts must be considered when assessing the temporal and spatial
dimensions of transforming energy systems to low-carbon technologies, and
policies to promote them. Moreover, inferences from past changes cannot
simply be extrapolated into the future because of the changing contexts for
energy-related technology developments.

For example, some capacities for energy-related innovations have potentially strengthened over time. Scientific investigation and experience-based
learning accumulate knowledge over time, including better methodologies
for research and experience-based learning. There is also an expansion of
scientific and engineering skills through growing investments in education.
But better ideas appear to be getting harder to find, needing progressively more
intensive research efforts to produce over time, both within and across sectors
and technologies (Bloom et al., 2019). Despite potentially diminishing returns
to research inputs, such R&D (research and development) efforts advanced
many low-carbon technologies in recent decades, and some are now deploying
commercially in markets (see Chapters 3 and 4). Encouragingly, low-carbon
technology concepts that could serve many—but not necessarily all—current
energy, energy-service and material demands exist at least at a pre-commercial
stage (Chapter 3). Indeed, some technologies have existed for a century or
more, such as electrolysers and fuel cells for production and use of hydrogen.
But their development and deployment remain limited, serving only commercial niches. More such concepts are needed to respond fully to the challenges
of transforming current energy systems to low-carbon technologies, enable

testing of alternative low-carbon technologies in initial markets and allow for market selection of the ones best suited to customer needs.

Other contexts for energy-related innovations and change have arguably become more challenging over time. Energy 'system' starting points for the first two industrial revolutions were decentralized and small-scale, and based on the capture and use of traditional energy resources that were inherently local. For example, low-energy-density fuels like feed, wood and dung, and conversion technologies like draught animals, have high transport costs relative to their value in providing energy and energy services. Deploying new technologies into these market contexts offered many early adopters clear net expected economic benefits. In contrast, starting points for transformation to low-carbon energy technologies are highly complex modern energy systems which emerged because of significant complementarities among technologies and scale economies. Some arise from network effects, as with railroads, paved public roads, and transmission and distribution networks for electric power and natural gas. Others are external to individual firms and households, like transport vehicles having the same drivetrain technologies and using standardized transport fuels. These systems of interrelated technologies, infrastructures, institutions and societal norms took one-half to one century to mature. But once established, they can lock-in some incumbent technologies by raising the costs to individual firms and households of switching to alternatives (Arthur, 1989; Unruh, 2000; Seto et al., 2016). Such lock-in effects are inherent to energy systems of interdependent technologies and infrastructures, which create high costs for transforming systems, at least in the early stages of adopting alternative low-carbon technologies. However, some low-carbon fuels and technologies afford 'drop in' or 'bolt-on' alternatives, such as sustainable biofuels and carbon dioxide capture, use and storage, which largely avoid such switching costs.

While recognizing that contexts for transforming current energy systems differ from those of past energy innovations, the experiences of the historic transformations and more recent technological innovations highlight several temporal and spatial characteristics of change. This process of change can be portrayed as consisting of four stylized stages: (1) development of new technology prototypes, primarily through investments in R&D; (2) demonstration of successful new prototypes at a commercial scale; (3) their early deployment and commercialization in initial markets where they are refined and adapted to customer needs; and (4) widespread diffusion across markets, sectors and countries. To give a sense of timescales for energy-related technologies, the median times for a sample of recent electricity-generation technologies to traverse their stages of research, development and demonstration (stages 1 and 2) and market deployment and commercialization in their initial market (stage 3) are 24 years and 18 years, respectively (Figure 1.1; Gross et al., 2018).

Timescales for these two stages for energy end-use technologies and consumer products are shorter: 12 years and 15 years, respectively.

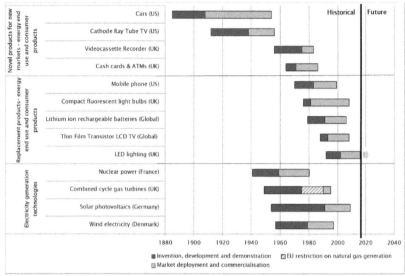

Notes: The figure groups innovation timescales by technology characteristics and category: novel energy end use and consumer products (new products with entirely new markets); replacement energy end use and consumer products (new technologies which replace but perform a similar function to an existing product); and electricity-generation technologies. Invention, development and demonstration are innovation activities prior to commercial availability. Market deployment and commercialization span the period from commercial introduction to a 20 per cent share of the potential market.
Source: Reprinted from R. Gross et al. (2018), How long does innovation and commercialisation in the energy sector take? Historical case studies of the timescales from innovation to widespread commercialisation in energy supply and end use technology, *Energy Policy*, 123, 682–99 (CC BY 4.0). https://doi.org/10.1016/j.enpol.2018.08.061.

Figure 1.1   Energy technology innovations and initial market deployment timelines

The least well researched and understood product life-cycle stage is 'market deployment and commercialization', but it is one that is likely to play a key role in transforming energy systems, so it is important to take a closer look at available evidence. In studies using a range of definitions and measures for this product life-cycle stage, a consistent finding is that new technologies which substitute easily for existing ones achieve market commercialization more quickly than novel technologies in initial markets (Grübler et al., 1999; Bento and Wilson, 2016; Sovacool, 2016; Bento et al., 2018). Technologies that are long-lived and components of interconnected technologies and infrastructures tend to take the longest to diffuse, but notable exceptions are jet engines and nuclear power, which government interventions helped to accelerate (motivated in part by military and energy security aims) (Grübler et al., 1999; Bento et al., 2018). Government interventions such as public procurement can thus accelerate the pace of commercialization of new technologies. There is also some evidence that the market deployment and commercial stage for new technologies deploying in subsequent markets advanced more quickly than in initial markets, although the size of the effect is not large (Wilson and Grübler, 2011; Bento et al., 2018). This finding points to some potential knowledge spillovers from initial to subsequent markets and benefits from investments in absorptive capacities for new technologies in subsequent markets, such as developing countries.

Drawing together available evidence on the stages of innovation and new technology diffusion highlights several stylized facts about their overall pace from invention to market maturity. Those energy-related technologies associated with a relatively slow overall pace in reaching product maturity have several common technology characteristics or deployment conditions (Grübler et al., 2016). They are: (1) multiple changes in technologies, infrastructures, social norms and institutions and large-scale investments needs; (2) extensive development and early-stage deployment of new technologies that must be tested in and adapted to market contexts; (3) low immediate individual adaptation benefits for firms and households, and complex coordination issues between centralized (regulatory) and decentralized (firm and household) decisions. More rapid transformations tend to have converse characteristics of slow transformations (Sovacool, 2016). They involve: (1) new technologies that easily substitute for incumbent ones, (2) alternative technologies that were previously used in other markets and can be readily adapted to new markets, and (3) new technologies that offer significant and multiple benefits to individual adopters. Potential low-carbon technologies and their deployment environments feature a range of such characteristics. Their eventual mix would likely reflect in part societal choices on timescales for change, with tighter timetables directing low-carbon technology development and deployment towards those that have characteristics associated with more rapid energy transformations.

CONCLUSION

Sparked by an industrial enlightenment and propelled by successive waves of energy-related innovations, human capabilities to capture energy on a vast scale and use it well have soared, transforming societies—both positively and negatively. Modern energy systems lifted standards to unprecedented heights in advanced industrial countries, and they make vital contributions to growing prosperity in developing countries. Current energy systems and fossil fuels are deeply woven through the fabric of modern economic activities and household lives. But mounting cumulative emissions of carbon dioxide from fossil fuels and other greenhouse gas emissions are confronting societies with the risks of climate change. To transform current energy systems, most firms and households must invest in low-carbon alternatives, but such changes raise challenging economic, social and political issues. They include the vast scale of the change in terms of new investments, lock-in of some incumbent technologies in investment decisions of individual firms and households, and distributional impacts within and across societies. Incumbent technologies also benefit from un-costed carbon dioxide emissions, a market failure that must be corrected. Markets have an important role in enabling these changes to take place, as do government policies and energy-reform strategies (Parts II and III).

Historic energy transformations and more recent technology developments provide useful insights into potential market-based dynamics of developing, commercializing and widely diffusing new low-carbon technologies on a global scale. At a systems level, energy transformations are slow-moving, spanning a century or more. Within systems, some changes are indeed slow, like those to infrastructures and energy networks, which consist primarily of long-lived and interdependent technologies. Other changes are relatively quick, like light-emitting diode technology that substitutes easily for compact fluorescent and incandescent lightbulbs. Most potential changes to low-carbon energy technologies fall somewhere in-between these two examples. Moreover, tighter timetables for transforming energy would likely tilt technologies choices towards those with characteristics that facilitate their more rapid deployment and widespread diffusion, at least in the first wave. It is thus important to take a close look at current energy systems and the useful services and materials they provide, along with foreseeable alternative low-carbon technologies that could maintain and grow their provision in advanced industrialized and developing countries (Chapters 2 and 3).

NOTES

1. The terms 'advanced industrialized' and 'developing' countries draw broad distinctions among countries, their level of technological development, industrialization and material living standards. This distinction is similar to—but not always precisely the same as—the International Energy Agency (IEA) country classification system for analysis of global energy use of Organisation for Economic Co-operation and Development (OECD) member countries and non-OECD countries. It is also similar to the 1997 Kyoto Protocol Annex II Parties (then OECD member countries), which assumed obligations to provide financial resources to developing countries for climate change mitigation and adaptation purposes.
2. Adapted from the definition of energy in the Oxford English Dictionaries. Retrieved from https://en.oxforddictionaries.com/definition/energy.

3. The International System of Units (SI), Laboratoire National de Métrologie et d'Essais. Retrieved from www.french-metrology.com/en/si/units-measurement .asp.
4. FAOStat. Retrieved from http://fenixservices.fao.org/faostat/static/documents/ CountryProfile/pdf/syb_5000.pdf.
5. IEA Data & Statistics and OECD Data. Retrieved from www.iea.org/data -and-statistics/data-tables and https://data.oecd.org, respectively. Figures exclude non-energy use of fossil-fuel resources for production of chemicals and other materials.
6. This section draws on extensive research on the agricultural revolution, human settlement and societal development around the world from several disciplines— archaeology, geography, history and energy. See, for example, Diamond (1997), Kander et al. (2013), Harari (2015), Scott (2017) and Smil (2017).
7. Historical national GDP data are from the Maddison Project Database Version 2018, University of Groningen. Retrieved from www.rug.nl/ggdc/historicalde velopment/maddison/releases/maddison-project-database-2018. See Bolt et al. (2018) on the database methodology.
8. This section and the next one draw on extensive research on the first and second industrial revolutions from several disciplines—history, economics and energy. See, for example, Mokyr (2002, 2009), R.C. Allen (2009), Kander et al. (2013), Jacob (2014), Gordon (2016) and Smil (2017).
9. Maddison Project Database Version 2018, op. cit.
10. This factor is the energy intensity of real GDP growth; that is, the per cent change in energy use divided by the per cent change in real GDP.
11. Maddison Project Database Version 2018, op. cit.
12. US energy figures are linear interpolations of data from Smil (2017), p. 458.
13. See Smil (2017, p. 333) on the efficiency of turbofan jet engines. The air passenger kilometres figures are World Bank statistics retrieved from https://data.worldbank .org/indicator/IS.AIR.PSGR.

2. Useful energy and the climate

Energy is demanded for the useful services and materials that it enables. These benefits from energy flow through the energy-using equipment and appliances in buildings, road and other transport vehicles, and industrial processes for making useful materials like steel, cement, plastics and chemicals from raw and recycled materials. The demand for energy is thus derived from the demand for other things. They include thermally comfortable, well-lit and productive building space, passenger- and freight-transport services, and useful materials for making and maintaining buildings and infrastructures, equipment, vehicles and other goods. To transform energy systems, while maintaining living standards that benefit from current energy use and creating opportunities for sustainable growth, it is necessary to change technologies in much of the energy-using and energy-producing capital stock. It is also important to use current and alternative technologies, energy and materials much more efficiently.

Transforming energy is particularly challenging in advanced industrialized countries as well as China and Russia with industries and infrastructures shaped by their formerly planned economies. These countries have the most extensively developed energy systems, with economic livelihoods and household lives heavily dependent on existing technologies and fossil fuels. Comfortable, modern buildings, extensive infrastructures and ease of transport are central to their household, community and economic activities. Some economic specializations of these countries also rest on their energy systems and fossil fuels. At the same time, specializations of these countries in innovation, industrial production and manufacturing are central to development and commercialization of alternative low-carbon technologies and renewable energy resources that are necessary to cut carbon dioxide and other greenhouse gas emissions. The structural economic changes and distributional impacts to which they give rise are thus inherent to energy transformations. To sustain these transformations politically, these societal impacts must be addressed together with the drive for economically efficient change.

In addition, technology systems that distribute useful energy to customers from where it is captured and converted from primary energy resources form networks, which give rise to network effects that influence the costs of transforming energy systems. For example, the wires and pipes that transmit and distribute electric power and processed natural gas from their production to customers are centralized networks and natural monopolies. Their geographical reach is shaped by the spatial density of stationary energy demands from buildings and factories, as well as any service obligations required by governments (Farsi et al., 2008). Decentralized distribution systems for liquid fuels benefit from indirect network effects and scale economies from geographically clustered customers with technologies that use the same fuels. These fuels are used for space heating in buildings and process heat in industries, as well as by transport customers. Similarly, existing electric power systems and new end-use technologies like battery electric vehicles give rise to new indirect network effects, but with early adopters facing high costs for switching to new technologies (Yu et al., 2016). In addition, customers are familiar with the existing technologies and energy supplies, which can anchor preferences on currently available choices (Kahneman, 2011, pp. 119–28). These scale economies, especially from network and demonstration effects, along with behavioural barriers to change, create inertia in existing systems and lock-in of some incumbent technologies in choices of individual firms and households, at least in the early stages of transforming energy systems.

It is therefore important to take a close look at current energy systems, the useful energy services and materials they provide, and the household and economic activities they support. This assessment sets the social, economic and technological contexts for transforming modern energy systems and the elimination of net emissions of carbon dioxide from them. An examination of low-carbon technology alternatives that could enable continued provision of the useful energy services and materials provided by current energy systems, but without the net emissions of carbon dioxide from fossil-fuel use and industrial processes, follows in the next chapter.

DERIVED ENERGY DEMANDS AND DELIVERED ENERGY

Energy demands derive largely from the need for the services and materials which require significant amounts of energy to produce. Energy services take many forms. In buildings, they include heating and cooling for their occupants' thermal comfort, and lumens of artificial lighting to facilitate work and household activities. Appliances, office machines and industrial equipment use energy to do useful work in households, commercial buildings and factories. In transport, passenger-km and tonnes-km measures are services in terms of distances travelled by people and goods over roads, rail or water or through the air. Material demands arise largely from the need for buildings and infrastructures but also for equipment, vehicles and consumer goods. The derived demand for energy depends in turn on the efficiency with which the energy-using capital stock coverts it into the useful services and materials required. For example, in buildings, both heating and cooling technology technologies together with the thermal efficiency of building fabrics in controlling heat flows jointly determine how much energy is needed to provide their occupants with the shelter and thermal comfort they demand. Similarly, in transport, the fuel efficiency of passenger transport is shaped by engine efficiency, vehicle weight and aerodynamics, as well as the number of passengers per vehicle, which determine the amount of energy used per passenger-km demanded.

The most significant factor in shaping demands for energy services and materials is economic development and per capita income. Basic comparisons of current residential and commercial building space, ownership of appliances, passenger and freight transport services and material stocks across advanced industrialized and developing countries reveal much about the relationship between income and demand for energy services and materials.[1] In particular, cross-country and time series data for advanced industrialized countries indicate that as per capita incomes rise over time, especially as countries grow through income levels in the range of $5,000 to $30,000 per capita, the spending of additional income on additional benefits from residential space and appliances as well as transport is high. But at higher per capita income levels, many of these demands increase much less with additional income, and increasingly plateau (Fouquet, 2014; Krausmann et al., 2017). Other factors that influence demand for energy services and materials include climates and seasons, the extent of urbanization and population density, and country land size. In addition, economic specialization in production among countries shapes energy use. For example, some countries, such as China, Japan and South Korea, specialize in producing primary materials, which are relatively energy-intensive to produce. Others, like the United States, specialize in less energy-intensive sectors, such as services.

For energy to be useful in producing energy-related services and materials, it must be available to customers when and where it is demanded. This ready availability of energy can involve some customer storage, like fuel tanks for vehicles and buildings that store liquid fuels for use as needed for transport and heating services. But self-storage is expensive and limited in scale. Most energy storage at scale is undertaken through timing of production of fossil-fuel resources, which are stored naturally underground and at scale within fuel production and distribution networks. For example, wholesale terminals and retail stores hold road transport fuels in large tanks distributed across the supply network to meet customer demands when and where they arise. Coal power generators store coal at plants for producing electricity when demanded, while natural gas networks store gas in depleted gas fields and salt caverns for serving seasonal heating and electricity demands.

The storage costs of fossil fuels are low because of their high energy densities and ease of storage in their chemical forms. In contrast, storage costs for electricity are high because it cannot be stored easily—almost all electricity must be used as it is produced. Most energy storage for electricity systems is as coal at power stations or within natural gas networks, or as water behind a dam at hydro-electric plants for use to meet electric power demands as they arise. Some produced power is also stored by converting it into another type of energy for storage, and then converted back into electricity when needed. Pumped hydroelectric storage is the least expensive and most extensively used way of storing electrical energy once produced. However, the cost of storing energy via pumped-hydro storage is several orders of magnitude greater per GJ than that of natural gas in a depleted gas reservoir or salt cavern (FERC, 2004; Lazard, 2019a).

As with storage, energy transport costs vary significantly across different types of energy carriers, with fuels having lower costs because of their higher energy densities and chemical form. For example, crude oil and oil products, which are the most energy-dense fuels by volume, are transported primarily by oil pipelines and tankers and have the lowest transport costs per km-GJ (Saadi et al., 2018). In comparison, natural gas is transported primarily by pipelines and liquified natural gas tankers and its transport costs per km-GJ are about double that of crude oil per km-GJ, largely because of its lower energy density by volume owing to its gaseous form. The costs of transmitting and distributing electricity are several times those for natural gas because wires are a more expensive and less efficient conduit for energy than natural gas pipelines.

To gain a sense of the value added to energy by energy storage and distribution, consider network charges in the European Union (EU), where regulation of electricity and natural gas transmission and distribution networks promotes their efficient operation and investment. Cost-reflective charges per unit of delivered energy to EU retail customers in 2018 were $17/megawatt-hour (MWh) for electric power and $7/MWh for natural gas.[2] For electricity, this amount includes cost-reflective charges for transmission, distribution and system-balancing services. For natural gas, it covers transmission and distribution costs, with storage remunerated in part by seasonal variation in wholesale gas prices. Natural gas storage builds in summer when demand and prices are low, and falls in winter when demand and prices are high. There are also costs incurred by energy suppliers in managing customer relationships, like meter reading and billing. In comparison, distribution, storage and retailing costs for transport fuels (diesel and petrol) in the EU in 2015 were about $12/MWh, with widely distributed retail sites accounting for much of this cost.[3] The storage and distribution costs of oil products are relatively low.

Overall, timely delivery to final customers costs about two-thirds of the wholesale cost of electricity per MWh, and half the EU import cost of natural gas in 2018. Delivery costs are about 20 per cent of the wholesale cost of refined oil products with a crude oil price of around $50 per barrel. The value of delivering energy when and where needed to customers is thus a significant share of the total value of energy supplied to them, especially for electricity and natural gas. The lower share for oil products reflects their higher wholesale price per unit of energy they provide. This wholesale price premium for oil products flows from their high energy density and ease of storage in transport vehicles that must carry with them their energy stores. For energy to be useful it must be available for use when and where customers demand, whether in stationary applications or transport.

COMFORTABLE, WELL-LIT AND PRODUCTIVE BUILDINGS

For shelter, comfort and productivity, people spend most of their time in buildings, especially in advanced industrialized countries (around 90 per cent), and buildings make up most of the physical capital stock in these countries.[4] Modern buildings in the main are comfortable and well-lit places for household life and work and for producing and consuming services. In advanced industrialized and developing countries (for which data are available), the residential floor space per capita averages are about 50 square metres and 24 m^2, respectively.[5] This space increases with per capita incomes, especially strongly over middle-income levels, and reflects variation in not only the increasing size of residences, but also decreasing occupancy with smaller family sizes. Advanced industrialized countries, with their relatively large share of services in economic activity, also have much of the world's commercial floor space. This commercial space is used primarily for retail, education, healthcare, hotel and restaurant services as well as administrative offices and warehouses.

Energy use is shaped significantly by building size—the space that needs to be heated or cooled to a comfortable and healthy temperature, and lit when natural lighting is insufficient. Space heating, cooling and ventilation as well as lighting account for about half of total energy use in buildings. Much of this is in advanced industrialized countries as well as Russia and China, where there are relatively large building stocks, cool temperate climates and large seasonal demands for space heating. Most commercial buildings and dwellings also need water heating, including for occupant hygiene and other purposes, and this accounts for about 10 per cent of building energy use. Productive activities within buildings associated with work, services and household life make up the balance of energy use.

Thermally Comfortable

The demand for space heating and cooling in buildings reflects preferences for levels of indoor thermal comfort that are similar to outdoor temperatures of East Africa, where humans first evolved (Just et al., 2019). Moreover, if outdoor temperatures become either very hot or cold, the maintenance of comfortable indoor temperatures becomes significant for human health and wellbeing (Fowler et al., 2015; Kenny et al., 2019). But most buildings do not maintain indoor temperatures at desired levels because they are imperfectly insulated and sealed—many older buildings poorly so.

If an indoor temperature is higher than that outdoors, building fabrics tend to conduct heat to the outside and cool the interior, and vice versa. Most heat transfers affecting buildings are through walls, windows, ceilings and floors that make up most of their external surface areas (MacKay, 2008, pp. 50–4; Rez, 2017, pp. 15–41). Buildings typically have walls with air-filled space between the interior surface and exterior cladding, because air is an insulator and creates a moisture barrier. Moreover, this wall cavity in modern buildings is typically filled in part with insulating materials that further reduce heat transfers. Buildings also need ventilation to maintain indoor air quality and control humidity, and this ventilation can affect interior temperatures. As air moves deliberately through opened windows and mechanical ventilation systems or unintentionally through gaps in building fabrics, heat can flow from buildings.

Heating, cooling and ventilation systems for buildings counter these heat flows to maintain comfortable indoor temperatures. The greater the differences between the desired indoor and actual outdoor air temperatures, and poorer the thermal performance of building fabrics, the more work demanded of these systems. Buildings also absorb energy from the Sun as infrared radiation, heating building fabrics and interiors—sometimes desired, other times not. They can be designed to manage heat from solar radiation to help lessen the demand for heating and cooling. This includes passive solar heating in cooler climates—orienting building windows and heat sinks like concrete floors towards to the Sun to absorb solar radiation—and window shadings and light-coloured roofs in warmer climates to reject the Sun's energy.

A boiler is the main technology used to provide space and water heating in buildings, especially where seasonal demand for heating is significant. Most boilers use natural gas, liquified petroleum gas (propane and butane) or fuel oil. Use of coal for heating has been largely phased out because of pollution affecting local air quality, as well as the costs of onsite coal storage and handling. Modern condensing boilers, which recover heat that would otherwise be wasted in flue gases, are 90–95 per cent efficient in converting fuel into useful heat. Non-condensing boilers are 70–80 per cent efficient (UNEP, 2017, p. 25).

Electric heating is the other widely used heating technology. Electric resistance heating is nearly 100 per cent efficient in converting electricity into useful heat. But because conventional thermal generation of electricity is less than 50 per cent efficient in converting thermal fuels into electricity, fuels tend to be used to produce heat directly in buildings rather than indirectly via electricity. Electric heating in advanced industrialized regions tends to be used primarily where low-cost power is available from hydroelectric sources (Canada, Norway, Sweden) or existing nuclear plants (France), or where heating needs are limited (Mediterranean climates). Electric resistance heating systems are less expensive to purchase and install than boilers and hot-water radiators. However, delivered electricity generated from coal or natural gas is more expensive than delivered heating fuels. Electric resistance heating, therefore, tends to be relatively less expensive than fuel boilers and hot-water radiators where demand for space heating and electricity costs are low, and vice versa.

In warmer climates, air conditioning is increasingly used to maintain comfortable indoor temperatures for work and household life. Air conditioners work by forcing a liquid refrigerant to convert into gas inside an interior-facing coil, a physical process (phase transformation) that cools the refrigerant and coils that contain it (Rez, 2017, pp. 29–36). An electric fan blows inside air over the cold coils to cool the air and warm the refrigerant. To repeat this process, an electric compressor then compresses the gaseous refrigerant within outside-facing coils to convert it back into a liquid, a process which heats the refrigerant and the coils that contain it. An outside-facing heat exchanger conducts heat away from the refrigerant. The cooler compressed liquid is then allowed to expand into the inside-facing coils and convert again into a gas, repeating the cycle. New air conditioners are 400–600 per cent efficient in using electricity to pump heat outside buildings, although the average efficiency of all air conditioners currently in use is around 300 per cent (UNEP, 2017, p. 25).

The average lifespan of building fabrics in advanced industrialized and developing countries ranges from 40 to 120 years, while that of heating, ventilation and air-conditioning equipment is 15 to 25 years (IEA and UNEP, 2013, p. 36). Therefore, the overall thermal performance of the building stock is largely determined by wide variation in past building practices and technologies rather than current ones. This is especially the case in advanced industrialized countries with cooler temperate climates, where shares of older buildings with thermally inefficient and poorly sealed building fabrics in the total building stock are large.

The amount of work demanded of heating, cooling and ventilation systems is determined not only by building stock characteristics but also the climate in which buildings are located. A measure of the relationship between climate and demand for thermal comfort is the annual heating and cooling degree days in a given location. This measure is based on an assumed desired indoor temperature, typically in the range of 16–21°C. A heating degree day is the difference between the desired indoor temperature and average daily outside temperature if this difference is positive. For example, on a given day and location, if the desired indoor temperature of 18°C and average outside temperature over the day is 16°C, this day is measured as a 2°C-day; if the outside temperature is 20°C, the heating degree day is nil. This calculation is made for each day of the year and totalled for a measure of annual heating degree days in a given location. A similar measurement is made for cooling degree days. Systematic worldwide measurement of heating and cooling degree days points to significant demands for space heating in temperate climates with significant seasonal temperature variation—especially Eurasia, Europe, Japan, North America, Northern China and South Korea—and space cooling demands in warmer climates (Mourshed, 2016; Atalla et al., 2018).

But the demand for heating and cooling in residential and commercial buildings ultimately depends on human behaviour—actual rather than assumed demands for thermal comfort—and effective control of heating and cooling systems to align these demands with system operation. Thermostats and radiator valves serve to regulate these systems so that they provide the desired room temperatures when occupied. Digital technologies also help to control heating, cooling and ventilation systems by zone or room and time of day to help align system operation with actual demands for thermal comfort. Available evidence on household heating suggests that zonal controls that heat individual spaces to different temperatures at different times can save energy compared with simpler manual controls—provided that they are appropriately commissioned (Lomas et al., 2018). But evidence also suggests that such systems can be difficult to use effectively.

Well-Lit Buildings

Interior light in buildings is fundamental to their productivity, and natural sunlight through windows is its best source—it is free and supports human health. The amount of light is measured in lumens, and on a clear day the amount of diffuse daylight is about 11,000 lumens per m² at the Earth's surface (NOAO, 2015). In a building, this light in an area close to a window might reduce to about 1000 lumens per m², and in the interior of the room it might fall to 50 lumens per m². In a household, the amount of light needed for normal household activities ranges from 100–300 lumens per m². In commercial buildings, the light needed for many activities is higher—in the range of 300–500 lumens per m², and for some specialized activities, such as in hospital operating theatre, it can be 1000 lumens per m² or more.

Electrical lighting provides light when sunlight is insufficient, and recent advances in lighting technology have substantially improved its quality and energy efficiency. Current light-emitting diode (LED) bulbs produce about 120 lumens per watt (W) of electricity.[6] The lumens produced by a lightbulb is the amount of light measured at its source, so 3000 lumens from bare lightbulbs would be needed to illuminate a 12 m² room with 250 lumens of light. This is the amount of light produced by two 100 W incandescent lightbulbs, a 23 W compact fluorescent bulb (CFL) and a 13 W LED bulb. Because of their longer lifespan and higher energy efficiency, LED bulbs provide more cost-effective lighting in most contexts than CFL bulbs, even though LED bulbs are more expensive to buy. Traditional incandescent lightbulbs are largely obsolete.

As with heating, actual demand for electric lighting in dwellings and commercial buildings depends on human behaviour. A manual electrical switch typically controls lighting, though people are more likely to turn on lights when entering a dark room than turn them off when leaving a well-lit one. Digital controls linked to motion detectors or lighting schedules help to align better lighting with demand.

Productive Dwellings and Commercial Buildings

Household life consists of many energy-using activities that contribute to the health and wellbeing of occupants, and the one that uses the most energy is heating water for hygiene and cooking. In dwellings with boilers, they typically provide both water and space heating. The main alternative is an electric immersion heater in a hot-water storage tank. Where incomes allow, households use electric refrigerators and freezers for storing food, and natural gas or electric stoves and ovens for cooking. There are of course a wide range of electrical kitchen appliances, including microwave ovens, electric kettles, and so forth. Households use electric dishwashers, washing machines, dryers and vacuum cleaners for cleaning, and televisions, telephones and computers for entertainment and communication as well as study and work. The extent of household energy use for these purposes depends on access to electric power, especially in South Asia and Africa, where access is not yet universal; per capita incomes; and the number of household occupants (IEA, 2017a; IEA 2017b, pp. 39–44).

In some developing countries, especially in South Asia and Africa, the main energy-using appliance in dwellings is a stove, which is used for cooking, heating water and, if needed, space heating. The main fuel source for a traditional stove is traditional or commercial biomass: wood, charcoal and dung. The efficiency of these stoves is typically poor, with only about 20 per cent of the heat produced from burning the fuel directed to cooking and heating water (IEA, 2017b, pp. 57–63). Moreover, these stoves are a significant source of indoor air pollution and a cause of respiratory illness. In contrast, modern liquified petroleum gas, natural gas or electric resistance stoves are about 50 per cent efficient in directing heat for cooking (and electric induction stoves about 70 per cent efficient).

Energy use for productive activities in commercial buildings reflects the range of economic activities undertaken within them. As for dwellings, it includes water heating to support occupant hygiene and other activities. It also includes electrical equipment in commercial offices and educational buildings, like computing, audio-visual and printing equipment for analysing and communicating information. Energy-intensive activities in hospitals and other medical buildings include operating and imaging, as well as laboratory analysis and sterilization processes (Morgenstern et al., 2016). While data on actual equipment energy use in commercial buildings is limited, it can nevertheless be expected that, for example, hospitals use more energy per m^2, including for thermal comfort to promote patient health, than do commercial offices.

Adding Up Building Energy Uses and Supplies

Energy use in buildings varies significantly with per capita income, building size (m^2 of floor space) and occupants per dwelling, as well as with regional variations in climates. Among advanced industrialized countries and Russia, total energy use in dwellings and commercial buildings ranges from 60–47 GJ per capita (GJ/pop) in the United States/Canada and Russia to 30 GJ/pop and Australia/New Zealand and Japan/South Korea (Table 2.1a). Total per capita building energy use in Europe is 37 GJ/pop. Key factors behind this variation are extensive use of dwelling and commercial building floor space, equipment and appliances in North America, and high space-heating demand in Russia per m^2 of floor space due to its cold climate. Energy use in the Asia-Pacific countries is lower because of warmer climates and, in Japan and South Korea, intensive use of building floor space. Europe too uses building floor space relatively intensively, but has relatively high space-heating demands, reflecting its cooler climate and older building stock.

Electricity and natural gas networks provide much of the energy supplies to buildings in these countries. Natural gas distribution networks have their most extensive coverage in North America, Europe and Russia, reflecting in part available domestic natural gas resources. Where available, natural gas tends to be used for space and water heating and cooking in buildings, although in the Nordic countries and Russia many buildings are served by district heating networks for their space and water heating. The heat source for these networks is typically heat from thermal generation of electricity and industrial processes that would otherwise be wasted. For buildings that are not served by natural gas or district heating networks, oil products (fuel oil and liquified petroleum gas) and biomass are the main energy supplies for space and water heating. Also, coal still sees limited use in buildings in Europe and Russia. All buildings in these countries are served by electricity distribution networks to meet

Table 2.1a   *Activities in buildings and energy per capita (energy GJ/pop unless noted otherwise; other indicators as noted, 2017)*

Advanced industrialized countries and Russia	United States/ Canada	Australia/ New Zealand	Europe	Japan/ South Korea	Russia
Activity indicators for buildings					
Per capita income (2017 Int'l $)	58,759	47,558	42,788	40,177	26,079
Population (millions)	361.5	29.4	446.1	178.2	144.5
Occupants per dwelling (number)	2.7	2.9	2.3	2.5	2.2
Dwelling floor space per capita (m²)	68	55	40	37	26
Dwellings with access to electricity (%)	100	100	100	100	100
Services share in total value added (%)	76	67	65	64	57
Commercial floor space per capita (m²)	25	19	15	20	6
Heating degree days (°C-day)	2496	763	2624	1922	4676
Energy use in dwellings					
Thermal comfort and lighting (GJ per m²)	0.27	0.14	0.39	0.16	…
Hot water, cooking and appliances (GJ per dwelling)	43	27	19	27	…
Total dwelling use (GJ/pop)	33	17	24	16	36
Energy use in commercial buildings					
Thermal comfort and lighting (GJ per m²)	0.42	…	0.51	0.33	…
Hot water, equipment and appliances (GJ/ pop)	16	…	5	8	…
Total commercial use (GJ/pop)	27	13	13	14	11
Total building energy use	**60**	**30**	**37**	**30**	**47**
Energy supplies to buildings					
Networks					
Electricity	30	18	12	13	8
Natural gas	23	7	13	7	16
Heat	0	0	3	1	18
Decentralized					
Oil products	4	2	4	6	3
Biomass	2	2	3	<1	1
Coal	0	0	1	<1	1

Notes: Data are for 2017 or most recent year available, except heating degree days, which are averages of the most recent five years. Figures for country groups are weighted averages by country populations, allowing for any missing observations. Missing observations are denoted by an ellipsis.

Sources: Data for activity indicators for buildings and their energy uses and supplies are from the International Energy Agency (IEA) Energy Efficiency Indicators Database 2020 (June), extended version. Supplemental data for activity indicators are from IEA (2013) and Atalla et al. (2018), and for energy uses and supplies from IEA Data & Statistics, retrieved from www.iea .org/data-and-statistics/data-tables. Data for per capita incomes, share of services in total value added, and share of dwelling with access to electricity are from the World Bank, retrieved from https://data.worldbank.org/indicator.

demands for lighting, space cooling, ventilation, services from appliances and equipment and, in some countries and regions, for space and water heating.

Among selected developing countries and regions (Brazil, China, India and Mexico as well as Africa), activities in buildings and related energy-service demands are much less energy-intensive than in advanced industrialized countries and Russia. The main derived demands for energy in dwellings are for hot water and cooking. These demands are particularly high in Africa because of the poor energy efficiency of traditional stoves (Table 2.1b). There is little energy demand for thermal comfort in residences because relatively warm climates provide much of this comfort, although demand for space cooling can be expected to rise along with per capita incomes in these countries. However, China has both an extensively developed housing stock, with floor space per capita on a par with Japan and South Korea, and significant seasonal heating demands in northern China. Commercial building floor space per capita in these developing countries remains underdeveloped, although that in Mexico and China is approaching two-thirds of the European level. Nevertheless, the provision of services in commercial buildings remains much less energy-intensive on a per capita basis than in Europe.

Electricity is the main modern energy supplied to buildings in developing countries, especially in urban areas where pre-capita incomes and energy demands are relatively high and centralized distribution networks cost-effective. However, coverage of electricity networks remains incomplete, especially in sub-Saharan Africa. With rising incomes, electricity-using appliances become more affordable, which in turn supports expected electricity demand and grid expansions. However, extending centralized grids to some rural areas remains relatively costly because of low demand loads. The falling cost of renewable power, especially solar photovoltaics (PV), makes decentralized electricity generation and micro-grids an increasingly viable option, particularly in sub-Saharan Africa and South Asia. Traditional biomass like wood and charcoal remains an important fuel source for developing country households, especially in sub-Saharan Africa and South Asia. Where incomes allow and better technologies become available, modern stoves using cleaner cooking fuels such as liquified petroleum gas and biogas offer households the potential to maintain and develop their household activities, while reducing the harm to health from traditional stoves, biomass use and indoor air pollution.

Table 2.1b   Activities in buildings and energy per capita (energy in GJ/ pop; other indicators as noted, 2017)

Selected developing countries	Mexico	Brazil	China	India	Africa
Activity indicators for buildings					
Per capita income (2017 Int'l $)	19,795	14,520	14,302	6185	4231
Population (millions)	124.8	207.8	1386	1338	1243
Occupants per residence (no.)	3.7	3.3	3.3	4.5	...
Share of dwellings with electricity (%)	100	100	100	93	53
Dwelling floor space per capita (m^2)	20	22	36	12	...
Services share in value added (%)	60	63	62	49	52
Commercial floor space per capita (m^2)	9	2	9	1	...
Heating degree days (°C-day)	710	118	2260	280	...
Energy use in buildings					
Total use in dwellings	6	5	10	5	11
Total use in commercial buildings	1	2	3	1	1
Total building energy use	7	7	13	6	12
Energy supplies to buildings					
Networks					
Electricity	3	5	4	1	1
Natural gas	<1	0	1	0	0
Heat	0	0	1	0	0
Decentralised					
Oil products	2	1	2	1	1
Biomass	2	1	2	4	10
Coal	0	0	2	<1	<1

Notes: As in Table 2.1a.
Sources: As in Table 2.1a.

EASE OF TRANSPORT

Transport is central to modern societies and economies. Efficient transport of goods and people both expands the geographical scope of markets and promotes economic specialization and scale in production among firms and regions. This specialization and scaling of economic activity and its clustering around transport infrastructure support the high productivity of modern economies. This influence of transport costs on the location of modern economic activities was seen originally in the siting of factories and towns during the first industrial revolution, when factories and towns that supported them were located close to natural resources and waterways for transporting raw materials and finished goods (Mokyr, 2009, pp. 209–12). Then ships were the most efficient transport mode and ground transport was rudimentary, but railroads transformed transport as industrialization advanced and reshaped the location of cities and industrial activity (Krugman, 1991, pp. 23–33).

In modern economies, the effects of transport costs, specialization, scale economies and clustering of economic activities are seen in urban agglomerations that bring together spatially markets for labour and products, and facilitate the spread of knowledge. They are supported by extensive road networks that enable flexible travel routes and vehicle operator choice as well as rail networks that provide low-cost mass passenger and freight transport. Road networks make possible a wider spatial distribution of economic activity than water, rail or air transport alone. But their integration in dense multi-modal transport networks helps intensify agglomeration benefits, even as freight and passenger transport costs fall. These benefits are reflected in the relatively high productivity of urban areas, even after allowing for the tendency of cities to attract more productive employees and firms. (Krugman, 1991, pp. 59–67; Glaeser and Maré, 2001; Combes et al., 2008; Combes et al., 2012).

Passenger Transport

Passenger transport demands reflect all motivations for travel, including (1) work-related, (2) household shopping, education, health care and other errands, and (3) leisure and other activities. In the United States, these three reasons for travel account for 22, 33 and 45 per cent of total annual passenger-km of households, respectively (McGuckin and Fucci, 2018, p. 16). Other advanced industrialized countries show similar patterns in the reasons for household travel (Schäfer, 2000). Moreover, an important characteristic of overall demand for passenger transport services is the relative invariance of the average amount of time that people spend travelling to variations across countries in average per capita income levels. There are on average across country populations relatively fixed time budgets for travel of one to two hours per day—about 5 per cent of total household time budgets (Schäfer et al., 2009, pp. 29–35). This average is broadly constant at an aggregate level across advanced industrialized and developing countries, but at disaggregated levels there is considerable variation. The underlying reasons for this apparent constancy at an aggregate level are not well understood, but it could reflect increasing disutility of spending more time travelling, with varying 'annoyance' levels across people (Mokhtarian and Chen, 2004). Greater passenger travel distances in advanced industrialized countries are largely enabled by affordable access to private automobiles and their convenience, as well as rapid air and rail transport (Dargay et al., 2007). In developing countries, people spend on average the same amount of time travelling as those in advanced industrialized countries but go much less distance.

In advanced industrialized countries, average annual per capita domestic passenger travel across all domestic transport modes is about 17,800 passenger-km per year.[7] Domestic passenger transport demand is higher than this average in North America as well as Australia and New Zealand, with their widespread access to private cars, low population densities and large land areas (Fulton et al., 2009). Passenger transport demand is lower than this average in Europe as well as Japan and South Korea, with their less widespread private automobile ownership and higher population densities, as well as access to extensive passenger rail networks. Among selected developing countries, average annual per capita domestic passenger travel across all transport modes is about 6000 passenger-km per year. Among these countries, passenger-km travelled varies with per capita income levels and the extent of affordable access to a private motorized road transport, including two- and three-wheeled vehicles and buses, which account for significant shares of total passenger-km travelled. Rail and air travel are the other main domestic modes of passenger travel in advanced industrialized and developing countries.

The amount of energy use for passenger transport reflects not only distance and mode of travel but also vehicle efficiency and occupancy, which are shaped by several factors. Three determine vehicle efficiency: that of its engine or motor in converting fuel or electricity into motion, vehicle mass (resistance to acceleration) and resistance encountered to motion from air and rolling surface or water.

Internal combustion, turbofan (jet) and turboprop engines are the most widely used prime movers in transport. Gasoline engines are about 35 per cent efficient at their optimal rotating speeds in converting heat from fuel combustion into rotary motion, and diesel engines about 40 per cent efficient (Rez, 2017, pp. 133–41). Internal combustion engines achieve peak efficiency within narrow ranges of rotating speeds. In stop-and-go traffic, these engines operate relatively inefficiently despite being paired to transmissions with a range of gearings to help manage optimal engine speed. Large marine diesel engines achieve peak efficiencies of up to 50 per cent and can operate near their peak on open water (Rez, 2017, p. 179). The thermal efficiency of turbofan or turboprop engines in airplanes is in the range of 60–80 per cent at cruising speeds, but much lower during take-offs (Rez, 2017, p. 171). Electric motors, used primarily in electric railroad locomotives and increasingly in automobiles, are around 90 per cent efficient in converting electricity into motion and they can maintain their efficiency over a wide range of rotating speeds (Rez, 2017, pp. 142–4). However, the charging and discharging of batteries in battery electric vehicles entails some efficiency losses (15 per cent;

MacKay, 2008, p. 261). The electricity distribution via catenary wires or third rails for trains also carries some efficiency losses (5–10 per cent).

For road vehicles, acceleration resistance from vehicle mass and resistance to motion from the rolling surface and air are the other factors that determine their overall energy efficiency. Road vehicle size is a key determinant of its acceleration, rolling and air resistance, with large sports-utility vehicles having more mass and less aerodynamic shapes than smaller vehicles. Two- and three-wheeled vehicles are significantly lighter than automobiles but provide less passenger safety and comfort. In motorway driving at speed, air resistance is much greater than rolling resistance and most energy is being used to displace air, while in 'stop-and-go' city driving most energy is used to overcome acceleration and rolling resistance (Rez, 2017, pp. 127–33).

For trains, the rolling resistance of a steel wheel on a steel rail is about one-third that of a tire on a paved road surface and, while they have a much larger mass than cars, trains operate on controlled rail networks that manage train journeys and speeds (Rez, 2017, p. 149). The small frontal area of a long, thin train also helps to minimize air resistance at speed. Electric trains are by far the more efficient mode of ground transportation, albeit less convenient than private automobiles in some contexts.

The resistance to motion encountered by ships and airplanes is similar to that of surface transport vehicles. At cruising speeds, ships and airplanes use most of their energy to overcome water and air resistance. For ships, this resistance is slightly greater than that for trains, and for airplanes less (Rez, 2017, pp. 171 and 182).

Private automobile occupancy is determined by preferences, especially for time and convenience. The average number of passengers in an automobile in advanced industrialized countries is 1.4 persons, where there is a revealed high willingness to pay for the convenience of private road transport.[8] This includes leaving automobiles unused most of the time and operating them when used at much less than full capacity on average. Public transport vehicles typically operate commercially with an incentive to optimize occupancy through choices of routes and schedules.

Given average occupancy rates and current vehicle fleet efficiencies, passenger trains and buses are the most efficient surface transport modes, using about 0.3 MJ to 0.7 MJ per passenger-km.[9] Two- and three-wheeled vehicles also use about 0.7 MJ per passenger-km. The European automobile fleet averages about 1.6 MJ per passenger-km, and that in Japan and South Korea 1.9 MJ per passenger-km. The North America fleet averages about 2.1 MJ per passenger-km, reflecting a larger share of large vehicles.

Goods Transport

Demand for freight transport is derived from the production of goods and the location of their customers. This demand is shaped by the degree of specialization within supply chains and locations of primary commodity and materials production and component manufactures. The main types of freight are bulk materials (ores, minerals, fossil fuels, bulk chemicals and agricultural commodities), general merchandise (food and clothing) and specialized products (automobiles, refined fuels and chemicals). Small package delivery also helps to complete the journey of goods from producers to customers.

Freight transport modes are adapted to different freight types. Bulk commodities are transported in vast quantities, so low transport costs are key to choices of transport mode. They tend to be transported by ships, trains and heavy-duty trucks, and for some commodities pipelines. Manufactured goods tend to be higher value and more diverse than bulk commodities. For these goods, logistics costs can be as important as transport costs. Rail freight and trucks are key transport modes for them, along with intermodal facilities. Specialized products share some characteristics with bulk commodities, but tend to require specialized ship, rail and truck transport equipment and air freight to reduce risks in transport. Their relatively high value tends to require more specialized transport services. For package delivery services, logistics costs can be more important than transport costs. Air freight and trucks, including medium- and light-duty vehicles, are the main transport modes for packages.

In advanced industrialized and developing countries, total freight services per capita average about 38,000 and 10,000 tonne-km per year, respectively.[10] This difference reflects primarily higher per capita incomes in developed regions, with their greater consumption of goods and investment in physical capital. It also reflects comparative advantages in production. Trucks provide about 70 per cent of domestic freight transport in Europe and about 60 per cent in North America and Australia. The latter reflects their production of bulk commodities and inland transport by rail. In developing countries and regions, freight transport is largely provided by trucks; however, China and Russia have extensive rail networks that handle more than half of their domestic freight transport.

The main factors that influence vehicle efficiency in transporting goods are similar to those for transporting passengers: engine efficiency, vehicle and payload weight and resistance to motion. The primary drivers and engine efficiencies of trucks, freight trains and ships are similar to those for passenger vehicle equivalents, as are the sources of resistance to motion. On averages, ships and trains use about 0.1 MJ to 0.2 MJ per tonne-km.[11] Heavy-duty trucks require about 1.5 MJ to 2.5 MJ per tonne-km, while light-duty delivery vehicles use about 17 MJ per tonne-km because their payloads are light relative to vehicle weight.

Adding Up Energy Uses in Domestic Transport

The demand for passenger transport increases significantly with per capita income and inversely with population density. In countries with higher average per capita income there is a tendency towards greater automobile ownership and more intensive use of them to meet most passenger travel demands (Table 2.2a). This is especially so in the United States, Canada, Australia and to a lesser extent Russia, with their relatively low population densities and large land areas. Air travel is the second most intensively used mode of domestic passenger transport in North America, Australia and New Zealand given their land size and low population densities, while in Europe, Japan and South Korea rail networks are used relatively intensively. In advanced industrialized countries and Russia, average annual passenger-km per capita is mostly in the range of 11,000–16,000; however, passenger transport demand in North America is about double that of other industrialized countries. Average annual per capita energy use for passenger transport in advanced industrialized countries ranges from 57 GJ/pop in North America to 17 GJ/pop in Europe and 18 GJ/pop in Japan and South Korea. Australia and New Zealand lie near the mid-point of this range.

Demand for freight transport is driven by specializations in goods production as well as customer locations. Among advanced industrialized counties and Russia, the United States, Canada, Australia and Russia specialize in primary commodity production (Hausmann et al., 2014). This activity gives rise to significant domestic bulk transport demands, and they are served primarily by rail in these countries, as well as domestic shipping in North America and Australia (Table 2.2a). However, trucks use the most energy per tonne-km and account for most energy use for freight transport, providing services for a range of commodities and goods. Total annual energy use for freight transport per capita is just over 20 GJ/pop in the United States and Canada as well as Australia and New Zealand, and about half that in Europe, Japan and South Korea.

In selected developing countries and regions, average annual passenger-kms travelled are significantly less than those in advanced industrialized countries— in Latin America and China about two-thirds the level in Europe and India about one-third (Table 2.2b). Also, automobiles provide a smaller share of this transport services, especially in countries with lower average per capita incomes. Domestic freight transport services in Latin America are at similar per capita levels as in Europe; however, China stands out for its relatively intensive use of domestic freight transport in its production and supply activities,

Table 2.2a   Domestic transport activities and energy per capita (energy in GJ/pop; other indicators as noted, 2017)

Advanced industrialized countries and Russia	United States/ Canada	Australia/ New Zealand	Europe	Japan/ South Korea	Russia
Activity indicators for transport					
Population density (pop/km²)	33	5	105	449	9
Automobile ownership per 100 population	75	59	52	52	32
Per capita passenger-km – road	24,815	12,815	10,619	7214	12,017e
Per capita passenger-km – air	3668	1652	202	600	650
Per capita passenger-km – rail	179	602	1174	2917	852
Total per capita passenger-km	28,662	15,069	11,995	10,731	13,519e
Per capita tonne-km – road	9268	8236	4304	1828	1751
Per capita tonne-km – rail	6347	14,650	819	168	17,255
Per capita tonne-km – ship	2654	3798	558	1273	465
Total per capita tonne-km	18,269	26,684	5681	3269	19,471
Energy use for passenger transport					
Per capita energy use – road	49	26	16	16	…
Per capita energy use – air	7	5	<1	1	…
Per capita energy use – rail	<1	<1	<1	<1	…
Total energy use per capita	57	31	17	18	…
Energy use for freight transport					
Per capita energy use – road	18	19	9	5	…
Per capita energy use – rail	2	2	<1	<1	…
Per capita energy use – ship	<1	1	<1	1	…
Total energy use per capita	21	22	9	6	…
Total transport energy use	78	53	26	24	27
Total transport energy supply					
Oil products	72	53	24	24	17
Biofuels	5	0	1	0	0
Natural gas	<1	<1	<1	0	8
Electricity	<1	<1	1	<1	2

Notes: Data are for 2017 or most recent year available. Estimated road passenger-km (e) are based on actual and estimated vehicle stocks and parameters for passenger loads and vehicle-km per year from the IEA Mobility Model. Figures for country groups are weighted averages by country populations, allowing for missing observations. Missing observations are denoted by an ellipsis. Data include only domestic transport activities, including in countries in Europe (that is, intra-EU transport is not included). Energy-related data on international aviation and shipping are reported separately and not attributed to individual countries, including intra-EU transport.

Sources: Data for transport activity indicators and energy uses and supplies for transport are from the IEA Energy Efficiency Indicators Database 2020 (June), extended version. Supplemental data sources for activity indicators are ICAO (2017, Presentation of Statistical Results), the IEA Mobility Model Database, ITF (2019, Statistical Annex), and OICA vehicles-in-use statistics, retrieved from www.oica.net/category/vehicles-in-use. Supplemental data for energy uses and supplies for transport are from IEA Data & Statistics, retrieved from www.iea.org/data-and-statistics/data-tables. Data for country per capita incomes and population densities are from the World Bank, retrieved from https://data.worldbank.org/indicator.

reflecting in part its economic specializations in materials production and manufactured goods. Average annual energy use for freight and passenger transport ranges among these selected developing countries ranges from 17 GJ/pop in Latin America to less than 5 GJ/pop in India and Africa. China's transport energy use per capita lies near the midpoint of this range.

Transport fuels are mostly oil products—automotive and aviation gasoline, kerosene-type jet fuel, gas oil (diesel) and fuel oil for ships. They are produced by refineries that distil and chemically treat crude oil to yield a mix of lighter liquids and heavier residues. The liquid products from this distillation process are, from lightest to heaviest: liquified petroleum gases, naphtha (a chemical

Table 2.2b   *Domestic transport activities and energy per capita (energy in GJ/pop; other indicators as noted, 2017)*

Selected developing countries	Mexico	Brazil	China	India	Africa
Activity indicators for transport					
Population density (pop/km^2)	64	25	148	450	53
Automobile ownership per 100 population	24	17	13	2	3
Per capita passenger-km – road	8390e	8394	6190e	2548e	...
Per capita passenger-km – air	288	621	646	83	...
Per capita passenger-km – rail	12	179	961	868	...
Total per capita passenger-km	8690e	9194	7797e	3499e	...
Per capita tonne-km – road	2312	3662	4543	1821	...
Per capita tonne-km – rail	698	1807	1717	489	...
Per capita tonne-km – ship	325	1091	7515	3	...
Total per capita tonne-km	3335	6560	13,775	2313	...
Total transport energy use	**17**	**17**	**9**	**3**	**4**
Total transport energy supply					
Oil products	17	13	8	3	4
Biofuels	0	3	0	0	0
Natural gas	0	1	1	<1	0
Electricity	0	0	<1	0	0

Notes: As in Table 2.2a.
Sources: As in Table 2.2a.

feedstock and gasoline additive), gasoline, jet fuel, kerosene, gas oil, lubricating oil, fuel oil, greases and waxes. Solid residues include petroleum coke and bitumen. While oil products provide most energy for transport, biofuels, natural gas and increasingly electricity fulfil some of these energy demands. Biofuels like ethanol and biodiesel are typically more expensive than comparable oil products per unit of energy. Mandates for their use are motivated by decarbonization, domestic resource availability and energy-security objectives. Mandates in Brazil, China, the EU and the United States for blending biofuels with suitable oil products account for most of world biofuel demand. In addition, natural gas is used in vehicles with suitable engines and fuel storage tanks, primarily in China, Russia, Europe and North America, because the fuel burns more cleanly than diesel, helping to improve urban air quality, and costs less than diesel.

International Shipping and Aviation

International shipping accounts for the bulk of freight transport in terms of tonnes-km, and this reflects shipments of bulk commodities like iron ore and steel, bauxite and alumina, coal, fertilizers, grains, and wood. These commodities account for about 60 per cent of ocean freight (UNCTAD, 2019, pp. 4–6). Oil, natural gas and chemicals account for about 30 per cent, and manufactured goods the balance. The most intensively travelled international shipping routes are between the Asia-Pacific region on the one hand and Europe and North America on the other.

International air passenger travel accounts for two-thirds of total air passenger travel—both domestic and international (ICAO, 2017). The most intensively travelled international routes are within Asia-Pacific, North America and Europe and between these regions. Taken together international aviation and shipping account for about 4 per cent of the total world final energy demand (IEA, 2019a, pp. 40–42).

USEFUL MATERIALS FOR BUILDINGS, INFRASTRUCTURES AND GOODS

Building and maintaining buildings and infrastructures as well as producing equipment, vehicles, appliances and some other consumer goods use materials and energy intensively. Among the most widely used materials by mass are cement and concrete, iron and steel, wood, chemicals and plastics, glass and aluminium (Allwood and Cullen, 2015, pp. 11–26). These materials have a range of useful properties: concrete's adaptability and strength (when combined with steel), steel's rigidity and strength, wood's structural strength and renewability, plastics' adaptability and light weight, glass's transparency and aluminium's strength and light weight. The focus here, though, is on cement and concrete, iron and steel, and chemicals and plastics because they use the most energy to produce and emit significant amounts of carbon dioxide from their industrial processes.[12] These processes transform raw materials like metal ores, limestone and fossil hydrocarbon resources into these useful materials. Some processes also use recycled materials from goods at the end of their product lives, and they reduce substantially the amount of energy and emissions from producing new useful materials.

Cement and Concrete

Cement is a widely used building material and, when combined with aggregates like sand and crushed stone, is formed into concrete shapes. Concrete is a ceramic, stone-like material that is easily shaped and typically strengthened with steel reinforcing bars. This combination of steel and concrete is a universal building material. In advanced industrialized countries and regions, the stock of cement estimated to be in buildings and infrastructures appears to have plateaued in the range of 15 tonnes per capita (t/pop) in North America, 20 t/pop in Japan and 25 t/pop in Europe in 2014 (Cao et al., 2017). Differences in levels reflect in part different choices in building materials, with those in North America using more wood than in Europe. There is evidence that per capita cement stocks have begun to decline in Japan, Sweden and the United Kingdom as material efficiency improves. Among developing countries, China, the Middle East and Russia have relatively high and rising per capita stocks of cement, in the range of 10–20 t/pop, which reflects a focus on heavy industry in development. In other developing countries, the stock of cement is estimated to be in the range of 2–5 t/pop. The world average per capita stock of cement is around 8 t. Total world cement production is 4.1 billion tonnes per year with an average expected life across all uses of 50 years (Krausmann et al., 2017; USGS, 2019, p. 43). About two-thirds of current production is in China and India, as these large developing countries expand their building stocks and infrastructures.

The most energy- and emissions-intensive stage of the production process for Portland cement—the most widely used type—is heating a ground mixture of limestone (calcium carbonate), sand (silica) and clay in a rotating kiln to a very high temperature (1800–2000°C) (Rez, 2017, pp. 198–200). This process causes a chemical reaction that creates pellets of calcium silicates called clinker. The heat drives out carbon from the limestone, leaving quicklime, which interacts with the sand to yield the pellets. This chemical reaction emits substantial amounts of carbon dioxide. Total thermal fuels used in clinker production averages 3.2 GJ/t (IEA, 2020a). While most energy in producing cement is for heating kilns, some is also used to prepare raw materials and in grinding clinker into cement. Recycling of concrete at the end of its use into aggregates for road and building construction helps avoid quarrying for aggregates and using landfills for building waste, but recycling concrete to recover cement is not practical because of the energy this would require.

Iron and Steel

Steel makes up much of the physical capital stock—structural steel in buildings, factories and infrastructure and steel components in vehicles, equipment and appliances. About three-quarters of the steel stock is in buildings and infrastructures (Pauliuk et al., 2013; Krausmann et al., 2017). In advanced countries with long industrial histories, like Germany, the United Kingdom and the United States, in-use per capita steel stocks are estimated to be in the range of 11–16 t/pop (in 2010). There is also evidence that per capita stocks plateau in that range. In recently industrialized countries like South Korea, steel stocks are around 6 t/pop. In developing countries, estimated steel stocks are in the range of 6–10 t/pop in Russia and the Middle East, 3–6 t/pop in China and less than 1 t/pop in India and Africa. The world average per capita steel stock is about 4 t. Total current world steel production is 1.7 billion tonnes with an average expected life across all uses of 44 years (Krausmann et al., 2017; World Steel Association, 2019). Again, much of this current apparent use of new steel production is in developing countries.

   The most energy-intensive part of the steel production process is smelting iron ore in blast furnaces fuelled with coke, which is a carbon-rich fuel made from coal. These furnaces serve to remove oxygen and other impurities in iron ore (iron oxide) through high temperatures (up to 1800°C—the melting point of iron) and chemical reactions. In particular, carbon monoxide from burning coke interacts with oxygen in the iron ore to reduce it to molten pig iron and impurities (slag) that can be easily separated. Waste gases include carbon dioxide from this chemical reaction. The pig iron is further processed in oxygen furnaces to remove some of the remaining carbon in the iron to produce steel. This two-step process for producing crude steel accounts for about 70 per cent of total world steel production and uses around 25 GJ/t of energy, mostly coal and electricity (World Steel Association, 2019; IEA, 2020b). The other main method for producing crude steel uses recycled material and electric arc furnaces, and this process requires around 10 GJ/t of energy. The world average recycling rate for steel at end-use is 80 per cent. However, this recycled material accounts for only about 30 per cent of new steel production because of the small size of the in-use stock at the end of its product life relative to current steel demands (Krausmann et al., 2017).

Chemicals and Plastics

Chemicals and plastics have a wide range of uses. Some, like plastics, are for durable uses while others are for production inputs and consumption. The industry produces four main product groups that account for much of its output and energy use (Levi and Cullen, 2018). They include olefins such as ethylene and propylene, which are widely used in plastics and are made from petroleum and natural gas feedstocks. The estimated world in-use per capita stock of plastics is about 0.4 t/pop, with an average expected lifetime across all uses of six years (Krausmann et al., 2017; Wiedenhofer et al., 2019). The other chemical products are largely used for production inputs and consumption. Ammonia is widely used to make fertilizers such as urea for agricultural production. Ammonia is made primarily from natural gas, but also from coal. Aromatics and methanol are produced from petroleum and natural gas feedstocks and are used as fuel additives and in consumer products. The industrial processes for producing plastics and various types of chemicals are too varied to set out in detail here, but descriptions of these processes, their feedstock and energy requirements are readily available (see, for example, Levi and Cullen, 2018).

Global annual production of plastics and chemicals is more readily measured in terms of value added rather than physical quantities, and total production amounts to about 1 per cent of world GDP (Oxford Economics, 2019). The sector accounts for 15 per cent of total world oil demand and 9 per cent of natural gas demand for use as both feedstocks and fuels (IEA, 2020c). China, Europe, Japan, North America and South Korea specialize in production of plastics and chemicals, accounting for about 90 per cent of the industry's total value added. These countries and regions also account for much of the apparent consumption of plastics and chemicals.

Adding Up the Energy in Useful New Materials

Access to and the cost of primary commodities and energy resources, as well as other productive capabilities like technologies, shape the pattern of economic specialization across countries in those goods (and services) that are traded in international markets. For example, among the advanced industrialized countries, Russia, and selected developing countries, Australia and Russia specialize in producing materials from primary resources and use energy intensively in per capita terms for their production—around 45 GJ/pop primarily from natural gas, coal and electric power (Table 2.3a and b). Russia also uses a significant amount of waste heat from other industrial processes such as electricity generation. China, Japan and South Korea also specialize in producing primary materials, but they also produce higher value-added products like chemicals, vehicles, equipment and appliances that use these

Table 2.3a   Materials production and energy per capita (energy in GJ/
pop; other indicators as noted, 2017)

Advanced industrialized countries and Russia	United States/ Canada	Australia/ New Zealand	Europe	Japan/ South Korea	Russia
Activity indicators for industry					
Cement production per capita (t/pop)	0.3	0.3	0.3	0.7	0.4
Iron and steel production per capita (t/pop)	0.3	0.2	0.3	1.0	0.5
Chemicals share in total value added (%)	2	1	2	3	…
Energy use for materials production					
Cement and other non-metallic minerals	2	3	3	3	…
Iron and steel	4	5	5	16	…
Chemicals and plastics (including feedstocks)	10	7	5	7	…
Other materials					
Wood and paper	6	3	3	3	…
Copper and other base metals	5	16	6	17	…
Aluminium and other nonferrous metals	1	5	1	1	…
Total energy use for materials	**28**	**39**	**23**	**47**	**43**
Energy supply for producing materials					
Natural gas	12	15	5	4	11
Coal	4	14	7	28	7
Biomass and waste	3	2	2	1	<1
Oil products	1	3	1	3	5
Heat	1	0	1	1	12
Electricity	7	12	5	10	8

Notes: Data are for 2017 or most recent year available. Figures do not include fossil hydrocarbon resources used as material feedstocks. Figures for country groups are weighted averages by country populations, allowing for missing observations. Missing observations are denoted by an ellipsis.
Sources: Data for industrial activity indicators and energy uses and supplies for industry are from the IEA Energy Efficiency Indicators Database 2020 (June), extended version. Supplemental data sources for activity indicators are Oxford Economics (2019), USGS (2019) and World Steel Association (2019). The data for country per capita incomes are from the World Bank, retrieved from https://data.worldbank.org/indicator.

materials intensively (Hausmann et al., 2014). Their per capita energy use for materials production ranges from 30 GJ/pop in China to 47 GJ/pop in Japan and South Korea. Coal and electricity are the main sources of energy for production processes. North America and Europe are less specialized in primary materials, apart from chemicals and plastics production in the United States

Table 2.3b   *Materials production and energy per capita (energy in GJ/ pop; other indicators as noted, 2017)*

Selected developing countries	Mexico	Brazil	China	India	Africa
Activity indicators for industry					
Cement production per capita (t/pop)	0.4	0.3	1.7	0.2	...
Iron and steel production per capita (t/pop)	0.2	0.2	0.6	0.1	...
Chemicals share in total value added (%)	...	...	3	...	...
Total energy use for materials	**7**	**15**	**30**	**6**	**3**
Energy supply for producing materials					
Natural gas	4	2	2	<1	<1
Coal	1	5	16	3	<1
Biomass and waste	0	4	0	1	1
Oil products	1	2	2	1	1
Heat	0	0	2	0	0
Electricity	1	1	9	1	1

Notes: As in Table 2.3a.
Sources: As in Table 2.3a.

and Europe. Most developing countries and regions are less specialized in producing materials.

ENERGY FOR GENERATING ELECTRIC POWER AND PRODUCING FUELS

In addition to being an input into producing useful energy services and material, energy is the main input into producing useful energy carriers such as electric power and oil products. For example, conventional thermal power generation plants produce 76 per cent of the world's electricity.[13] They use primarily coal and natural gas to produce steam to rotate steam turbines and generate electric power (68 per cent), although oil, biomass and material wastes are also used as fuels in these plants (8 per cent). Some thermal plants also use fissile materials to produce steam (13 per cent). The average efficiency of conventional and nuclear thermal plants in converting these primary energy resources into useful electric power and heat is about 40 per cent. The rest of the primary energy dissipates into the atmosphere as waste heat from this energy conversion. Hydro-electric (7 per cent) and variable renewables such as wind turbines and solar PV (4 per cent) round out the generation mix. Energy is also a significant input into the thermal distillation of crude oil into refined oil products. The overall energy efficiency of oil refineries is in the range of 90–94 per cent depending on the quality of crude oil inputs ('heavy' versus 'light' feedstocks) and type of distillation and chemical treatment process used in refining, such as hydro-processing (Han et al., 2015). In other words, the energy conversion losses are in the range of 6–10 per cent.

Across countries, factors that shape the mix of thermal fuels and renewable energy resources used to produce electric power include access to the primary energy resources and their generation technologies, security of energy supplies and affordability (Table 2.4). Countries in the Asia-Pacific region use coal relatively intensively as the primary energy source for electric power generation (60–80 per cent share), owing to widespread coal resources in the region and its relatively low transport costs. In Russia and Mexico, natural gas is used relatively intensively (about 60 per cent), reflecting abundant domestic and regional resources, and in Brazil hydro-electric power contributes almost 50 per cent of the primary energy for electric power generation. Systems in North America and Europe use a diversified mix of primary energy resources. Also, most systems have small but growing shares of variable renewable generation from wind and solar PV, in the range of 2–10 per cent. Their expansion reflects environmental impacts associated with fossil-fuel use, including local air pollution, especially in urban areas, and risks of climate change (IEA, 2016; pp. 20–22; Prüss-Ustün et al., 2016, pp. 46–66; IPCC, 2018).

CURRENT ENERGY SYSTEMS AND THE EARTH'S CLIMATE

Fossil-fuel use is a major contributor to altering the composition of the Earth's atmosphere, its surface temperature and climate, as are industrial processes, agriculture practices and land-use changes. The Earth receives solar irradiance from the Sun, and much of this is converted into heat as it is absorbed by the atmosphere, clouds and surface (land and oceans). The Earth radiates some of this energy as infrared radiation back into space, but the atmosphere also traps some and reradiates it back to the surface. The gases in the atmosphere that allow solar irradiance to pass through to heat the Earth's surface, but trap some of the heat being radiated away from it, are so-called greenhouse gases. Without them, the Earth's average surface temperature would be much lower than it is—a freezing minus 18°C rather than a temperate 15°C (Wolfson, 2016, pp. 320–22). The main naturally occurring gases are water vapour and carbon dioxide, which cycle through the Earth's systems, and had reached a natural balance before the industrial revolutions.

The balance has been substantially disrupted by human activities, including the combustion of fossil fuels, industrial process emissions, and agricultural and land-use practices on an ever-increasing scale since the first and especially the second industrial revolutions. These activities raised the atmospheric concentration of carbon dioxide to about 410 parts per million (ppm) by volume

Table 2.4   *Electric power and heat generation and energy per capita (energy in GJ/pop, 2017)*

Advanced industrialized countries and Russia	United States/ Canada	Australia/ New Zealand	Europe	Japan/ South Korea	Russia
Electric power and heat use	50	36	32	34	63
Energy to produce electricity and heat	109	92	63	75	98
Coal	37	55	16	31	17
Crude oil and oil products	1	2	2	4	2
Natural gas	28	19	12	22	58
Nuclear	28	0	20	11	15
Hydro-electric	7	5	2	2	5
Wind and solar PV	5	9	5	2	0
Biomass and waste	3	2	6	3	1
Selected developing countries	Mexico	Brazil	China	India	Africa
Electric power and heat use	9	10	20	4	2
Energy to produce electricity and heat	22	15	40	11	5
Coal	3	1	31	8	2
Crude oil and oil products	3	1	0	<1	1
Natural gas	12	3	2	<1	2
Nuclear	1	1	2	<1	<1
Hydro-electric	1	6	3	<1	<1
Wind and solar PV	1	1	1	0	0
Biomass and waste	<1	2	1	1	0

Notes: Data are for 2017 or most recent year available. Figures for country groups are weighted averages by country populations, allowing for any missing observations. The difference between electric power and heat use and energy to produce electricity and heat is primarily the energy conversion losses in generating electric power and transmission losses in its transport.
Sources: Energy data are from IEA Data & Statistics, retrieved from www.iea.org/data-and-statistics/data-tables. The data for country per capita incomes are from the World Bank, retrieved from https://data.worldbank.org/indicator.

in 2018 from an estimated 280 ppm average in the centuries before industrialization began—an estimated 50 per cent increase.[14] Most of the carbon dioxide emitted from human activity has occurred in the last half-century (Figure 2.1). Coal use and land-use changes account for about 33 and 30 per cent, respectively, of cumulative emissions from 1870 to 2018, with oil (25 per cent), natural gas (10 per cent) and other emission sources (2 per cent) accounting for the remainder.[15] The rising concentrations of carbon dioxide, methane and other gases from human activity are causing the Earth's climate to change significantly. The estimated rise in average global surface temperature since the start of industrialization is around 1°C, with wide regional variation (IPCC, 2018 and 2021).

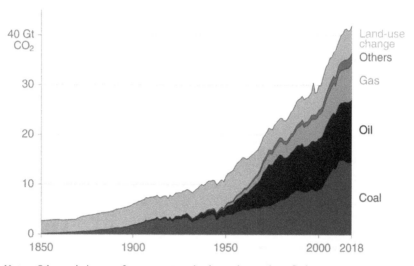

Note: Other emissions are from cement production and natural gas flaring.
Source: Reprinted from Global Carbon Project, Carbon Budget 2019 (CC BY 4.0). Retrieved from www.globalcarbonproject.org/carbonbudget.

Figure 2.1   *Annual global CO_2 emissions from fossil fuels, cement and land use (in giga-tonnes of carbon dioxide ($GtCO_2$))*

Fossil carbon dioxide emissions from the world's energy systems and industrial processes were an estimated 37 $GtCO_2$ in 2017 (Muntean et al., 2018). Advanced industrialized countries, China and Russia accounted for about two-thirds of these emissions, with China, the United States, the EU-28 and Russia emitting 10.9, 5.1, 3.6 and 1.8 $GtCO_2$, respectively (Figure 2.2). Among other developing countries, India is the most significant emitter (2.5 $GtCO_2$). In per capita terms, though, carbon dioxide emissions reflect the patterns of per capita energy use and fossil fuel mix outlined above. Per capita emissions are relatively high in North America, Australia, Japan and South Korea—in the range of 10–17 tCO_2/pop. China and the 28 EU member countries have per capita emissions of 7–8 tCO_2/pop. Per capita emissions in India are much lower, at 1.8 tCO_2/pop. Among the main emitting sectors, electric power generation, industrial materials production, transport and buildings directly contribute 38, 22, 22 and 9 per cent of emissions, respectively. The technology uses that emit much of the fossil carbon dioxide are conventional thermal electricity generation plants, blast furnaces, cement kilns, refineries and chemical plants, industrial boilers, transport vehicles and boilers for space and water heating in buildings.

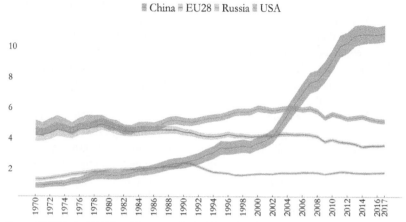

■ China ■ EU28 ■ Russia ■ USA

Note: EU28 is the 28 EU member countries in 2017. The primary data source for energy and industrial process emissions is the Emissions Database for Global Atmospheric Research (EDGARv5.0). For data on all greenhouse gas emissions, including emissions from land use changes, see Olivier and Peters (2018).
Source: Reprinted from M. Muntean et al. (2018), *Fossil CO₂ Emissions of All World Countries—2018 Report*, Luxembourg: Publications Office of the European Union (p. 9). https://doi.org/10.2760/30158.

Figure 2.2   *Annual fossil CO₂ emissions from energy and industry by major emitting countries (in giga-tonnes of carbon dioxide (GtCO₂), with shaded uncertainty bands)*

CONCLUSION

To cut carbon dioxide emissions, the firms and households that use fossil fuels; own the emitting electric power and industrial plants, transport vehicles and buildings; and consume the energy services and materials must do so more efficiently and cleanly. They must also make timely investments in alternative low-carbon technologies and energy supplies. Most of these firms and households are in advanced industrialized countries, Russia, China and, to a lesser extent, India, where current energy systems are extensively developed. Those in other developing countries must also increasingly focus their investments on low-carbon technologies, renewable resources and sustainable growth, especially as these alternatives become increasingly cost-effective. Moreover, it is becoming more apparent what are feasible and desirable low-carbon alternatives to current technologies and traditional use of fossil fuels. These alternatives emerged from investment in energy-related innovations over recent decades and more recent public and private investments in market creation for lower- and low-carbon technologies, especially those that generate electric power from ubiquitous solar irradiance and wind resources. These renewable generation technologies are increasingly competitive with conventional thermal generation technologies, even without costing carbon dioxide emissions. Similar developments are seen in lithium-ion battery packs and battery electric vehicles, which are increasingly competitive with internal-combustion-engine vehicles, especially in Europe, where transport fuel taxes are high.

But these efforts—important as they are—remain incomplete and partial. A much more comprehensive set of alternative technologies would be needed to achieve the goals of the Paris Agreement, and those low-carbon technologies that are increasingly commercialized would need to deploy and diffuse much more widely and rapidly across countries. The next chapter takes stock of low-carbon technology alternatives across the main energy-using and energy-producing sectors.

NOTES

1. On panel data relationships (across countries and over time) between average per capita income levels and: (1) residential floor space, see IEA (2017a, pp. 3–5); (2) appliance ownership, see Dobbs et al. (2012, pp. 23–32) and IEA (2017a, pp. 5–7); and (3) passenger-km and tonne-km, see ITF (2015) and Hellebrandt and Mauro (2016, pp. 111–6).
2. ACER (2019a). Report data retrieved from https://aegis.acer.europa.eu/chest. Figures are converted into USD at the 2018 average exchange rate.
3. Fuels Europe (2020). Fuel price breakdown. Retrieved from www.fuelseurope.eu/knowledge/refining-in-europe/economics-of-refining/fuel-price-breakdown. The 2015 figure is based on a wholesale petrol price of €0.53/l, which includes costs of crude oil, refining (€0.11/l) and distribution (€0.11/l) but excludes fuel taxes. The distribution cost is converted into USD at the 2015 average exchange rate. The average Brent crude oil price in 2015 was $52/bbl.
4. See Robinson and Gershuny (2013) for estimates of how and where people spend time. Capital stock figures are from OECD national balance sheet data. Retrieved from https://stats.oecd.org/index.aspx?DataSetCode=SNA_TABLE9B.
5. IEA (2013) and IEA Energy Efficiency Indicators Database 2020 (June Edition), extended version. Developing country figure does not include Africa.
6. See MacKay (2008, pp. 57–9). The efficiency of LED and other lightbulbs is updated to current standards. LED bulbs produce about 117 lumens per W (230 voltage), compact fluorescent bulbs produce about 68 and incandescent bulbs 15.
7. Energy Efficiency Indicators Database 2020 (June), extended version, and IEA, Mobility Model Database.
8. IEA Energy Efficiency Indicators Database 2020 (June Edition), extended version.

9. IEA Energy Efficiency Indicators Database 2020 (June Edition), extended version. These figures are 2017 averages for the vehicle fleets in advanced industrialized countries.
10. IEA Energy Efficiency Indicators Database 2020 (June Edition), extended version, and IEA Mobility Model Database.
11. IEA Energy Efficiency Indicators Database 2020 (June Edition), extended version, and IEA Mobility Model Database.
12. Allwood and Cullen (2015, p. 26). See also Rez (2017, pp. 185–214) on industrial processes for producing these and other materials.
13. IEA World Energy Balances. Retrieved from www.iea.org/data-and-statistics/data-tables.
14. The annual average atmospheric concentration of carbon dioxide reached 407 parts per million (ppm) in 2018, up from an estimated 280 ppm in 1750 (Joos and Spahni, 2008). The measure of the 2018 average annual atmospheric concentration of carbon dioxide is from the US National Oceanic and Atmospheric Administration, Earth System Research Laboratory. Retrieved from www.esrl.noaa.gov/gmd/ccgg/trends.
15. Global Carbon Project, Supplemental data of Global Carbon Budget 2019 (Version 1.0) [Data set], https://doi.org/10.18160/gcp-2019.

3. Net zero emissions and low-carbon alternatives

Near elimination of carbon dioxide emissions, including human use of carbon-containing resources in the Earth's interior and surface, is necessary to stabilize the atmospheric concentration of carbon dioxide and the Earth's surface temperatures at any given level. This eventual need has been foreseen for some time, although timescales and technologies for achieving this outcome were highly uncertain (Hoffert et al., 1998; Matthews and Caldeira, 2008). Energy systems with net zero emissions of fossil carbon dioxide would be very different from current, fossil-fuel-based systems. They would need to capture non-fossil primary energy resources on a scale comparable to current use of fossil fuels. They would also need technologies that convert available low-carbon energy into the useful services and materials that societies demand. Any residual carbon dioxide emissions from fossil fuel use and industrial processes would need to be either captured and physically stored or removed from the atmosphere by increasing natural absorption of carbon dioxide though land-use changes such as reforestation or using technologies to capture and remove carbon dioxide from the air.

The core of a low-carbon energy system is increasingly clear with ongoing advances of low-carbon alternatives. This core is low-carbon electric power and large-scale increases in electrification of activities, such as surface transport and thermal comfort in buildings. Low-carbon power can be generated at a global scale using abundantly and widely available renewable solar, wind and water primary resources, as well as fissile materials. None of these power-generation technologies emit fossil carbon dioxide, though they remain more expensive in terms of their overall system costs than dispatchable thermal generation of electric power using fossil fuels. Nevertheless, the costs of low-carbon alternatives are falling significantly, as are those of complementary flexibility technologies necessary for balancing supply and demand profiles for electric power. Moreover, electric power is clean to use and versatile, but costly to distribute and difficult to store, though storage costs are declining rapidly.

In addition, since not all energy uses have the potential to be cost-effectively electrified with foreseeable technologies, low-carbon fuels would be necessary too. These fuels include hydrogen and advanced biofuels. When burned in an engine or oxidized in a fuel cell, hydrogen produces only energy and water. Burning biofuels emits into the atmosphere carbon dioxide that was recently absorbed from it by plant photosynthesis, creating a cycle that can be repeated without increasing the concentration of carbon dioxide in the atmosphere. But bioresources would need to be sustainably produced to avoid competition with food production and land-use changes that could release carbon stored in plants and soils into the atmosphere as carbon dioxide. This sustainability constraint limits the potential scale of bioresources available for use in energy systems and industrial processes. Low-carbon fuels also include non-traditional fossil-fuel use, including its combination with carbon dioxide capture, use and storage (CCUS) technologies and increased natural carbon stores, such as afforestation and reforestation measures.

Energy end-use technologies would need to change too. They include heating equipment in buildings, boilers, furnaces and turbines for industrial plants and vehicle engines that use fossil fuels. They would need to be adapted to use low-carbon fuels or manage any residual fossil carbon dioxide emissions or be replaced with alternatives that use low-carbon electricity. Such changes would maintain benefits from energy use such as thermal comfort in buildings, ease of transport and useful materials, while eliminating net emissions of carbon dioxide. Moreover, the capital stock that uses energy is widely dispersed among firms and households—much more so than the capital stock that produces and supplies energy. For example, there are currently about two billion dwellings—though not all have access to modern energy nor use fossil fuels—and one billion passenger cars in the world—almost all of which use fossil fuels. There are thousands of industrial plants that use fossil fuels across energy-intensive sectors and many countries. The vast number of firms and households that participate in energy systems, in both demand and supply, together with information requirements and uncertainties around most investments in alternatives, point to a central role for markets in change. They would need, though, supporting government policies to address multiple market imperfections.

Before turning to the knowledge, capabilities and government policies that advance and guide low-carbon technologies and renewable resources through markets—the subject of subsequent chapters—it is useful to describe the alternatives that could form low-carbon energy systems. This description is illustrative of what alternatives could emerge from the processes of innovation and commercialization of renewable resources and low-carbon technologies. It is neither a prescription of which ones should be successful nor a prediction of future successes. It is also important to examine why it is necessary for energy systems to have net zero emissions of carbon dioxide from fossil resources and why this outcome is time-bound if climate change is to be limited. Such a clear endpoint sharpens the focus on cost-effective pathways to achieving this outcome.

WHY TIME-BOUND COMMITMENTS TO NET ZERO EMISSIONS OF CARBON DIOXIDE?

Carbon dioxide emissions from human activities destabilize the Earth's climate because they add a prominent greenhouse gas to the atmosphere (NASA, 2020). In addition to carbon dioxide, naturally occurring and human-caused greenhouse gases include water vapour, methane, nitrous oxide and fluorinated gases. Earth-system modelling estimates that removals from the atmosphere of carbon dioxide emitted from human activities by natural 'sinks' and its storage in carbon reservoirs in the oceans, land and Earth's interior are partial and slow relative to the pace of these emissions and their influence on the climate. Most of the Earth's carbon is stored in its interior, and the rest is in the atmosphere, oceans, plants and soils.

Carbon flows between these reservoirs through natural exchanges that make up the Earth's carbon cycle, which has both a 'slow track' and 'fast track' (Archer, 2010, pp. 1–20; Ciais et al., 2013). Carbon reservoirs in the slow track are in the Earth's interior. They include limestone (mostly calcium carbonate, $CaCO_3$) and other carbon-containing rocks as well as fossil hydrocarbon resources. Natural turnovers of carbon stored in these geological reservoirs are estimated to span tens of thousands to millions of years. Geological processes such as the water cycle, weathering of rocks by wind and water, tectonic plate movements and volcanic eruptions drive these turnovers of carbon. Carbon reservoirs in the fast track are in the atmosphere and oceans and on land. In the atmosphere, carbon is largely in the form of carbon dioxide (CO_2) and methane (CH_4). In oceans, carbon takes several forms—carbonic acid (H_2CO_3), carbon dioxide-consuming marine life like plankton and corals, and calcium carbonate sediments on ocean floors. On land, the vegetation, soils and permafrost contain organic (carbon-containing) compounds and matter. Natural turnover of carbon in these reservoirs is estimated to range from decades to millennia. Life on the Earth (the biosphere), carbonic acid in rainfall and carbon dioxide exchanges between the atmosphere and ocean surface waters drive this turnover. Natural interactions between the geological processes in the slow track and the biological, chemical and geological processes in the fast track are judged to be small and relatively stable (Ciais et al., 2013).

However, among other things, extraction of fossil fuels from the Earth's interior and their combustion transfer, by human activity, a large amount of carbon in a relatively short period of time from the slow track of the carbon cycle to the atmosphere as carbon dioxide. Once in the fast track of the cycle, land vegetation, plankton and corals work to remove some of this added carbon dioxide and store it as increased biomass, as do oceans, albeit with a consequence of increasing ocean acidification that inhibits marine life (Friedlingstein et al., 2019). But these processes are estimated to remove just over half of human-caused emissions (Figure 3.1). Moreover, the geological processes that store carbon in the Earth's interior are negligibly slow relative to the pace of carbon dioxide emissions from human activities and climate change.

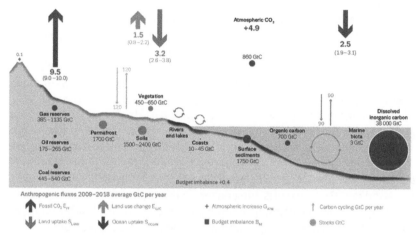

Notes: Schematic representation of the overall disruption to the global carbon cycle caused by human activities, averaged globally for the decade 2009–2018. See legends for the corresponding arrows and units. The uncertainty in the atmospheric CO_2 growth rate is very small (±0.02 giga-tonnes of cumulative carbon emissions (GtC) per year) and is neglected for the figure. The human disruptions occur on top of an active carbon cycle, with fluxes and stocks represented in the background and taken from Ciais et al. (2013) for all numbers, with the ocean gross fluxes updated to 90 GtC per year to account for the increase in atmospheric CO_2 since publication, and except for the carbon stocks in coasts, which are from a literature review of coastal marine sediments.
Source: Reprinted from P. Friedlingstein et al. (2019), Global carbon budget 2019, *Earth System Science Data*, 11, 1783–838 (CC BY 4.0). https://doi.org/10.5194/essd-11-1783-2019.

Figure 3.1   *The global carbon cycle and its fluxes from human activities (annual average in giga-tonnes of carbon (GtC), 2009–18)*

Of the estimated 615 GtC (±80 GtC) from 1870 to 2017 from burning fossil fuels and human changes to land use such as deforestation, natural carbon sinks on land and in oceans are estimated to have removed around 30 and 25 per cent, respectively (Le Quéré et al., 2018). The estimated increase in carbon in the atmosphere is 250 GtC (±5 GtC) (about 920 Gt of carbon dioxide $(GtCO_2)$).[1] The cumulative airborne fraction of the carbon emissions from this past human activity is thus about 40 per cent, with evidence that this share is increasing gradually over time (Ciais et al., 2013, Collins et al., 2013). The UN Intergovernmental Panel on Climate Change (IPCC) assesses unequivocally that emissions from burning fossil fuels and land-use change are the dominant causes of the almost 50 per cent increase in atmospheric carbon from the estimated pre-industrial level of about 595 GtC (IPCC, 2014a and 2021).[2]

Earth-system modelling yields two important results for assessing the average surface temperature response to carbon dioxide emissions. One is that the global average surface temperature response to a carbon dioxide emission reaches a peak after an initial adjustment period of about a decade and remains broadly stable for more than a century thereafter (Matthews and Caldeira, 2008; Solomon et al., 2009; Ricke and Caldeira, 2014). This constancy over time reflects two offsetting processes in the models—the gradual transfer of heat from the atmosphere to deep ocean waters through their mixing with surface waters and the gradual absorption of carbon dioxide from the atmosphere by the biosphere and oceans. In the initial adjustment period, relatively rapid heat transfers from the atmosphere to surface ocean waters dominate. The second result is that this temperature response does not depend on the atmospheric concentration of carbon dioxide at the time of an emission (Matthews et al. 2009; MacDougal and Friedlingstein, 2015). Again, this stability reflects two offsetting factors in the models. One is a diminishing warming effect from increasing atmospheric concentrations of carbon dioxide, and the other is an increasing airborne fraction of cumulative emissions as the biosphere and oceans become less effective in absorbing them. These Earth-system modelling results mean that the projected temperature response to a carbon dioxide emission is approximately independent of time and its atmospheric concentration, at least over the next century (Collins et al., 2013; Van der Ploeg, 2018; Dietz and Venmans, 2019).

These two apparently stable Earth-system processes thus simplify for modelling to a stable relationship between cumulative emissions and surface temperature response, which is useful for policy making (M.R. Allen et al., 2009; Matthews et al., 2009; Zickfeld et al., 2009). This relationship—the transient climate response to cumulative carbon emissions (TCRE)—is defined as the average surface temperature change in response to 1000 GtC of carbon dioxide emissions from human activities (Collins et al., 2013). This amount of cumulative emissions from human activities is expected to give rise to an approximate doubling of atmospheric carbon dioxide, given the cumulative airborne faction of emissions so far and its gradually increasing trend. The projected temperature change associated with the cumulative emission is measured for time windows centred on the beginning of the sustained emission of carbon dioxide (the start of the second industrial revolution) and the time of the doubling of its atmospheric concentration. While the TCRE focuses on carbon dioxide emissions and temperature change, it is important to emphasize that other greenhouse gas emissions such as methane, and other aspects of climate change such as extreme weather events and sea-level rise, are also important for policy making.

The IPCC assesses that the TCRE is likely in the range of 0.8–2.5°C per 1,000 GtC (Collins et al., 2013). This range reflects uncertainties around both the carbon cycle and average surface temperature responses to atmospheric carbon dioxide, but it does not include highly uncertain feedbacks like potential methane releases from thawing permafrost. Nevertheless, the measure provides a useful standardized gauge of how complex Earth-system models project the climate response to cumulative carbon dioxide emissions from human activities. The TCRE can also be compared to the observed relationship between cumulative emissions and average surface temperature change. For example, the estimated cumulative carbon dioxide emissions from human activities between 1876 and 2010 and average surface temperature change over this period are 525 GtC (±75 GtC) and 0.87°C (likely range 0.75–0.99°C), respectively (M.R. Allen et al., 2018; Le Quéré et al., 2018; Rogelj et al., 2018a).[3] In making such assessments of observed cumulative emissions and temperature outcomes it is also important to allow for other temporary factors that can affect measured temperatures, such as volcanic eruptions (Krakatoa in 1883) and variations in solar irradiance.[4]

If the TCRE is positive and approximately constant, net zero emissions of carbon dioxide from human activities are necessary for stabilization of surface temperatures. Moreover, given the IPCC confidence range for the TCRE, it would be necessary to keep cumulative emissions from the start of industrialization below 1000 GtC to limit the likely change in surface temperatures to less than 2°C (Collins et al., 2013; Rogelj et al., 2018a).[5] An IPCC allowance for the net contribution to warming of other greenhouse gases and aerosols lowers this carbon budget somewhat, although the balance of their climate effects is relatively uncertain. The IPCC estimates that the remaining carbon budgets from the beginning of 2018 for two temperature change outcomes—1.5°C and 2°C—are likely to be 115 GtC and 320 GtC, using the 67th percentile of the estimated TCRE (Rogelj et al., 2018a). These estimated carbon budgets take account of other likely greenhouse gas emissions and aerosols. The IPCC assigns medium confidence to these budgets because of uncertainties around the TCRE, other greenhouse gas emissions and omitted feedback effects such as permafrost thawing.

The IPCC sets out two simple linear emission pathways and illustrative timescales for achieving net zero emissions while remaining within the two carbon budgets. They use constant absolute emissions cuts from the beginning of 2017 to the point in time when the carbon budget would be exhausted. Given estimated emissions of 11 GtC (±1 GtC) in 2017 and a remaining carbon budget of 115 GtC, consistent with the goal of limiting temperature change to 1.5°C, the timescale for elimination of net emissions of carbon dioxide from human activities would be about 20 years (by the year 2037). For a remaining carbon budget of 320 GtC, the necessary timescale for achieving net zero emissions would be about 60 years (by the year 2077). But emissions continued to rise after 2017, so the illustrative timescales have narrowed from these linear IPCC examples. There is in addition a range of plausible non-linear pathways for eliminating net emissions of carbon dioxide from human activities, including ones that allow for 'negative' emissions from removing and storing safely atmospheric carbon dioxide (Riahi et al., 2017; Rogelj et al., 2018a). Their timescales for achieving net zero emissions vary. Nevertheless, all share the same end point for energy systems—net zero emissions of carbon dioxide from human use of carbon-containing resources extracted from the Earth's interior. It is therefore important to take a close look at what energy systems and industrial processes with net zero emissions could look like.

RENEWABLE RESOURCES AND LOW-CARBON TECHNOLOGIES: THE 'HARD' CHALLENGES

It is increasingly likely that an energy system with net zero emissions of carbon dioxide would rely heavily on vast amounts of low-cost, low-carbon electricity generated from widely available renewable energy resources such as solar, wind, bioenergy and hydropower (Bruckner et al., 2014; S.J. Davis et al., 2018). But these renewable energy resources give rise to some difficult challenges for energy systems. Electric power systems would need considerable flexibility to cost-effectively manage time-varying and uncertain differences between electricity demand and generation from variable renewable energy resources like solar irradiance and wind. If these challenges could be addressed, there would be substantial potential to adopt electricity-using alternatives to fossil-fuel-using vehicles, boilers and industrial processes. But not all energy services and industrial processes could be feasibly electrified with foreseeable technologies, and some would need low-carbon fuels for their energy supplies. They include aviation, shipping and long-distance heavy-duty surface transport vehicles, as well as production of materials like steel, cement, plastics and chemicals. Achieving sufficient flexibility in the electric power system to manage variable renewable supplies would also likely need low-carbon fuels. These activities within energy systems would need new energy-using technologies and industrial processes, new energy-storage technologies and low-carbon fuels such as bioenergy and hydrogen (including ammonia, NH_3). Carbon-management technologies like CCUS as well as carbon removal from the atmosphere through reforestation and direct air capture could also be needed to reach net zero emissions from energy and industrial processes. Figure 3.2 provides a schematic of such an energy system.

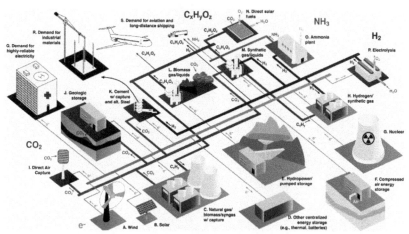

Note: (A to S) indicate the dominant role of specific technologies and processes; A–H, electricity generation and transmission; I–K, carbon management; L–N, hydrocarbon production and transport; O, ammonia production and transport; P, hydrogen production and transport; and Q–S, end-uses of energy and materials.
Source: Reprinted from S.J. Davis et al. (2018), Net zero emission energy systems, *Science*, 360(6396), eaas9793, with permission of the American Association for the Advancement of Science. https://doi.org/10.1126/science.aas9793.

Figure 3.2   *Schematic of an energy system with net zero emissions of carbon dioxide*

To envision more fully what a future low-carbon energy system could look like, it is necessary to examine closely potential alternatives for various energy services, industrial processes, low-carbon fuels and flexibility in electric power systems. Some alternatives face substantial technological and cost barriers that would need to be overcome (Figure 3.3). For example, fossil fuels provide much of the energy storage necessary for varying electricity generation with demand variations and maintaining overall system balance, especially across weeks and seasons but also within days. Alternatives from integrating heating and electric power systems, developing energy-storage technologies and low-carbon fuels could provide balancing capacity in systems where variable renewable generation has a significant share of the total. Alternative processes for smelting metal ores using hydrogen or carbon-free electrolysis, or managing carbon dioxide from current processes, would be necessary to eliminate their net emissions. Low-carbon fuels like hydrogen or advanced biofuels for heavy transport would also be necessary. But most of these alternatives

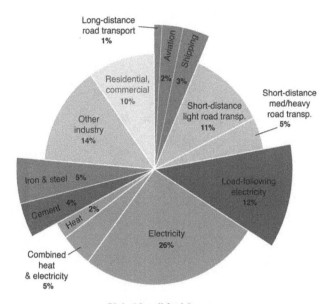

**Global fossil fuel &
industry emissions, 2014**

Notes: Estimates of CO_2 emissions related to different energy services and materials production, highlighting those that will be the most difficult to decarbonize with longer pie wedges. The sectoral breakdowns shown for 2014 are based on data from the International Energy Agency (IEA) and EDGAR 4.3 databases, with estimated total carbon dioxide emissions of 33.9 $GtCO_2$ and difficult-to-eliminate emissions of 9.2 $GtCO_2$. In 2017, estimated global emissions based on the EDGAR 5.0 database were 1.5 per cent above their 2014 level, though their sector composition would have remained largely unchanged. 'Load-following electricity' refers to current demand profiles for electric power. The highlighted iron and steel and cement emissions are those related to the dominant industrial processes only. Long-distance road transport is primarily heavy-duty road freight. Residential and commercial emissions are only those produced directly by firms and households.
Source: Reprinted from S.J. Davis et al. (2018), Net zero emission energy systems, *Science*, 360(6396), eaas9793, with permission of the American Association for the Advancement of Science. https://doi.org/10.1126/science.aas9793.

Figure 3.3   *Difficult-to-eliminate emissions from current energy systems (share of total global emissions in per cent; emissions in $GtCO_2$)*

are just conjectures rather than investment projects that transform energy systems. They are relatively immature, unproven at the necessary scale and costly compared with incumbent technologies. In contrast, other alternatives are increasingly competitive with incumbent technologies, such as wind and solar photovoltaic (PV) generation of electric power, battery electric light-duty vehicles and electric heat pumps for buildings, though these technologies too appeared difficult and costly a decade or two ago.

The technology alternatives set out here for low-carbon electricity and fuels as well as energy end-uses are those that have been or are being demonstrated in large-scale pilot projects, or they are already in use commercially, including in different applications. Tight timescales for achieving net zero emissions from energy would tend to tilt alternative technology choices towards those that are closer to being commercially available in markets and substitute more readily for current technologies (Chapter 1). However, rival technologies could well dominate low-carbon energy systems in the long run.

Variable Renewable Generation and Flexibility Technologies in Electricity Systems

There are two compelling reasons why low-carbon electric power would form the core of a future low-carbon energy system. One is that renewable resources like solar irradiance and wind are widely available around the world, and these technologies could be deployed on a scale that could fulfil expected final energy demands assuming that these demands could all be met with electric power. A second is that the costs of these once relatively expensive technologies have fallen substantially with increasing deployment and use, benefiting from knowledge gained through experience and the increasing scale of their manufacture. They are increasingly competitive with existing electric power generation technologies—a trend with further potential (Farmer and Lafond, 2016; Lafond et al., 2018).

The global technical potential for renewable electricity far exceeds current and expected future total final energy demand. The potential for solar PV generation is three to 30 times the current total global annual final energy use of about 360 exajoules (EJ), excluding 40 EJ of non-energy uses of fossil hydrocarbons (Arvizu et al., 2011). That of onshore and offshore wind generation is in the range of 85 EJ to 580 EJ and hydropower is about 50 EJ (Kumar et al., 2011; Wiser et al., 2011). Much of the uncertainty about the technical potential for solar PV and onshore wind generation is around the availability of land for them, and for offshore wind the availability of sites in shallow waters. Variations in assumptions about the future technical performance of energy-conversion technologies are less significant. However, estimates for offshore wind do not include deep waters and floating offshore wind technologies that could vastly expand its technical potential.[6]

Estimates of the global technical potential for all primary bioenergy resources, which could be used for a range of purposes, from electricity generation and heat to chemicals and liquid transport fuels, vary widely—from about 50 EJ to 150 EJ (Chum et al., 2011; IEA, 2017c). These estimates assume that bioenergy crops do not displace food, fodder, fibre and wood production from agricultural lands and forests. Key uncertainties arise around the availability of agricultural and forest residues and biogenic municipal wastes, scope for dedicated bioenergy production through productivity gains in existing land use, and extent and productivity of marginal lands used for bioenergy production.

The overall physical availabilities of solar, wind, hydropower and bioenergy resources, as well as geothermal and tidal, which are more locationally specific, are thus unlikely to limit the feasibility and scale of using these primary energy resources to produce useful electric power and potentially low-carbon fuels. Their challenges lie elsewhere. One is cost, but for solar PV and wind-turbine technologies this barrier is increasingly being overcome. A second is their integration into current energy systems.

The costs of renewable electricity generation technologies have fallen significantly since governments began major policy supports for their deployment from the mid-2000s—initially in Europe and subsequently in the United States, Japan, China and India (Aklin and Urpelainen, 2018, pp. 36–42). The levelized cost of electricity generation (LCOE) provides a measure of its combined investment and operating costs. This cost is the average, time-discounted cost of electric power per megawatt hour (MWh) assuming that a plant operates over its expected lifetime at a load factor consistent with its availability and economic dispatch of plants in the electricity system. The global weighted-average levelized cost of new utility-scale solar PV and onshore wind projects fell to about \$68/MWh and \$53/MWh in 2019, respectively, from \$378/MWh and \$86/MWh in 2010 (Figure 3.4). Auction results for future deployment of these technologies point to further cost reductions in 2021 to \$39/MWh and \$43/MWh, respectively. The global weighted average of the levelized cost of electricity of offshore wind fell to \$115/MWh in 2019 from \$161/MWh in 2010. In comparison, the mid-range levelized cost of electricity from thermal generation plant ranges was \$56/MWh in 2019 for new combined-cycle gas turbine plants and \$175/MWh in 2019 for natural gas peaking plants[7] operating at an expected 10 per cent capacity (Lazard, 2019b). That for new coal plants lies within this range.

The full cost of any new electricity-generation technology, however, involves not only the investment and operating costs of the technology itself, but also the costs of integrating the generation capacity into the electricity system to which it is added (Joskow, 2011; Borenstein, 2012). There are several characteristics of electric power generated from variable renewable energy resources that impose significant integration costs, although all generation technologies have such associated costs (Pérez-Arriaga, 2011; Ueckedt et al., 2013; Hirth et al., 2015).

Firstly, electric power produced from these variable resources cannot be adjusted to the extent that output of dispatchable natural gas or coal-fired

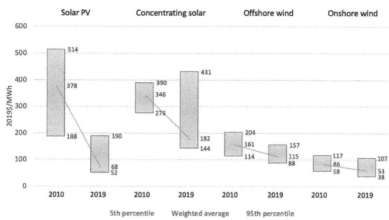

Notes: The lines join the global weighted-average LCOEs in 2010 and 2019 derived from international data on individual plants commissioned in each year. The project-level LCOE is calculated with a real weighted-average cost of capital of 7.5 per cent for Organisation for Economic Co-operation and Development (OECD) countries and China and 10 per cent for the rest of the world. The range bands for each technology and year represent the 5th and 95th percentile bands for renewable projects.
Source: Data are from IRENA (2020).

Figure 3.4   *Global levelized costs of newly commissioned renewable power generation plants (capacity-weighted average and range in 2019$/MWh, 2010 and 2019)*

generation plants can be varied with demand. Moreover, because the supply of variable renewable energy does not necessarily match closely demand for electricity, and its storage is expensive, integration of variable energy into the system gives rise to so-called profile costs. These costs are reflected in the tendency for electricity generated from variable renewable energy to attract relatively low wholesale electricity market prices because their supplies—but not necessarily demand—are positively correlated with resource availability, especially as their share in generation increases. Given their low operating (marginal) costs and resulting priority in economic plant dispatch, generation from thermal plants with relatively high operating costs declines and sets the wholesale price less often. Indeed, if electricity production from variable renewable energy exceeds demand, its supply can be curtailed or electricity prices can turn negative to pay for increased demand to balance the system. Profile costs also arise from the operating and physical depreciation costs of more frequent ramping and cycling of thermal plants needed to balance overall system supply and demand as the share of renewables in generation rises.

Secondly, most trading of electric power—the matching of expected supply and demand—takes place well in advance of the real-time physical balancing of the electric power market by the system operator. Because power from variable renewable energy is uncertain until produced, any differences between forecasted and actual electricity generation must be balanced at short notice, giving rise to real-time system-balancing costs. Electricity system operators use a mix of balancing and ancillary services provided by generators and users of electricity to maintain system reliability (Chapter 6).

Thirdly, the supply of variable renewable energy is location-specific. Good generation sites—affordable land with abundant renewable resources for solar PV and onshore wind—are often located away from cities where most electricity demands arise. Such distances make necessary costly extensions to transmission grids, especially for offshore wind with its high relatively costs per kilometre for transmission infrastructure.

While integration costs of variable renewables' generation depend on the existing electricity system to which they are added and how the system changes over time, it is possible to make a few general points about them (Hirth et al., 2015). Integration costs, especially profile costs, tend to increase with the share of each variable renewable energy in the generation mix. This cost can be managed in part through technological and geographical diversity, including interconnections between electricity systems. Also, some variable resources have more constant profiles than others. In particular, offshore wind tends to be more constant than onshore wind, which is more constant than solar PV. The correlation of variable renewable energy resources with demand can also be significant, and a positive correlation can lead to negative profile costs, such as intraday variation in solar PV generation and demand for thermal comfort from air conditioning. System flexibility from demand management, energy storage and flexible generation can also reduce profile costs of variable renewable energy. For example, where feasible, reservoir and pumped hydroelectric power can be an important source of emission-free, flexible generation. With respect to locational factors, solar PV can typically be located closer to electricity demand than onshore or offshore wind with lower grid extension costs. Decentralized solar PV can also help avoid grid reinforcement costs as additional electricity demands are added to the system.

Available evidence finds that profile costs account for much of the system-integration costs of variable renewable energy into otherwise unchanged electricity systems dominated by conventional thermal generation (Hirth et al., 2015; OECD/NEA, 2019, pp. 19–20). For example, with the share of wind in total electricity generation in the 30–40 per cent range, the modelled incremental profile costs of additional wind capacity in current electricity systems are in the range of \$36–50/MWh.[8] Profile costs for solar PV generation tend to be larger because its generation is more concentrated in fewer hours than that of wind, unless its generation correlates well with intraday variation in electricity demand. Estimates of short-term balancing costs that arise from uncertainties around forecasted variable renewable resource availability are relatively small, around $4/MWh. Grid-extension costs for onshore wind and solar PV are in the range of $5–15/MWh (OECD/NEA, 2019, p. 62). Therefore, incremental system-integration costs for variable renewables at generation shares of 30–40 per cent in existing power systems with no other changes are projected at about $40–65/MWh. Similarly, a recent OECD/ Nuclear Energy Agency (NEA) study projected system-integration costs at a 50 per cent share of variable renewables (25 per cent each for onshore wind and solar) in an otherwise unchanged conventional power system to be about $50/MWh. These integration costs are in addition to the investment and operating costs of wind and solar PV generation.

The key to bringing down integration costs of variable renewable energy is reducing profile and balancing costs through lower-cost flexibility technologies that have net zero emissions (Strbac and Aunedi, 2016). Even in electric power systems with well-diversified renewable resources, a highly reliable system with a high share of electric power from variable renewables would require substantial amounts of dispatchable supply (generation or storage) or demand management (Kroposki, 2017). But as the share of variable renewable energy in generation rises, the share of flexible resources in the total would decline, so minimizing the capital costs of flexibility is key to cost-effective integration of variable renewables (Hirth and Steckel, 2016).

Electricity demand management is particularly attractive as a source of flexibility because it has relatively low capital costs. In current electricity systems dominated by thermal generation plants, demand management has long been used for load reduction to manage unplanned plant outages and peak electricity demands. Typically, these services have been provided directly by large industrial and commercial customers or indirectly by demand aggregators. They modify the timing of electricity use to take advantage of periods of low wholesale electricity prices and avoid high prices. Large firms that buy electric power in the wholesale market can choose to modify electricity demands in this way to the extent that it is cost-effective for them to do so. Demand aggregators engage with groups of smaller customers on ways to manage their electric power demands (and self-generation) to take advantage of variation of wholesale price variations.

Recent advances in electricity metering, communication and control technologies expand the range of technically feasible demand management options, including shifted heating and cooling cycles in buildings, automatic scheduling of water heating and appliances, and smart charging of electric vehicles. The potential for demand management to reduce electricity demand to flatten their peaks and align them more closely with variable renewable energy with current technologies is estimated to be in the range of 7–14 per cent of peak electricity demand in the United States and Europe (Gils, 2014; Bronski et al., 2015). There is also significant potential for demand management in developing countries with rapidly growing electric power systems (E. Hale et al., 2018). Reducing demand peaks improves the capital efficiency of the electricity system, by avoiding its expansion to serve peak demands that occur infrequently.

Low-cost thermal-storage technologies for industrial processes and buildings offer similar low-cost flexibility as demand management. For example, low-cost electricity when in excess supply from variable renewables could be used to heat firebricks to high temperatures needed for industrial processes, which could then be discharged for industrial processes when electricity prices are high (Forsberg et al., 2017). The capital cost of clay firebricks is a small fraction (<10 per cent) of those for battery and pumped hydroelectric storage facilities. In buildings, hot-water storage tanks are already widely used in buildings to store heat and provide hot water when needed. But it is also feasible to store heat produced from electricity in a range of low-cost materials at much higher energy densities than water (Imperial College, 2018, p. 32). They include materials like paraffin that change phases between a solid and liquid at relatively low temperatures. These materials release heat at predictable temperatures as they cool and change from a liquid to solid. This heat can be used reliably for space heating.

Flexibility in electricity systems can also be achieved by temporarily storing energy from electric power in another form for future conversion back into electricity. There are several such storage technologies serving a range of system needs, including arbitrage of the time profile of electric power supply and demand as well as primary and secondary responses for system balancing (Figure 3.5). The technologies that can store the most energy and discharge the most electric power over the longest period of time are pumped hydroelectric and compressed-air storage systems. They store electric power as kinetic energy and then convert it back into power when needed. Flywheels and batteries can provide instantaneous and short-run responses to unexpected changes in supply and demand, but their energy storage and power capacities are more limited than pumped hydro and compressed-air storage.

Pumped hydroelectric systems are a mature and widely used technology, although their scope can be limited by societal and environmental concerns around large dams. Compressed-air systems store energy in underground reservoirs like natural salt caverns but are relatively immature, with large two-scale demonstration projects operating in Germany and the United States. The next generation of this technology is being designed to store both compressed air and heat from the compression process to boost overall energy efficiency. The levelized cost of electricity discharged from newly built pumped hydroelectric

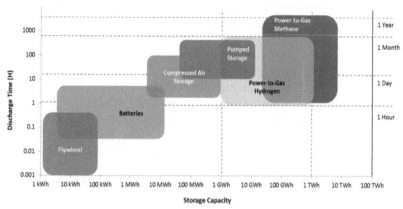

Notes: Discharge time is the amount of time a storage technology can maintain its output. Storage capacity is the amount of power an energy-storage device can discharge.
Source: Reprinted from J. Moore and B. Shabini (2016), A critical study of stationary energy storage policies in Australia in an international context: The role of hydrogen and battery technologies, *Energies*, 9(9), 674 (CC BY 4.0). https://doi.org/10.3390/en9090674.

Figure 3.5   *Discharge times and storage capacities of power storage technologies (in time and GWh)*

and compressed-air storage is in the range of $220–350/MWh, assuming that the systems are cycled at least weekly and a cost of electric power to pump the water or compress the air of $50/MWh (Schmidt et al., 2019).

Flywheels and rechargeable batteries have much higher capital costs per energy-storage capacity than pumped hydroelectric or compressed-air storage. To be cost-effective, they need to be cycled frequently to provide balancing and ancillary services for the system. For example, the levelized costs of electricity discharged from flywheels and lithium-ion batteries is currently in the range of $250–300/MWh when cycled daily (Schmidt et al., 2019). In comparison, the levelized cost of a natural gas peaking plant is in the range of $150–200/MWh at a 10 per cent load factor (Lazard, 2019b). Given current cost trends, the levelized costs of electric power from flywheels and lithium-ion batteries are projected to fall to the range in which they would be competitive with natural gas peaking plants. These projected cost declines are especially strong for lithium-ion battery packs, with their increasingly widespread use in transport and stationary applications. Battery technologies based on alternative chemistries and cheaper materials also hold potential for lower costs. For example, while flow batteries have lower energy and power densities than lithium-ion batteries, they are potentially suitable for stationary applications. New flow chemistries using relatively abundant, low-cost materials like sulphur could significantly boost their cost-effectiveness and the range of system applications to which they could be applied (Li et al., 2017).[9]

There is also potential for electric vehicle batteries to provide balance services for electricity systems, increasing battery utilization and the energy services they provide (both transport and system balancing). However, it remains to be established whether the charging infrastructure, communication and control systems could be cost-effective and whether vehicle-owner consent for diminished vehicle control and potentially shorter battery life in exchange for payments could be manageable and widely acceptable at grid-relevant scales (Davis et al., 2018). The key trade-offs are among efficiency gains and cost saving from more intensive battery cycling, battery charging, communication and payment infrastructure costs, and potential disruption and depreciation costs incurred by electric vehicle owners.

While storing electricity as thermal, kinetic and chemical energy can provide balancing services to electric power systems over seconds to weeks, large seasonal variations in electricity demand would have to be met through storage of low-carbon fuels, especially where seasonal demands for space heating in buildings are increasingly met with electricity. Fuels are the most cost-effective energy store for large amounts of energy over long periods of time because of their chemical form and high energy density. The technologies for converting low-carbon fuels into electricity are conventional thermal generation technologies that currently use fossil fuels such as combustion turbines suitably adapted to the combustion characteristics of low-carbon fuels. Low-carbon electricity can also be converted into low-carbon fuels; for example, using electrolysis of water to produce hydrogen.

Actual integration costs for variable renewable energy resources such as wind and solar PV therefore depend on the rate of progress in developing and deploying flexible energy resources on electric power systems. Some flexibility resources are currently available and cost-effective, such as demand management and system interconnections to diversify across variable resources. Others require further system integration and technological advances to become cost-effective, such as heat and battery storage. Sustained and rapid progress in developing these flexibility resources would be key to keeping integration costs of variable renewables low as their share in total generation increases. For example, there is significant potential to keep these integration costs below \$20/MWh for onshore wind in the UK electric power system even as its share in total generation increases above 50 per cent through demand management, system interconnection and storage (Figure 3.6). Key to achieving this outcome are well-designed wholesale markets for electric power that create market incentives for investment in both variable renewable generation and flexible resources (Chapter 6).

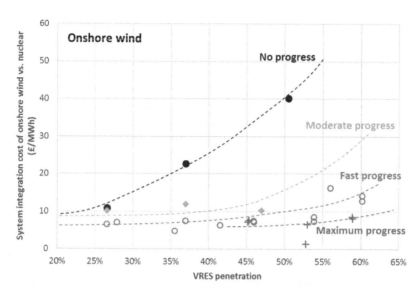

Notes: System-integration costs include profile, balancing and grid costs, and those for nuclear power are deemed to be small. Variable renewable energy supply (VRES) penetration is the modelled share of onshore wind in total UK power system generation. No progress on system flexibility assumes only current thermal generation technologies provide system flexibility. Moderate to maximum progress in increasing system flexibility refers to the pace at which least-cost alternatives in terms of interconnections with other power systems, demand-side response and storage resources are deployed in the system alongside the increasing share of onshore wind generation. The average exchange rate in 2015 was $1.53/£1 and cumulative US consumer price inflation from 2016 to 2019 was 7.9 per cent, so the vertical axis in 2019$ is 0, 16, 33, 50, 66, 82 and 99.
Source: Reprinted from G. Strbac and M. Aunedi (2016), *Whole-system Cost of Variable Renewables in Future GB Electricity System*, London: Imperial College (p. 12). https://doi.org/ 10.13140/RG.2.2.24965.55523.

Figure 3.6   *UK system-integration costs for onshore wind, its generation share and flexibility (in £/MW/h and per cent share of wind in total generation)*

Low-Carbon Fuels

There are two primary routes to low-carbon fuels. One relies on sustainable bioenergy resources for solid, liquid and gaseous hydrocarbon fuels that contain no fossil carbon and do not cause carbon dioxide emissions from direct or indirect land-use changes. The other relies on low-carbon hydrogen or hydrogen-based fuels like ammonia that are carbon-free fuels for combustion in turbines and engines or oxidation in fuel cells.

In contrast to energy-dense fossil hydrocarbon resources created through the forces of geological pressure and heat over millions of years, primary bioenergy resources have relatively low energy densities. These resources are thus costly to collect and transport for processing and use relative to the energy they contain. This factor significantly shapes their production and supply. The main sustainable bioenergy feedstocks are solid biogenic municipal wastes that are collected from households and firms, and agricultural and forest residues, as well as energy crops grown on marginal lands and algae in water.

An example of a current cost-effective use of biomass resources is the burning of forest residues for heat in pulp and paper production. This biomass is a low-cost by-product of pulp and paper production. Biomass and biogenic wastes are also being increasingly used for thermal generation of electricity and co-generation of heat and power. The global average levelized cost of electric power from new biomass and waste generation plants in 2019 was $66/MWh, which is at the upper end of the levelized cost range for new a combined-cycle gas turbine (Lazard, 2019b; IRENA, 2020, p. 13). About 4 per cent of the total primary energy used to produce electric power and commercial heat in 2017 was biomass and wastes.[10] However, it is important to emphasize that some combusted wastes emit fossil carbon dioxide, such as plastics made from fossil hydrocarbon resources. For combustion of solid municipal wastes to be free of fossil carbon dioxide emissions, plastics would need to be made from renewable resources.

The main technology for producing biogas—a mix of methane and carbon dioxide—is anaerobic digestion of wet biomass. This is a mature technology that uses bacteria and other micro-organisms to break down organic matter without light and in oxygen-deprived conditions. Feedstocks include animal and agricultural residues, biogenic solid waste and sewage from households, firms and industrial plants. Energy crops too are used, including being blended with less energy-dense feedstocks. But because the energy density of most feedstocks is low, anaerobic digestors are often sited close to their feedstocks and small in scale. The cost of biogas is largely determined by feedstock collection and transport costs and the size of digesters. If upgraded to remove the carbon dioxide and other impurities to produce biomethane, its current cost is in the range of $10–17/GJ, compared with that of natural gas in 2018 in the range of $3–9/GJ (IEA, 2020d, p. 38). Being mature technologies, the scope for cost reductions in anaerobic digestion and biogas purification is limited to primarily incremental process and logistical efficiencies.

An alternative technology for producing biogas is partial oxidation of dry biomass using high-temperature heat in an oxygen-controlled environment. This pyrolysis process, which is also used for coal gasification, produces syngas, a hydrogen-rich gas that also includes carbon monoxide and carbon dioxide. An advantage of biomass gasification is that it uses a wider range of feedstocks than anaerobic digesters, including forest residues and other cellulosic plant residues. However, partial oxidation of biomass is a less well-developed and more energy-intensive technology than anaerobic digesters for converting biomass into a useful fuel. There are also significant technical barriers still associated with this technology (Ericsson, 2017).

Liquid biofuels are mainly ethanol from energy crops like sugar cane and corn, and biodiesel and synthetic aviation fuel from oil seeds and increasingly waste oils. These biofuels currently account for about 4 per cent of transport fuels worldwide. They are relatively costly compared to the gasoline, diesel and kerosene jet fuel for which they substitute and are typically blended with fossil fuels under policy mandates. Blending up to 10 per cent ethanol with gasoline is cost-effective, though, because it substitutes for relatively more expensive gasoline additives such as ethers that are used to improve combustion. However, first-generation biofuels are associated with carbon emissions from production processes and potential direct and indirect land-use changes, including competition with food production and other existing land uses.

Advanced biofuels seek to overcome these land-use challenges by developing conversion processes that use a wider range of feedstocks like cellulosic residues from agricultural and forest production and crops that can be grown on marginal lands. The technology pathways that extract sugars from cellulosic residues and convert them into ethanol and methanol have seen some progress. The pathways to advanced biodiesel and synthetic aviation fuels have proved, however, more difficult to advance. For example, it is feasible to integrate biomass gasification with a Fischer–Tropsch process to converts syngas into synthetic lubricants and fuels. However, these technologies have proved difficult to scale beyond small demonstration projects, and remain very costly. Advanced biofuels—ethanol and biodiesel—are currently \$4–40/GJ more expensive than the fossil fuels for which they substitute (IEA Bioenergy, 2020, p. 57).

Biorefineries are industrial plants that integrate pyrolysis, gasification, and other thermochemical and biochemical processes to produce both material feedstocks and fuels. They are for bioresources what oil refineries and chemical plants are for crude oil. There are, for example, more than 800 biorefineries in the European Union (EU), of which more than 500 produce bio-based chemicals, 360 liquid biofuels and 140 bio-based composites and fibres (multi-product facilities are counted more than once) (Parisi, 2018). Of those facilities, almost 200 are integrated biorefineries that combine the production of bio-based materials and energy. The location of most biorefineries shows correspondence with existing chemical clusters and ports, with a significant concentration in Belgium and the Netherlands. Agricultural resources are the feedstock source used by most biorefineries in all EU countries. Feedstock logistics, conversion technology integration and multiple products are potential routes to greater cost-effectiveness of bioresource feedstocks and their conversion technologies.

Hydrogen is a carbon-free fuel that is already widely used as a feedstock in industrial processes such as crude oil refining and producing ammonia and methanol. It is mainly produced via steam reforming of natural gas, which reacts methane with water in the presence of catalysts at high temperature and pressure. Some hydrogen is also produced by coal gasification, mainly in China. Steam reforming is a mature technology, but one that emits fossil carbon dioxide as a by-product. The levelized cost of hydrogen produced in this way is around $8/GJ in the United States, Russia and Middle East, where the natural gas feedstock is abundantly available and relatively low-cost (IEA, 2019b, p. 42). Adding CCUS to a standalone steam methane reformer to manage the carbon dioxide emissions adds about 50 per cent to the levelized cost of hydrogen. Other potential technologies for converting natural gas into hydrogen include auto-thermal reforming (a combination of steam reforming and partial oxidation) and methane-splitting at a very high temperature. The former process yields a relatively pure stream of carbon dioxide to facilitate its capture and storage or use. The latter produces hydrogen and solid carbon rather than carbon dioxide emissions. These technologies are at the demonstration project and research and development stages, respectively.

An altogether different way to produce low-carbon hydrogen is electrolysis of water using low-carbon electricity. Most electrolysis uses alkaline electrolytes and metal catalysts to produce hydrogen, a well-established technology used in some chemical industries. It is 60–70 per cent efficient in converting electric power into chemical energy as hydrogen. At a grid-provided cost of electric power of $40/MWh, the levelized cost of hydrogen from alkaline electrolysis is in the range $17–25/GJ, assuming that electrolysers are operated at capacity (IEA, 2019b, p. 47). To reach the current cost level of hydrogen from steam methane reformers with carbon dioxide capture and storage, there would need to be significant technical improvement in electrolysers, lower costs from an increased scale of deployment and potentially dedicated renewable or nuclear generation to supply electric power (and potentially heat for high-temperature electrolysis). For example, two demonstration projects for producing electrolytic hydrogen from offshore wind generation are at the planning stages in the Netherlands and United Kingdom.[11]

Alternative electrolysis technologies to alkaline systems are polymer electrolyte membrane (PEM) and solid oxide electrolyser cell (SOEC) electrolysers. Two advantages of PEM electrolysers are they operate flexibly and efficiently around their design capacity and they produce hydrogen at high pressure without additional compression. A disadvantage is their relatively high cost because they need expensive materials like platinum and have a relatively short operating life. SOEC electrolysers use steam for electrolysis, and low-cost materials like ceramics and nickel. They need a low-carbon energy for both high-temperature heat and electricity generation. It is also possible to operate an SOEC in reverse mode as a fuel cell converting hydrogen back into electricity and providing electricity-grid balancing services. However, this electrolyser technology is immature and its materials degrade relatively quickly because of very high operating temperatures.

Ammonia is also a carbon-free fuel, which is already widely used in fertilizer production. It is produced primarily by using steam methane reforming of natural gas to yield hydrogen and combining the hydrogen with nitrogen from the atmosphere using the Haber–Bosch process. It is feasible to produce low-carbon ammonia using either steam reforming of natural gas with CCUS or electrolytic hydrogen. Production of ammonia is more expensive than that of hydrogen because of the added cost of the Haber–Bosch process.

Potential advantages of low-carbon ammonia over hydrogen as a fuel are the relatively low pressure at which it condenses into a liquid at ambient temperature, its somewhat higher energy density by volume and its use in suitably adapted internal combustion engines and turbines. These properties facilitate ammonia's bulk transportation and storage, as well as potential use as a transport fuel in conventional vehicles. However, combustion of ammonia must be carefully controlled to avoid emitting nitrogen oxide, a pollutant that interacts with sunlight and causes smog (Kobayashi et al., 2019). Also, for use in most fuel cells, ammonia must be stripped of its nitrogen, which requires energy and reduces efficiency.

Low-Carbon Fuels and Better Batteries for Heavy Transport

Heavy-duty trucks, rail locomotives, ships and airplanes have substantial passenger, payload and distance requirements. If they carry their energy with them, these vehicles need energy carriers with high densities by volume and weight. Batteries have relatively low energy densities because in part they must contain within them all their chemical reactants and products. In contrast, fuels that are combusted or oxidized in fuel cells use air and then vent their exhaust gases or steam (S.J. Davis et al., 2018). As chemical energy stores, fuels thus have significant inherent advantages over batteries in terms of energy density.

Not all fuels, however, have comparable energy densities. The most energy-dense fuels by volume are oil products such as gasoline, jet fuel and diesel—in the range of 34 MJ/l to 38 MJ/l. The energy density of liquid ammonia and compressed hydrogen (700 bar) are 15 MJ/l and 10 MJ/l, respectively. If hydrogen is used in a fuel cell to produce electricity and drive electric motors, the energy efficiency of the drivetrain is about 50 per cent greater than that of an internal combustion engine, so a fuel-cell electric truck requires only two-thirds the energy of an internal-combustion-engine truck to deliver the same tonne-km service (IEA, 2019b, Annex). The fuel storage volume of hydrogen is thus about 2.5 times that of diesel for comparable service delivery. In addition, a hydrogen-fuel-cell heavy-duty truck can be refuelled in 10 to 15 minutes, which would be much faster than a comparable battery electric vehicle could be recharged with prospective technologies. Similar considerations would point towards using hydrogen fuel cells to power trains for railway lines that are not cost-effective to electrify with catenary wires (Staffell et al., 2019). A city in China (Foshan City) and region in Germany (Lower Saxony) are among the first to operate hydrogen-fuel-cell trams and trains.[12]

The overall transport service cost per tonne-km of a fuel-cell electric heavy-duty truck is currently more than twice that of a comparable truck with an internal combustion engine, including the payload sacrifice to accommodate hydrogen fuel storage (Hunter and Penev, 2019). Key technical and performance challenges with hydrogen fuel cells are their durability and cost of the fuel cells, hydrogen storage tanks and refuelling infrastructure. First developed by NASA, PEM fuel cells for transport applications use relatively expensive and not very durable platinum-group metals for catalysts. Innovation efforts focus in part on reducing these material costs and improving durability (Hunter and Penev, 2019). There is also significant potential to reduce costs given experience curves of stationary fuel cells in Europe, Japan, South Korea and the United States (Staffell et al., 2019; Weidner et al., 2019). Hydrogen storage tanks, which are made from expensive composite materials, are also a significant cost component, but have perhaps less scope for future cost reductions than fuel cells. In addition, the rollout of hydrogen refuelling infrastructure is necessary for the uptake of hydrogen-fuel-cell vehicles. As with existing refuelling stations for oil products and new charging networks for battery electric vehicles, there are significant external network effects with this infrastructure. The first nationwide network of hydrogen refuelling stations is being developed in Germany though a public–private partnership.[13] There is thus long-run technical potential for hydrogen-fuel-cell heavy-duty trucks to achieve total costs of ownership comparable to conventional diesel vehicles (Hunter and Penev, 2019).

Other alternatives for heavy transport are rechargeable batteries that achieve much higher energy-storage densities than those made with lithium ion. At least two alternative battery chemistries—lithium-sulphur and lithium-air— have the potential to provide such advances (IEA, 2020e, p. 114). They are currently at the small prototype and concept stages, respectively. In addition, supercapacitors, which combine properties of electrochemical batteries and capacitors, are an alternative to batteries. Capacitors store energy as static electricity on the surface of a material rather than in chemical form. Supercapacitors using metal organic framework materials are at the small prototype stage and can achieve significantly higher energy-storage densities than lithium-ion batteries, and faster charging times.[14] Graphene nanotubes are also a potentially attractive material for supercapacitors (Castle and Hendry, 2020). This alternative is at the research stage (Sammed et al., 2020).

For ships, hydrogen fuels cells are an alternative, as are low-carbon ammonia and methanol for use in conventional internal combustion engines. The latter two fuels have the advantage that they could be used in existing marine diesel engines with few modifications, though ammonia's combustion must be carefully controlled. The relatively low energy density of these low carbon fuels means more space for fuel storage and less for cargo or passengers relative to existing fuels. For aviation, advanced biofuels or synthetic aviation fuels with relatively high energy densities appear more suitable than other low-carbon fuels. These low-carbon hydrocarbon fuels would be largely drop-in alternatives for existing jet engines. Alternative propulsion technologies for commercial airplanes are feasible, albeit over multi-decadal time horizons.

Low-Carbon Industrial Processes and Materials

There are two basic ways to eliminate or manage carbon dioxide emissions from industrial processes like production of iron and steel, cement, plastics and chemicals (Energy Transitions Commission, 2018, p. 61; Material Economics, 2019, pp. 32–5). One is to transform industrial processes to use low-carbon electric power, hydrogen, bioresources and recycled materials as alternatives to fossil hydrocarbon resources for energy, chemical reactants and material feedstocks. There are a range of feasible alternative processes for producing useful materials, and a few are highlighted here that are at the stage of large-scale demonstration projects. The other way to net eliminate emissions is to manage any emissions from residual fossil-fuel use and industrial processes by applying CCUS technologies to some existing processes and increasing natural carbon sinks such as reforestation. Both new industrial processes and carbon dioxide management could be necessary in sectors to eliminate their net emissions of carbon dioxide. However, the highlighted alternative would raise material costs substantially—in the range of 20–115 per cent (Material Economics, 2019, p. 10).

For iron and steel production, an alternative process to smelting iron ore using coke to remove oxygen and other impurities is direct reduction of the ore using hydrogen as a heat source and chemical reactant. This process is similar to the existing direct reduction of iron ore using natural gas, but substituting hydrogen. As with the existing process, the resulting sponge iron is further processed in electric arc furnaces, usually together with steel scrap, to produce new material. The processes involved in preparing the iron ore for direct reduction could also be electrified. Both direct reduction of iron ore using hydrogen and recycling steel using electric arc furnaces substantially eliminate carbon dioxide emissions from new steel production. The hydrogen direct-reduction process is being piloted in a large-scale public–private demonstration project in Sweden.[15] An alternative approach to low-carbon steel would be to apply CCUS technologies to existing integrated steel production facilities. This technology could remove about 90 per cent of carbon dioxide emissions from steel production, so additional carbon dioxide removal through co-firing blast furnaces with coke and charcoal or increasing natural carbon sinks would be necessary to achieve net zero emissions. But to achieve this capture rate, existing facilities would need substantial restructuring to concentrate emission streams. Another alternative production process would be low-carbon electrolysis of iron ore, as with smelting bauxite in aluminium production.[16] However, this alternative for iron ore appears only a long-run prospect.

For cement, the key decarbonization challenge arises from the significant energy and process emissions from the calcination of limestone in clinker production. Alternatives involve CCUS on either existing production processes or restructured ones that separate energy and process emissions. This restructuring could involve either electrification of kilns used to produce clinker or combusting fossil fuels with pure oxygen to yield a stream of carbon dioxide for capture. Estimated carbon dioxide capture rates are around 95 per cent for existing production processes and more than 99 per cent for restructured ones (Material Economics, 2019, p. 175). A large-scale demonstration project of carbon capture and storage from cement is being implemented in Norway.[17] In addition, the lime (calcium oxide) in surface concrete is chemically unstable and absorbs some carbon dioxide from the atmosphere (Xi et al., 2016). New production processes use this property of cement to strengthen the concrete product and reduce its net emissions.[18] However, this absorption effect alone is insufficient to manage fully process emissions from cement production.

For plastics and chemicals, the decarbonization challenges relate to energy and process emissions in their production and for plastics their disposal by incineration at the end of product lives. Many plastics and chemicals are produced from fossil hydrocarbon feedstocks—naphtha from crude oil and ethane from natural gas. Steam crackers transform these feedstocks into high-value chemicals such as olefins (ethylene and propylene) and aromatics (benzene and toluene) using high temperatures, steam and catalysts to promote the required chemical reactions. Chemicals such as ammonia are produced by steam methane reforming of natural gas to produce hydrogen, and the Haber–Bosch process that combines it with nitrogen from the atmosphere. An alternative to eliminate net emissions from these processes would be to fit carbon dioxide capture and storage technologies to existing refineries, steam crackers and steam reformers to manage their emissions. However, this approach would leave unmanaged emissions from incineration of plastics at the end of product lives.

A renewables-based alternative would be to use a range of biomass feed-stocks to produce bio-methanol, bio-ethanol and bio-naphtha using conversion processes such as anaerobic digestion and gasification of biomass for hydro-carbons. As with biomethane production, challenges with these alternative approaches would be scale in gathering feedstocks and their processing into high-value chemicals. These considerations would point to closer integration of plastics with other bioresource-based sectors such as agriculture, food, forestry, pulp and paper for residues and wastes as well as municipal biogenic waste systems. Innovations in recycling of plastics at the end of product lives could also add significantly to feedstocks for new production. This bioresource-based alternative for plastics could use low-carbon electricity for industrial process heat. Similarly, low-carbon electricity could be used to produce hydrogen via electrolysis for ammonia production.

'EASILY' TRANSFORMED BUILDINGS AND LIGHT-DUTY VEHICLES

Buildings and light-duty road vehicles are relatively easy to decarbonize compared to heavy industry and commercial transport and some flexible resources for electric power systems. Their transformation is 'easy' because in part the energy services they provide can largely be electrified with existing technologies for electric heating and drivetrains. Moreover, these alternative technologies are becoming increasing competitive with incumbent ones, albeit with significant government early-deployment support (Chapter 4).

Electric resistance heating is almost 100 per cent efficient in converting electricity into heat, and widely used in buildings in warmer climates where heating demands are low; however, heat pumps are much more efficient. They convert electricity into useful heat by taking available heat in the air or ground, and 'pump' it inside a building. Their operation is similar to that of air conditioning—but in reverse—and is a mature technology. They can achieve efficiencies of 300 per cent or more in converting electricity into useful heat by using solar thermal energy from the air or ground. While ground-source heat pumps are relatively effective in producing heat, they are also expensive because they require extensive ground works or bore holes to tap into the relatively stable heat in the ground. Air-source heat pumps are much less expensive than ground-source ones, but their efficiency declines as the gap widens between outside air and desired indoor temperatures. In cooler cli-mates, air-source heat pumps can require high standards of building thermal efficiency and potentially a back-up heating technology for cold snaps (outside air temperatures below a few degrees C) such as electric resistance heating or a boiler that uses a low-carbon fuel. In countries and regions with existing natural-gas distribution networks, better building insulation and hybrid heating systems are alternative technology combinations to fossil-fuel boilers and thermally inefficient buildings (Imperial College, 2018, 12–16).

Low-carbon heating fuels are an alternative or complement to electric space heating. Biomass boilers and stoves are a standalone solution, particularly in rural areas where biomass supplies are locally available and local air quality issues are manageable. However, in urban and suburban areas boilers using a low-carbon gaseous fuel like biomethane, synthetic methane or hydrogen could be more suitable, especially if these fuels could be distributed to buildings through existing or cost-effectively modified gas distribution networks. They could also be more suitable if poor thermal efficiency of existing buildings is difficult to rectify. District heating systems that pipe to buildings hot water produced from low-carbon energy are also an alternative in urban areas. The cost-effective pathway to low-carbon buildings, though, would depend on the context—the current building stock, infrastructure and feasible adaptations to low-carbon technologies and fuels (Imperial College, 2018, 23–6).

The electrification of light- and medium-duty vehicles using battery electric drivetrains is underway owing in part to their relative light payloads. Electric vehicles offer several potential advantages over those with internal combustion engines (Sperling, 2018). They have better performance, improve local air quality and avoid carbon dioxide emissions when they are charged with low-carbon electricity or fuelled with low-carbon hydrogen. They are simpler to manufacture, with fewer moving parts, and much more energy-efficient to operate than vehicles with internal combustion engines. They allow more flexible designs by placing modular batteries within vehicle frames and motors near wheels and by eschewing bulky engine blocks, radiators and drivetrains.

Their drawbacks are high costs and, for battery electric vehicles, limited range and access to charging points and lengthy battery charging times. However, the costs of batteries, which are the most expensive component of battery electric vehicles, are falling significantly with increasing scale of production. On current cost-reduction trajectories, in the second half of the 2020s the still relatively high cost of battery electric vehicles is expected to be offset in the United States by saving on fuel costs (in present value terms) (Lutsey and Nicholas, 2019). This customer breakeven point is set to occur earlier in Europe because of higher fuel taxes. At the same time, the energy density of lithium-ion batteries is expected to continue to rise, helping to extend vehicle range. Development of direct-current rapid charging technologies is also cutting charging times.

CONCLUSION

This assessment of alternative technologies identifies feasible ways to maintain current energy services and useful materials with low-carbon technologies that cut deeply current fossil carbon dioxide emissions and manage those that cannot be eliminated. The highlighted technologies are those that are plausible alternatives to current technologies over the next several decades. With recent advances in renewable-generation and battery-storage technologies, it is increasingly clear that low-carbon electric power and the electrification of much of surface transport and heating for buildings will be central to low-carbon energy systems. However, substantial technological uncertainties remain around hard-to-electrify and hard-to-decarbonize activities. The full range of low-carbon technologies that prevail in the long run should be those that withstand the test of competitive markets. This approach requires innovation and commercialization capabilities that span a wide range of technologies and sectors, recognizing that there are significant technological adjacencies among them. It also requires comprehensive and coherent market-oriented policies that address multiple market imperfections that hold back innovations and commercialization of low-carbon technologies. This includes a role for industrial polices, especially in countries initiating technological disruptions and sectors where returns on upfront investments in innovation and market creation are attenuated by market structures, cost complementarities among low-carbon alternatives and potentially time-inconsistent government policies (Chapter 4). It also includes emissions pricing appropriately calibrated to climate goals and effectively sequenced with industrial policies (Chapter 5).

Not all low-carbon alternatives are necessarily more expensive in the long run than the incumbent technologies and fossil fuels they would replace in terms of energy services provided. The cost of delivered electricity in systems dominated by variable renewables energy like wind and solar PV in the long run could be about the same as that of electric power from systems dominated by conventional thermal generation systems. This would require sustained progresses in technological innovation in renewable generation and flexibility technologies and their cost-effective integration, supported by well-designed electric power markets (Chapter 6). Moreover, some electricity-using technologies are inherently more efficient than the thermal ones for which they could substitute—heat pumps for condensing boilers and electric motors for internal combustion engines, helping to keep down the costs of delivered energy services. However, some low-carbon alternatives would be more expensive in terms of the energy services and useful materials they yield. This appears likely for some aspects of commercial transport and heavy industry, pointing to the long-run importance of emissions pricing (Chapter 5) and policies to increase the efficiency with which low-carbon fuels and materials would be used (Chapter 7).

NOTES

1. 1 GtC equals 3.67 $GtCO_2$.
2. 1 ppm of carbon dioxide in the atmosphere is equivalent in mass to 2.13 GtC.
3. The estimated change in average surface temperatures is the difference between the period averages of 1850 to 1900 and 2006 to 2015, which are centred on 1876 and 2010, respectively. Instrumental measurements of surface temperatures are widely available from 1850.
4. The wide temperature measurement window around 1876 lessens the weight given to the temporary cooling effect of the Krakatoa eruption on measured surface temperatures.
5. These carbon budgets assume that residual warming is negligible after the budget constraint is reached and carbon dioxide emissions stop (Rogelj et al., 2018a).
6. See, for example, Equinor, www.equinor.com/en/what-we-do/floating-wind.html.
7. Peaking plants generate electric power over short time intervals to meet periodic peak power demands.
8. Hirth et al. (2015), based on a survey of more than 100 modelling studies of power systems in Europe and North America. The original estimate is €25/MWh to €35/MWh. Figures were converted using the 2014 average exchange rate of $1.33/€1 and cumulative US consumer price inflation to 2019.
9. A flow battery is an electrochemical cell that stores energy in chemicals dissolved in liquids contained within the system and separated by a membrane that allows ion exchanges.
10. IEA World Energy Balances. Retrieved from www.iea.org/data-and-statistics/data-tables.
11. Gasunie, www.gasunie.nl/en/news/europes-largest-green-hydrogen-project-starts-in-groningen, and Orsted, https://orsted.co.uk/media/newsroom/news/2020/02/gigastack-phase-2.
12. Alstrom, www.alstom.com/our-solutions/rolling-stock/coradia-ilint-worlds-1st-hydrogen-powered-train and Ballard, www.ballard.com/markets/rail.
13. H2 MOBILITY, https://h2.live/en/h2mobility.
14. See Lamborghini, www.lamborghini.com/en-en/news/hybrid-lamborghinis-super capacitor-technology-patented-mit.
15. See Hybrit, www.hybritdevelopment.com.
16. See Elysis, www.elysis.com/en#carbon-free-smelting.
17. See Norcem, www.norcem.no/en/CCS, and Northern Lights, https://northern lightsccs.com/en.
18. See Carbon Cure, www.carboncure.com.

PART II

Advancing and guiding low-carbon alternatives

4. Supporting innovation and early deployment of low-carbon alternatives

Eliminating net emissions of carbon dioxide and other greenhouse gases from energy systems requires transforming the current capital stock for energy supply and end-use. This transformation is underway and seen in aspects of energy systems now deemed relatively easy to transform: energy efficiency, renewable electricity generation, lithium-ion battery packs and battery electric vehicles. Just two decades ago, though, energy systems appeared largely difficult to change because of the high cost and limited availability of low-carbon alternatives, apart from some then easily accessible, low-cost energy-efficiency gains. Innovations, early deployment and, over time, widespread diffusion of low-carbon technologies are thus central to transforming energy systems. But it is important to examine why alternative low-carbon technologies have reached and disrupted some aspects of energy systems but not others, and what barriers confront the full range and extent of innovations and low-carbon-technology deployments necessary to eliminate net emissions from energy.

The source of innovation and technological advances is knowledge, which is accumulated through scientific investigation, its engineering application and learning from experience. Science-based innovations flow from successful research and development (R&D) activities that create knowledge and make possible new production processes and products (Jorgenson and Griliches, 1967). Patents are often used as a measure of the output of these investments in knowledge. In addition, there are experience-based innovations and improvements to engineering's use of science through which firms and their customers adopt, learn and adapt new technologies as they produce and use them (Arrow, 1962a). These learning processes are often described by experience curves for new technologies that measure improvement in technical performance, such as lower unit cost, in relation to their cumulative deployment or the passage of time. There are also significant interactions between science- and experience-based learning processes, with many technological advances produced jointly by them (Mowery et al., 2010).

Directions for innovation activities and technological advances take their lead from the contexts in which they take place, including changes in relative prices, increases in expected market size and shifts in government policies

(Hicks, 1932, pp. 124–5). For example, recent innovations in energy efficiency and vehicle electrification trace back at least in part to the oil price 'shocks' of the 1970s and 1980s. Nuclear safety was also then called into question by nuclear accidents at Three Mile Island and Chernobyl. In response to the oil price shocks and heighted energy-security concerns, governments in North America, Europe and Japan implemented policies to improve energy efficiency and increase support for energy R&D. Higher energy prices and government polices significantly oriented innovations as measured by patents towards greater energy efficiency in industry, buildings and vehicles as well as vehicle electrification, at least for a while. Government R&D supports for nuclear and renewable energy technologies that were then introduced, however, yielded limited innovation results, as electricity generators readily substituted away from new nuclear capacity and oil generation to coal and natural gas plants. But innovations in renewable power generation and battery electric vehicles subsequently accelerated with government policies that supported early investments in these technologies after the 2008–09 global financial and economic crisis, helping to create markets for them. For energy-related technologies, the real resource costs of investments in market creation for new technologies tend to be at least an order of magnitude larger than those in R&D (Lester and Hart, 2012, pp. 31–7). These costs are incurred in advancing new technologies to achieve the performance characteristics, scale and costs at which they become widely deployable within and across energy systems.

Investments in innovation, experience-based learning and market creation for new technologies, however, potentially confront market imperfections that can dull incentives to invest in them. Market returns to upfront investments in innovations and market creation for new technologies largely arise from Schumpeterian processes that create and manage market rents from newly differentiated products or created cost advantages (Aghion and Howitt, 1998, pp. 216–25). Some markets clearly support such differentiation and associated price-cost margins, such as those for consumer products like automobiles; others less so, such as those for electric power. Moreover, some market returns depend on government policies, such as emissions pricing and electricity market reforms. Associated with such policies are policy risks from swings in political sentiments and power as well as potential time inconsistencies in their application associated with their distributional impacts. Markets for future emissions and electric power prices that could be used to manage these risks are largely missing, especially over horizons for managing long-run investment risks. In addition, knowledge creation by one person or firm does not necessarily preclude use of this knowledge by others. Patents protect intellectual property rights, but they also formalize and codify knowledge which facilitate further learning and innovations. Investments in experience-based learning and scale economies in producing new technologies, especially those with potential for wide application and mass manufacturing, also create potential cost spillovers, particularly among complementary technologies that work together to lower overall system costs.

Such potential market imperfections affecting innovations and low-carbon-technology advances thus warrant careful assessment and, if appropriate, government policy responses. Some of these policy responses reside within the innovation systems of some countries, especially those that specialize in innovation. These systems consist of educational and research institutions that engage in basic research, fiscal incentives and financing instruments to support private R&D activities, and direct public funding of R&D. Key aspects of these innovation systems support R&D activities rather than specific technology developments, with the orientation of private R&D shaped by broader economic and policy contexts. Direct funding of R&D by governments, though, affords them the opportunity to direct R&D activities towards particular technology fields. Other government policies aim at directing private investment in innovation by expanding markets for low-carbon technologies. Emissions pricing is one such potential policy (Chapter 5). However, governments have tended to pursue sectoral policies that directly create markets for low-carbon technologies and thereby support industries that produce them, rather than economy-wide emissions pricing. This policy approach reflects in part missing markets for emissions pricing, its significant adverse distributional impacts and potential time inconsistencies in implementation (Chapter 5). These sectoral policies can also serve to address knowledge and cost spillovers from advancing low-carbon alternatives in energy systems, especially among complementary technologies in energy systems.

This chapter takes a close look at the factors in the broader economic and policy contexts that have directed or have the potential to guide innovations towards low-carbon technologies. It also examines the market imperfections and barriers that can hold back investments in innovation and early deployment of low-carbon technologies, along with government policies to overcome them, including market-creating and industry-supporting (industrial) policies.

FROM LOW-CARBON-TECHNOLOGY CONCEPTS TOWARDS ENERGY-SYSTEM-WIDE CHANGES

Accumulating knowledge sustains the scientific, engineering and commercial journey of new technologies from concept and prototype to demonstration and early deployment, and over time widespread deployment as those successful mature (Figure 4.1). In early conceptual stages of new technologies, relevant knowledge tends to accumulate primarily through R&D activities, although broader economic and policy contexts influence the direction of this innovation. Progress often involves much experimentation—trial and error—and any

advances are rarely a linear march of technological progress. This initial innovation stage includes engineering applications of scientific concepts to technology prototypes. Successful technology prototypes then become candidates
for demonstration at scale to assess prospects for commercial viability and, if
necessary, for external financing for demonstration, market entry and early
deployment. For example, a range of low-carbon-technology demonstration
projects are in implementation or being planned across several energy-supply
and end-use sectors (Chapter 3).

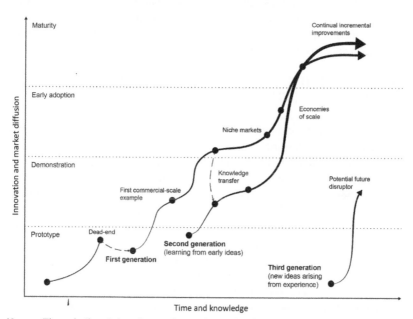

Notes: The main 'input' that advances innovations and technological progress is knowledge,
which accumulates over time through investments in R&D and experience-based learning.
Much knowledge accumulates through experimentation and trial and error, so the march of
technological progress is rarely linear, with significant feedback loops from experienced-based
learning to new R&D investments, as well as knowledge spillovers such as demonstration effects
of successful technological and commercial advances.
Source: Reprinted from IEA (2020), *Energy Technology Perspectives 2020: Special Report on
Clean Technology Innovation,* Paris: International Energy Agency (p. 23). Retrieved from www
.iea.org/reports/clean-energy-innovation. All rights reserved by the IEA.

Figure 4.1   *Innovation and technology diffusion stages, feedback loops
              and knowledge spillovers*

If prototypes scale successfully and technologies command financing, their
early market entry is typically characterized by cost or performance gaps with

incumbent technologies against which they compete. This was the experience of wind turbines and solar photovoltaic (PV) units as well as lithium-ion battery packs and battery electric light-duty vehicles as they entered markets over the past two decades (Chapter 3). Following market entry, these technologies matured steadily on the back of significant government early-deployment supports and are increasingly narrowing cost and performance gaps with incumbent technologies. Within the next decade these technologies are expected to become increasingly competitive with incumbent technologies given current rates of technical performance improvements. In fact, solar PV and onshore wind turbines are already cost-competitive where renewable resources are favourable, at least for generation shares over which their overall system costs are easily accommodated.

R&D ACTIVITIES, DIRECTED INNOVATION AND LOW-CARBON TECHNOLOGIES

Economic principles and evidence suggest that innovations by private firms tend to be directed towards where they are more profitably deployed. The profitability of an innovation would be expected to rise if the goods or services produced by the new technology command a higher price or if the technology uses more efficiently an input that becomes relatively more expensive (Acemoglu, 2002). Similarly, changes in policies and regulations and their 'shadow prices' are expected to induce innovations, and this effect is seen in innovation responses to a range of environmental policy changes (Popp et al., 2010; Popp, 2019). In addition, the larger the size of the market for the good or service produced by the innovation, the stronger the incentive to innovate. Such market influences are seen in the induced energy innovations following the oil price shocks of the 1970s and 1980s, and heightened nuclear safety concerns which curtailed investment in new nuclear capacity. The higher energy prices directed innovation as measured by the relative increase in high-value patents towards more energy-efficient industrial processes and vehicles as well as vehicle electrification (Popp, 2002; Crabb and Johnson, 2010; Aghion et al., 2016; and Figure 4.2). There is also evidence that appliance, building and vehicle efficiency standards, then introduced by governments in response to these developments, significantly boosted energy-efficiency innovations as measured by efficiency outcomes (Knittel, 2011; Noailly, 2012).

In the 2000s, amidst another oil price shock as well as the 2008–09 financial and economic crisis, climate concerns increasingly came to the fore of policy agendas with a strengthening of the economic case for urgent action to limit climate change (Stern, 2007, pp. xv–xix). Importantly, several major governments responded to the economic crisis by implementing 'green' fiscal stimulus measures, including direct government supports for renewable generation

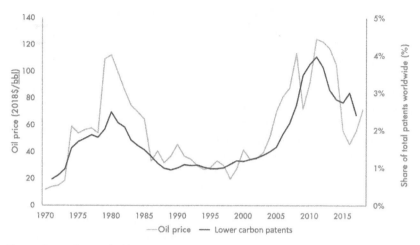

Figure 4.2   *Share of lower-carbon technology patents in total patents worldwide and oil prices (in per cent and 2018$ per barrel)*

that boosted demand for these technologies. The 2009 European Union (EU) Renewables Directive ushered in EU-wide feed-in tariffs for renewables, with EU countries clubbing together to support large-scale renewables deployments following initial programmes in Denmark and Germany (Newbery, 2016a; Aklin and Urpelainen, 2018, pp. 92–9). The US Energy Improvement Act (2008) and American Recovery and Reinvestment Act (2009) introduced and strengthened tax credits for renewable-generation investments (Council of Economic Advisors, 2016). China's 12th Five-Year Plan (2011–15) stepped up renewables deployment and further prioritized development of solar PV and wind-turbine manufacturing industries (Casey and Koleski, 2011). The Japanese government also introduced substantial renewable generation supports following the 2011 Fukushima nuclear accident (Skea et al., 2013). These measures to support early deployment of renewables, together with existing government R&D supports, were associated with a significant increase worldwide in patented innovations in solar PV, wind technologies and other related technologies (Bettencourt et al., 2013; Huenteler et al., 2016; Kavlak et al., 2018). At the same time, the decade of high oil prices from the mid-2000s and increasingly stringent vehicle emission (efficiency) standards likely encouraged further drivetrain and efficiency innovations in transport vehicles. This assumes the spurs to innovation of the previous oil price shocks and regulatory changes carried forward.

In addition to these direct impacts on innovation from policy changes, their potential influence on innovation extends beyond firms directly affected by them—not only across technologies but also regions and countries. For example, there were significant induced innovations at firm level between those in solar PV and wind turbines on the one hand and energy-storage technologies on the other, with innovations in one field being associated with more innovations in the other (Lazkano et al., 2017). There was a spur too to innovation in energy efficiency of conventional thermal generation technologies by innovations in their low-carbon-technology alternatives. Moreover, both domestic and foreign policies that support demand for low-carbon technologies were associated with significantly greater innovation by firms (Peters et al., 2012; Dechezleprêtre and Glachant, 2014). While the impact of domestic policies on innovation was greater than that of foreign ones, a larger overall size of foreign markets was associated with stronger impacts on innovation from early-deployment support policies. Similar geographical spillover effects on innovation are seen with state-level renewable portfolio standards in the United States (W. Fu et al., 2018).

There is in addition evidence from firm-level patent data that emission pricing policies, such as the EU Emissions Trading System (ETS), were associated with greater innovation as measured by patents. This effect can be seen by comparing a matched set of firms that are similar, but with one set having only plants just beneath the threshold for inclusion in the EU-ETS and the other having plants just above the threshold. Such policy threshold effects create 'natural experiments' for economic analysis. In this case, the regulated group of firms had significantly more low-carbon patents than the unregulated group, but because the group of regulated firms is small this effect was not large in the overall EU context (Calel and Dechezleprêtre, 2016). However, the analysis does not include potential innovations by upstream suppliers to the regulated firms or the downstream users of their products. These additional impacts on innovation were significant too, with an estimated overall policy impact in the United Kingdom of low-carbon-technology patents increasing by 20–30 per cent, albeit from a small base (Calel, 2020).

In this context, it is important to note the EU-ETS pricing of carbon dioxide emissions remained well below €20/tCO$_2$ from 2009 to 2017 (Bayer and Aklin, 2020). This low emissions allowance price was associated with an estimated

4 per cent cut in emissions from what would have otherwise occurred, largely from increased energy efficiency and switching to lower-carbon fossil fuels such as natural gas. But this policy instrument does not yet appear to be associated with strong innovation and low-carbon-technology deployment. Evidence from a survey of German businesses suggests that the coherence and credibility of EU decarbonization policies is important to innovative firms and the strength of their activities (Rogge and Schleich, 2018). This survey evidence also finds that renewable-generation deployment supports and coal phase-outs have been stronger influences on renewables innovation than EU emissions pricing policies. These survey results perhaps reflect past tensions among overlapping policies within the EU framework and a lengthy EU-ETS reform process. Reform was necessary in part because of its original inflexible design and the economic shocks arising from the 2008–09 global crisis and subsequent sovereign debt stresses in the euro area. With the introduction in 2019 of a market stability reserve and other measures, the EU-ETS proved more resilient to economic shocks associated with the novel coronavirus pandemic in 2020 than in the previous crisis.

Looking forward, the challenge is to identify energy-reform strategies that can strengthen the direction of innovation towards low-carbon technologies (Chapter 9). There are both direct and indirect mechanisms to achieve this outcome. For direct government funding of R&D activities, the allocation to low-carbon technologies could and should increase if the Paris Agreement goals are to be achieved. For private investment in R&D, the challenge is to strengthen expectations of larger markets for low-carbon technologies and the shifts in relative prices and costs that would support this outcome. Emissions pricing is one such policy, to the extent that it can be credibly implemented; heterodox policies such as market-creating and industry-supporting policies also have this effect. In the early deployment of low-carbon technologies so far, the latter have credibly expanded markets for them and induced significant innovations.

INVESTMENTS IN EXPERIENCE-BASED LEARNING AND SCALE ECONOMIES

Across the stages from early deployment to widespread diffusion of those innovations and new technologies that prove successful, much knowledge is produced through R&D as well as through by-products of production and use of new products. Uncovering their respective contributions is challenging, but empirical studies of innovation and technological performance improvement rates provide some insights. Technological performance is measured potentially along many dimensions. Perhaps the most notable is the number of transistors on a small integrated circuit chip. Moore's Law predicts that the number of transistors on an integrated processing chip increases at a constant rate over time (G.E. Moore, 1965). In fact, the prediction—largely correct so far—was that the number of transistors on a chip would double every two years, thus increasing exponentially over time, at least until technical barriers are encountered. Another commonly used principle for technical performance improvements is Wright's Law, originally applied to aircraft manufacturing, which states that improvements increase at a constant rate with cumulative production (Wright, 1936). In this case, the prediction was that the labour time required to manufacture an aeroplane would decline by 30 per cent for each doubling of its cumulative production.

In transforming energy systems, important technical performance characteristics are unit costs of technologies. When used as a measure of technical performance improvements, it is important to recognize that, in addition to experience-based learning, scale economies in manufacturing and deploying new technologies can reduce unit costs. Such economies arise when variable and fixed costs associated with given production capabilities rise more slowly than the units produced. They are, for example, particularly significant for manufactured low-carbon technologies that can be mass-produced.

Evidence across a range of technologies shows that the technical performance of many—not just integrated circuit chips—improves at a constant rate and thus grows exponentially over time (Nagy et al., 2013; Farmer and Lafond, 2016; Magee et al., 2016). But this correlation leaves unanswered the question of what is causing sustained rates of improvement. Many technologies experience sustained growth in both production over time and their technical performance, leaving Moore's Law and Wright's empirically equivalent in such cases (Magee et al., 2016; Lafond et al., 2018). Experience-based learning could thus be causing some technical improvements in products with steady growth rates in production. But other factors could be at work too.

There is in addition evidence that greater patenting activity is associated with higher technological performance improvements rates and that this activity explains around two-thirds of the variation in these learning rates (Benson and Magee, 2015; Magee et al., 2016, Kittner et al., 2017). This finding points to the significance of R&D investment in advancing technological performance, though patents can be jointly produced from scientific investigation and experience-based learning. Both R&D activities and experience-based learning appear significant contributors to technological performance improvements. It is therefore important in policy making to assume neither that such processes necessarily carry on indefinitely nor over-attribute technological performance improvements to experience-based learning (Thompson, 2012; Nordhaus, 2014). Equally, it is important not to ignore them because they are uncertain and 'hard' evidence about them is inherently limited in their early deployment stages (Akerlof, 2020).

Low-carbon technologies show wide variation in technical performance improvement rates. Experience curves relate the technical performance improvement rates of low-carbon technologies to their cumulative production and deployment, with learning rates typically expressed per doubling of their installed capacity. A widely used performance measure is the unit cost of installed energy capacity of a technology such as solar PV modules. The estimated learning rates for variable renewable generation technologies are 23 per cent for solar PV modules, 19 per cent for their inverters and 5 per cent for wind turbines (Figure 4.3). For energy-storage technologies, estimated learning rates are 19 and 16 per cent for lithium-ion battery packs for electric vehicles and utilities, respectively, and 13 per cent for vanadium-redux flow batteries. The learning rate for pumped hydroelectric systems, a mature technology, is minus 1 per cent. Estimated learning rates for proton-exchange membrane fuel cells in stationary applications and electrolysers—two related technologies—are similar at 16 and 18 per cent, respectively. However, the rate for solid oxide fuel cells (SOFC) is minus 2 per cent.

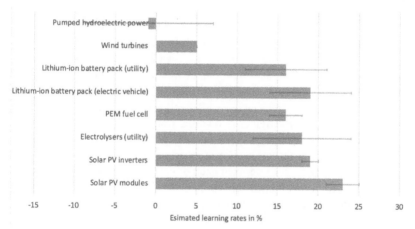

Notes: Estimated learning rates are based on linear regressions of technology prices per unit of nominal energy capacity and cumulative deployed nominal capacity for each technology. The learning rate uncertainty range is the 95 per cent standard error confidence interval where available.

Sources: Data are from Schmidt et al. (2018) and, for wind turbines, Lafond et al. (2018).

Figure 4.3   Estimated learning rates from experience curves for low-carbon technologies (in per cent per doubling of deployed capacity)

Characteristics of technologies with relatively high learning rates are that they are component technologies and mass-manufactured rather than complex capital investment projects (Huenteler et al., 2016). Solar PV modules and inverters, lithium-ion battery packs for electric vehicles and utility storage as well as electrolysers and fuel cells are manufactured goods with potential for standardization and mass production. However, not all such technologies necessarily achieve high learning rates, as the experience with SOFC shows. In contrast, large complex equipment and investment projects can show slower learning rates, as with wind turbines and pumped hydroelectric projects, or indeed significantly negative ones, as with second-generation nuclear power plants (Grübler, 2010; Escobar Rangel and Lévêque, 2015).

An in-depth study of technical performance improvements in solar PV modules from 1980 to 2012 provides further insights into their under-pinning factors (Kavlak et al., 2018). This study breaks down the overall cost of modules into cost components that are related to different performance-improvement mechanisms. They include public and private R&D investments, experience-based learning from manufacturing and scale economies in manufacturing arising from fixed costs (plant size). This detailed analysis of reductions in costs of the components of solar PV modules points to the significant contribution of public and private R&D investment, especially in the period 1980 to 2001, when the technology was deployed primarily in niche applications (Figure 4.4). This investment in knowledge delivered an estimated 60 per cent of total module cost reduction during this early stage of its development, primarily from increased PV cell and silicon usage efficiencies that flowed from R&D activities. However, from being a minor contributor in the first period, economies of scale became a significant one in the second, from 2001 to 2012. In this period, R&D investment and scale economies contributed an estimated 40–45 per cent each of cost reductions. Throughout the two time periods, the estimated contribution of experience-based learning was relatively small, about 10 per cent overall.

This study also examined how government policies influenced the three technical performance improvement mechanisms over the three decades. The analysis assumes that market-creating policies such as feed-in tariffs for electric power from renewables and renewable portfolio standards, catalysed private R&D investments, experience-based learning and scale economies. The respective contributions of public and private R&D are difficult to distinguish, but R&D expenditures were evenly split between public and private investments (Nemet and Kammen, 2007). If public and private R&D contributions are assumed to be evenly split as well, market-creating policies fostered about 60 per cent of cost reductions over both periods, 1980 to 2001 and 2001 to 2012 (Figure 4.5). In other words, the market-stimulating policies both directed significant private investment in R&D towards solar PV and

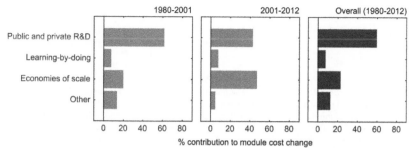

Notes: The three panels show the estimated contribution of public and private R&D, learning-by-doing, economies of scale and other factors to solar PV module cost decline in 1980–2001 (left), 2001–2012 (middle), and 1980–2012 (right). Those technology changes that required a lab setting or a nonroutine production activity (for example, experimental production line) are attributed as being caused by R&D. Improvements as a result of repeated routine manufacturing activity are attributed to learning-by-doing and changes that result from increases to the scale of the module manufacturing plant and from volume purchases of materials to economies of scale.
Source: Reprinted from G. Kavlak, J. McNerney and J.E. Trancik (2018), Evaluating the causes of cost reduction in photovoltaic modules. *Energy Policy*, 123, 700–710, with permission of Elsevier. https://doi.org/10.1016/j.enpol.2018.08.015.

Figure 4.4   *Estimated contributions of R&D, learning-by-doing and scale economies in the decline in solar PV model costs (in per cent, 1980–2012)*

drove down unit costs through significant scale economies and, to a lesser extent, experience-based learning (Bettencourt et al., 2013; Huenteler et al., 2016; Kavlak et al., 2018). Public R&D spending contributed the balance of unit cost reductions.

In principle, returns to upfront private investments in experience-based learning and scale economies arise from market rents created and managed by the Schumpeterian process of 'creative destruction'. This process consists of successive waves of disruptive new technologies reaching and penetrating markets. But for at least some low-carbon technologies, this engine of market dynamics and technological advances may sputter. Some knowledge accumulated from successful new technologies and their use can diffuse across firms, and some scale economies are external to firms, especially as complementary low-carbon technologies are integrated into energy systems. Both processes dull returns to the upfront investments in innovation and market creation. Moreover, returns to some low-carbon technologies can depend on long-run government policies, such as emissions pricing and electricity market reforms, which face potential time-inconsistency concerns (Chapters 5, 6 and 9).

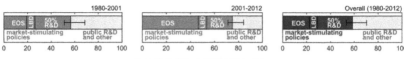

% contribution from market-stimulating policies

Notes: The three panels show the estimated contributions market-stimulating policies such as feed-in-tariffs and renewable portfolio standards to solar PV module cost reduction in 1980–2001 (left), 2001–2012 (middle), and 1980–2012 (right). R&D =research and development, LBD =learning-by-doing, EOS=economies of scale, Other=other mechanisms such as spillovers from other industries such as semiconductors. Scale economies, learning-by-doing, and private R&D were all catalysed by market-stimulating policies. Data does not allow for separate the effects of private and public R&D. To accommodate this, 50 per cent of total R&D contributions are added to the contributions of scale economies and learning-by-doing. This effectively assumes that 50 per cent of R&D improvements came from private R&D, broadly in line with the share of R&D expenditures in low-carbon energy. Uncertainty bars show the total contribution from market-stimulating policies that would result under alternate assignments to mechanisms, without accounting for uncertainty in the private R&D estimate.
Source: Reprinted from G. Kavlak, J. McNerney and J.E. Trancik (2018), Evaluating the causes of cost reduction in photovoltaic modules. *Energy Policy*, 123, 700–710, with permission of Elsevier. https://doi.org/10.1016/j.enpol.2018.08.015.

Figure 4.5   *Contribution of market-stimulating policies of solar PV module cost reductions (in per cent, 1980–2012)*

ACCUMULATING NEW CAPABILITIES AND KNOWLEDGE SPILLOVERS

Investments in knowledge, whether as R&D activities or as by-products of production and use of a new technology, face a general appropriability problem. Returns to successful innovations and technological advances not only accrue to the innovating firm but can also spill over to competing firms and those firms and households that purchase the innovative product (Arrow, 1962a and 1962b). The knowledge produced by an innovating firm can flow to rivals in the same industry and to firms in related industries through several channels. They include incomplete patent protection, an inability to keep innovations secret, and potential for reverse engineering and imitation. Benefits of innovation can also flow to other firms and households if the new product can only be imperfectly differentiated from its market alternatives so that the full value of the innovation cannot be reflected in its price. For example, product differentiation of automobiles affords more scope to capture value from investments in innovation and market creation than that of electric power, a relatively homogenous commodity. These appropriability problems could lead to underinvestment by private firms in innovation and early deployment of low-carbon technologies compared to a social optimum that accounts for both private returns and social value they create, in addition to any environmental externalities that are inadequately internalized.

A large number of studies across many countries and economic sectors find that the private returns to investments in R&D activities are usually higher than those to physical investments. Estimated private real rates of return on R&D investments fall largely within the range of 10–30 per cent, suggesting that firms are able to appropriate significant returns to R&D investment (Hall et al., 2010). Nevertheless, estimated social returns to R&D investment are usually substantially greater than private returns. The extent of such knowledge spillovers, though, varies widely across industries and countries. Across industries within the United States, they range from nil to 100 per cent (Hall et al., 2010). This wide range could reflect variations in the extent to which new products that benefit from the knowledge spillovers complement or substitute for the initial innovation. There is also evidence of international knowledge spillovers, which operate through international trade and investment as well as through the general diffusion of formalized knowledge through various media.

While investments in R&D yield expected rates of return above normal returns, they have two characteristics that distinguish them from ordinary investments that can constrain them (Hall and Lerner, 2010). Firstly, successful R&D activities create intangible assets from which future profits may be generated. These assets, though, are embedded partly in the knowledge and skills of firm employees, who may at some point separate from firms that funded the R&D activities. Moreover, because such assets are intangible they cannot normally be used to secure debt financing. Secondly, the variance of returns from investments in R&D across all sectors is much higher than that for other types of investment and heavily skewed towards relatively few highly successful R&D projects (Scherer and Harhoff, 2000). The combination of high uncertainty, which is difficult to manage through R&D sequencing (real options) and diversification, and intangible knowledge assets makes market investment and financing of R&D activities more difficult than for tangible investments. Evidence shows that, for larger firms, internal cash flows are a much more significant source of funding for R&D than external debt (B.H. Hall and Lerner, 2010). Smaller firms with less internal cash flow and access to financial markets can thus face constraints on their R&D activities due to financial market imperfections.

As a new technology reaches the market, learning from the experiences of producing and using it accumulates knowledge as a by-product of these other activities. If the accumulation occurs entirely within a firm, this dynamic scale economy is similar in effect to economies of scale that can arise from large fixed costs. If firms can finance cost-effective knowledge-accumulation and reach an efficient scale, there is no constraining market failure. However, if this accumulation of knowledge and capabilities from experience-based learning—or from R&D activities—occurs across interrelated firms (a supply chain) or in a workforce, it spills over across many firms and employees. In fact, there is a relatively high tendency for knowledge-intensive firms, innovation activities and knowledge workers to concentrate in geographical clusters—more so than for other economic activities—suggesting a role for physical proximity in such spillovers (Audretsch and Feldman, 2004). While it is possible to internalize these spillovers at least partially through commercial relationships and contractual agreements, negotiating and contracting costs bind the scope of such arrangements. To the extent that they cannot be internalized, experience-based learning faces a similar appropriability problem that affects R&D investments.

In addition, new technologies often require significant periods of formative production and deployment as they mature and attain the cost and performance characteristics that make them attractive to and widely understood by potential customers. Firms build on successful innovations and learn from the production and use of technologies to sustain improvements in technical performance and costs. At this stage of advancing technologies, customer engagement increasingly influences technological progress. Information flows from early technology adopters and potential followers also become more important. The underpinning processes of early technology deployment and increasing diffusion are thus information flows—learning from experience, adapting technologies and developing knowledge and capabilities of producers and customers (Geroski, 2000).

While it is not possible to directly observe such information flows, indirect evidence suggests that they could be important to the early deployment and increasing diffusion of new technologies. For example, there is a tendency for clustering of both technological innovation and early adoption within regions of countries (Baptista, 2000). Clusters of innovators and early adopters within a region tend to dominate initially, with slower diffusion of a new technology beyond an initial region to others in a country. This effect is stronger in early deployment of a technology and fades as it matures and diffuses more widely. This evidence is consistent with knowledge spillovers between producers and customers of a new technology and among their users, especially if experience-based knowledge remains informal and uncodified and physical proximity facilitates information flows.

The international diffusion of new technologies follows similar spatial patterns. But rather than through proximity, new technologies and their embodied knowledge flow through imports and foreign direct investments, as well as formally codified knowledge through various media channels (Keller, 2004). Innovations tend to diffuse from the country in which they originate and spread from more to less proximate countries over time (Comin and Hobijn, 2010). The pace of such diffusions has hastened over time, with more recent innovations across a range of technology fields and characteristics diffusing more quickly than those of the second industrial revolution. This could reflect both rapid growth in international trade and foreign investments—globalization— as well as growing capacities to absorb and adapt new technologies across countries (Mancusi, 2008). Moreover, the significance of geography in the diffusion of technologies tends to fade as technologies mature (Comin et al., 2012).

The evidence on knowledge spillovers from innovations and early deployment of low-carbon technologies is broadly in line with that for new technologies in general. Estimated spillovers from low-carbon-technology patents worldwide in electricity and transport indicate that they are significant and comparable in scope to those from information and computing technologies (Dechezleprêtre et al., 2017). Based on a measure of patent citations that takes account of both the forward citations of a patent and how frequently the forward citations are cited, low-carbon energy technologies also generate more knowledge spillovers than the fossil fuel-based technologies for which they substitute. Moreover, among US firms, low-carbon-technology patents are cited more frequently and have more general application across a range of technology fields than patents in other technology fields, except those for computers (Popp and Newell, 2012). Energy-related innovations also benefit significantly from knowledge spillovers from other sectors, as evidenced from patent citations. This spillover is significant from chemicals and electronics to energy but less so from information and communication technologies, at least up to that point in time (2012) (Nemet, 2012a).

Analysis of low-carbon-technology R&D also finds that, consistent with overall evidence on R&D returns, those on low-carbon-technology innovations are relatively high, highly variable and skewed towards a few successful innovations. Using citations of past patents in new patent applications as a measure of the value of the patents, three-quarters of US patents filed for low-carbon technologies such as wind, solar PV, nuclear, fuel cells and hybrid vehicles attract no—or only a few—forward citations (Popp et al., 2013). However, there is a select group of highly cited patents that account for most of the forward citations, and these successes are the culmination of several advances building one upon another.

The early deployment and increasing diffusion of low-carbon technologies also follows a spatial pattern similar to that of new technologies in general. Low-carbon-technology innovations occur more frequently in firms which have previously engaged in low-carbon innovation and are exposed to more low-carbon innovations in their home markets. Such effects on innovation at firm level are found with electric and hybrid drivetrain technologies for vehicles as well as renewable-power-generation and energy-storage technologies (Noailly and Smeets, 2015; Aghion et al., 2016; Lazkano et al., 2017). This suggests that firms' knowledge stocks, be they acquired from R&D skills or experience-based learning, are associated with knowledge spillovers from other firms or through labour markets. They also create 'path dependencies' in innovation among firms and locations. Moreover, early adopters of low-carbon technologies tend to be in the same country or region where innovations and low-carbon-technology production occur, although this locational association is shaped significantly by market-creating government policies (Chapter 8).

The wedge between private and social returns to innovations and the presence of imperfect market mechanisms to internalize knowledge spillovers among firms and their customers attest to their economic significance. Private returns to innovation are relatively high, but highly variable and heavily skewed towards a few successful innovations. They can be sufficient to support significant private investments in R&D, including in low-carbon technologies, when directed by market and policy developments and support by government policies such tax credits for R&D activities. Similarly, imperfect market arrangements, such as geographical clusters of innovative firms, commercial relationships and contracting among them and clusters of innovative firms and early adopters, can be sufficient to internalize knowledge spillovers. These arrangements impose market disciplines on upfront investments in innovation, experience-based learning and scale economies. Such disciplines help to ensure that these investments are privately and socially beneficial. However, they do not necessarily ensure that all socially beneficial upfront investments are undertaken by private firms. What is a sufficient condition in economics is often treated as a necessary condition in policy making, typically to conserve the role of markets and limit scope for government policy interventions and potential policy failures.

SWITCHING COSTS TO LOW-CARBON TECHNOLOGIES AND SPILLOVERS FROM INCREASING SCALE

In addition to market barriers from knowledge spillovers, there are several technical performance characteristics of existing energy systems which impose high switching costs to alternative technologies, at least initially. Energy service and material demands are largely shaped through sustained investments in long-lasting built environments, such as road layouts, rail networks, land-use patterns, buildings, factories and specializations in economic activities (Seto et al., 2016). Within this built environment of long-lived infrastructure, conventional thermal electric power-generation plants, oil refineries and other heavy industrial plants have relatively long expected asset lives (30–50 years), as well as some types of heavy transport vehicles such as ships and aeroplanes. Other energy-using technologies have somewhat shorter asset lives

(10–30 years), such as road transport vehicles and boilers for buildings. Given current investment trends in some incumbent technologies, there is a risk of some being scrapped prematurely before the end of their expected asset lives, which would add to future switching costs in those sectors (Vogt-Schilb et al., 2018). Such high costs could arise from the expected long asset lives of some current technologies and their incompatibility with emission-reduction goals of limiting the expected change in global surface temperatures to well below 2°C or 1.5°C. Investments in new coal-fired thermal generation plants would be an example of such a risk.

At the same time, customers can face various types of switching costs in adopting new technologies. These costs can vary with characteristics of firms and households that are potential technology adopters. They include the nature and vintage of their existing stocks of capital, skills and capabilities for adopting new technologies, and other input costs that they face. The characteristics of the new technology matter too, such as their degree of simplicity or complexity, and interdependencies with existing technologies (Chapter 1).

The existing capital stock is the basis from which energy systems are to transform, and the economic lifespans of these assets significantly affect the costs of change, with very long-lived assets tending to lock-in existing energy demands by increasing switching costs. For example, the configuration of existing industrial plants shapes the feasibility and costs of adopting low-carbon alternatives, as does the thermal efficiency of existing building fabrics. The location of dwellings, schools, commercial buildings and industrial plants together with transport infrastructures shape daily commuting choices. They create inertia in existing systems, at least until the long-lived assets are renewed or replaced. For example, technologies that are long-lived and components of interconnected technologies and infrastructures tend to take the longest to diffuse (Chapter 1). Notable exceptions, though, are jet engines and nuclear power plants, which government interventions helped to accelerate owing in part to military and security aims.

There are in addition significant network effects within existing energy systems and scale economies in manufacturing some component technologies for systems. For early adopters of alternative technologies, high switching costs can arise from network effects of existing energy infrastructures and scale economies (Arthur, 1989). For example, potential early adopters of low-carbon heating technologies in buildings and battery electric road vehicles face high costs. The utility derived by early adopters of battery electric or fuel-cell electric vehicles is constrained by limited recharging and refuelling infrastructures for these alternative technologies. If the stock of alternative vehicles expands together with their supporting infrastructures, the utility of these vehicles increases and lowers the switching costs of later adopters. But this new network effect requires standardization of component technologies in vehicles and energy supply which are often developed jointly by government and industry. More generally, just as cost complementarities among incumbent technologies in current energy systems lower overall system costs of delivered energy services, there are also significant cost complementarities among component technologies of low-carbon energy systems.

Early adopters of alternative technologies that have potential to be manufactured and deployed at a global scale also face relatively high unit costs before they reach mass production—costs which are allocated by markets among their producers and customers, or taxpayers via government policies. These cost characteristics of incumbent technologies in existing energy systems raise switching costs for early adopters, while investments by early adopters and innovating firms in market development lower costs for later adopters. Examples of low-carbon technologies that have potential to be manufactured at global scale include solar PV modules and inverters, battery packs for vehicles and stationary applications, electric motors, electrolysers and fuel cells. Other low-carbon technologies are more complex and less amenable to mass production, including wind turbines, pumped hydroelectric and compressed-air energy-storage facilities, and carbon dioxide capture and storage infrastructures. Large-scale nuclear power plants share these characteristics as well, but the R&D focus on small modular nuclear reactors aims to transform this technology to one involving modular manufacturing.

R&D SUPPORTS AND MARKET-CREATING POLICIES FOR LOW-CARBON TECHNOLOGIES

Policy challenges in advancing low-carbon technologies include addressing knowledge spillovers arising from R&D activities, innovation successes and experience-based learning from their early production and use. Economies of scale in producing some incumbent technologies and network externalities from their integration in current energy systems can also create barriers to the uptake of low-carbon-technology alternatives. Some aspects of policies to advance low-carbon technologies are the same as policies for innovation in general and attract a broad consensus in economics. For example, fiscal supports and financing programmes that help innovating firms internalize knowledge spillovers from R&D typically apply to all R&D activities rather than to specific technologies. Targeted government R&D spending on low-carbon technologies and financing programmes for small and medium-sized firms are also fundamental. Other policy aspects are less well settled. They include market-creating and industry-supporting (industrial) policies that support early adoption and production of low-carbon technologies. They also help direct private investments in innovation towards low-carbon technologies. The main alternative to market-creating and industrial policies for low-carbon technologies would be emissions pricing. The innovation and industrial policy responses to the challenges in advancing low-carbon technologies are considered in turn. Emissions pricing is the focus of the next chapter.

Innovation Policies

Evidence on low-carbon innovation as measured by patents points to significant knowledge spillovers from successful innovations and a role for government policy to internalize their value into R&D investment decisions. In theory, government spending to support R&D should equal the value of the knowledge spillovers, despite protections afforded by patents, so that firms take them into account in their innovation decisions (Goulder and Schneider, 1999).[1] But in practice, it is difficult to measure the size of these spillovers and to calibrate accordingly government support. A large number of empirical studies finds wide variation in the estimates of the private and social returns to R&D investments. Estimated private returns to R&D are typically in the range of 10–30 per cent, with a meta-study finding a median private return of 14 per cent (Hall et al., 2010; Ugur, 2016). Estimates of social returns to R&D are similarly wide ranging: 30–50 per cent, which is around double private returns (Hall et al., 2010). An economy-wide estimate of the social return to R&D investment for the United States is 30 per cent (Jones and Williams, 1998).

The returns to R&D, however, vary across industries and countries. The estimated increases in firm and industry value added from more R&D investment tend to be higher in R&D-intensive sectors, although the associated rates of return are not necessarily higher. Across countries, the returns to R&D vary with the level of technological development, with the productivity of R&D tending to be lower in technologically more advanced industrialized countries because of diminishing returns to scale. Subsequent countries benefit from knowledge spillovers from leading countries, depending on their capacity to adopt new technologies, which is developed through their domestic R&D investments (Mancusi, 2008). Thus, while available evidence provides broad indications of the value of knowledge spillovers and factors that influence them, it does not provide precisely prescribed amounts.

Governments financially support R&D activities in three main ways. They are: fiscal incentives for R&D, such as tax credits for private firms; financing programmes for small and medium-sized enterprises; and direct funding for basic research and R&D government research laboratories (Hepburn et al., 2017). Corporate tax credits against corporate taxes owed (and other tax benefits for R&D, like accelerated depreciation) are granted in many countries for corporate spending on activities that qualify as research as defined by government. These tax incentives help make marginal R&D projects in private firms more profitable and can encourage them to invest in innovations that are close to market (Hall and Van Reenen, 2000). However, they do not necessarily target the types of R&D with the largest knowledge spillovers, nor do they support technologies that are not close to market. Such targeting is difficult to calibrate and allocate in practice.

Evidence shows that tax incentives do significantly increase R&D spending by firms, and this in turn is associated with new patents that increase firms' market values (Bloom et al., 2002; Hall et al., 2005). However, direct evidence on the relationship between R&D tax credits and innovation outcomes is limited. For example, a study of low-carbon innovation worldwide from 1986 to 2005 finds a significant boost to energy-efficiency patents from R&D tax credits, but not to patents for electric vehicles, solar PV and wind technologies (Aghion et al., 2016). However, there is evidence that the pace of innovation in these technologies increased in the 2000s and beyond as governments stepped up their early-deployment support for them (Bettencourt et al., 2013; Huenteler et al., 2016; Kavlak et al., 2018).

Direct government grants and loans schemes with grant elements for private firms provide R&D support in ways that can both internalize knowledge spillovers and ease financial market constraints on R&D investment. Grants and loans are typically allocated through competitive mechanisms for evaluating project proposals, and these mechanisms can be used to help to direct supports by firm type and technology domain; for example, to smaller firms that may face financial market constraints or technologies with high knowledge spillovers. A study of a German grant scheme for R&D projects finds that, for matched sets of firms that received and did not receive grants, those that received one allocated more funding to R&D and the additional private and public R&D spending was associated with more patents (Czarnitzki and Hussinger, 2018). This effect was greater for small firms than for larger ones, which is suggestive of potential financial market constraints on R&D investment by smaller firms. A study of a US Department of Energy grant scheme for R&D projects by small firms finds similar effects on R&D spending and patenting outcomes (Howell, 2017). In addition, increased patenting and spillovers to technologically related firms are seen from a UK policy that increased R&D tax credits for small firms, which subsidized R&D investment and increased after-tax cash flows (Dechezleprêtre et al., 2019).

In addition to support for R&D investments by private firms, governments provide funding for basic research by universities and undertake direct research in government institutes. In general, university research tends to produce the more highly cited research output (Hepburn et al., 2017). But in fields related to low-carbon technologies, government-funded research in the United States, where there is a network of government energy-research institutes, is particularly impactful (Popp, 2017). Low-carbon-technology patents assigned to the US government are more highly cited than those from private firms, and the government patents are more likely to be cited in patents than similar articles from universities. These findings suggest that the US government energy-research institutes build effective links between basic research in universities and applied research in firms. There is also evidence that government research support effectively extends to demonstration projects at a commercial scale, helping to overcome barriers at this stage in advancing new technologies (Mowery et al., 2010; Weyant, 2011).

Heterodox Market-Creating and Industry-Supporting Policies

While economic principles and empirical evidence provide relatively clear guidance on the need for and effective forms of government support for R&D activities, the economic case for supporting early deployment of new technologies is less clear-cut. Most market-creating policies for low-carbon technologies subsidize early adopters to invest in low-carbon alternatives that face relatively high costs compared to those in incumbent ones. Such policies tend to be adopted in countries that specialize in innovations and manufacturing low-carbon technologies, with a view to indirectly benefiting firms that make upfront investments in innovations and market creation (Chapter 8). These market-creating policies are thus also domestic-industry-supporting. By supporting customer investments rather than those of producers, their design helps to maintain competition among firms producing low-carbon alternatives as a further spur to advancing these technologies.

An economic rationale for such government intervention based on market imperfections is dynamic external economies—similar to the economic case for tariff protection of infant industries (Cordon, 1974, pp. 257–64). Examples of such potential scale economies external to producing firms are knowledge spillovers among related producers and among end users that hasten technological advances and diffusion of new technologies. The geographical clustering of firms producing specific low-carbon technologies is suggestive of such spillovers. Information flows between producers and customers could also accelerate technological progress, as suggested by geographical clustering of innovating firms and early adopters of new technologies. In addition, inherent to all energy systems, whether current or low-carbon, are significant positive cost and profit spillovers among complementary technologies. These cost complementarities are fundamental to minimizing overall system costs of delivered low-carbon energy services and materials. There are also network effects among energy customers. However, firms making upfront investments in innovation and market creation are not necessarily able to internalize sufficiently these cost and profit spillovers through commercial relationships and expansions of firm scope. If not, there would be under-investment in them in competitive markets (Spence, 1976; Dixit and Stiglitz, 1977).

But as with knowledge spillovers from R&D and successful innovations, it is difficult to directly estimate the value of these positive cost spillovers from dynamic external economies. Available estimates for wind turbines and solar PV panels suggest that dynamic external economies are relatively small and that the value of accumulated experience-based knowledge depreciates relatively quickly (Nemet, 2012b; Kavlak et al., 2018). However, these assessments impose assumptions about how information flows among firms and what types of investments in knowledge contribute to different types of technical performance improvements. For example, an assessment of the solar PV module manufacturing industry found that R&D investments of firms significantly enhanced their capacities to absorb technologies and their embodied innovations from other firms rather than directly produced innovations that improved technical performance (Hoppmann, 2018). Nevertheless, it would appear that such external dynamic economies among firms are relatively small, certainly compared to the scale of supports provided by market-creating industrial policies for wind and solar PV.

The larger dynamic efficiency gains, especially for those low-carbon technologies which can be manufactured at a global scale, appear to arise from scale economies in manufacturing. These benefits can in principle be realized by manufacturing firms if they are able to differentiate the quality of their innovative products and attract current and future price premiums to remunerate the sunk costs of developing new technologies and markets. For example, the differentiation of battery electric vehicles as also being premium automobiles (for example, Tesla) allows product pricing that at least partially remunerates their relatively expensive battery electric drivetrains in their development and early deployment.[2] In contrast, liberalized wholesale markets for electric power appear less supportive of such a product-differentiation and innovation strategy. For example, liberalization of the UK wholesale electricity market was associated with an abrupt drop in the sector's R&D activities and innovation (Jamasb and Pollitt, 2011 and 2015). More generally, product characteristics and market structures influence the extent to which investments in innovation and market creation are remunerated in markets. Government-designed wholesale electric power markets and more traditionally organized power systems appear less remunerative of investments in these upfront investments than other markets (Newbery, 2018).

But just because in principle industrial policies for low-carbon technologies may be warranted in some contexts, this does not necessarily mean they should be pursued. Common criticisms of such policies are their ineffectiveness and vulnerability to rent-seeking by industry interests. The apparent success of such industrial policies for wind and solar PV as well as lithium-ion battery packs and electric vehicles offers anecdotal support for their potential effectiveness (Chapter 3). While these examples clearly suffer from selection bias, they nevertheless account for most of the real resources that governments directed towards early deployment of low-carbon technologies. The argument that these policies were misdirected would need to establish a plausible counterfactual argument of better low-carbon technologies in timescales consistent with long-run climate goals.

These low-carbon-technology anecdotes are consistent with broader empirical evidence on market-creating and industry-supporting policy interventions.[3] Evidence on the potential effectiveness of industry-supporting policies comes from evaluations of regionally targeted investment supports in manufacturing. These studies use variation in eligibility over time and threshold effects in eligibility at points in time to isolate policy impacts. Evaluations of place-based investment subsidies in Italy and the United Kingdom that aimed to promote regional growth through manufacturing investment show that they significantly increased investment and sustained growth in activities that benefited from them (Bernini and Pellegrini, 2011; Cerqua and Pellegrini, 2014; Criscuolo et al., 2019). Similar enduring impacts are seen from government procurements of aeroplanes on airframe manufacturing (Jaworski and Smyth, 2018). There is also evidence that investment subsidies in Chinese manufacturing sectors that were allocated in sectors exposed to international competition saw relatively strong increases in productivity over time (Aghion et al., 2015). Sectors receiving producer subsidies but protected by tariffs and sheltered from international competition performed less well. This finding points to the importance of industrial policy designs that maintain market competition in driving performance improvements and curbing rent-seeking.

The relationship among competition, innovation and productivity gains is among the more complex in economics. A Schumpeterian view sees a degree of transitory market power as necessary for innovation because this market power provides a return to investments in it. In contrast, a more orthodox view sees product market competition as the spur to innovation because market incumbents with some degree of market power have more to lose from disruptive change (Arrow, 1962b). But if both market incumbents and new entrants have available to them potentially disruptive innovations, the ability of incumbents to shape market developments is attenuated by new market entrants and potential competition (Gilbert and Newbury, 1982; Aghion and Howitt, 1998, p. 223–4). In these circumstances, market incumbents with some degree of market power could have a stronger incentive to innovate than would those operating in perfectly competitive markets, though the strength of this effect can be attenuated by organizational and technological switching costs faced by incumbents (March, 1991; Holmes et al., 2012). In the context of transforming energy systems, time-bound government commitments to net zero emissions, to the extent supported by comprehensive domestic energy-reform strategies to implement them, can have the effect of increasing potential competition from low-carbon alternatives. Such policy strategies involve government supports for R&D activities, including those to ease financing constraints faced by small and medium-sized firms, and investments in market creation for low-carbon alternatives which directly reinforce the climate goal. They in turn can direct investments towards innovations and market creation for low-carbon alternatives, strengthening competition in enabling energy-system transformations.

CONCLUSION

The journey from innovation towards widespread diffusion of low-carbon technologies within energy systems and across countries is necessary to achieve net zero emissions of carbon dioxide from energy, a transformation that has been underway for some time. This awakening of low-carbon technologies took its early directions from the oil price shocks of the 1970s and 1980s and safety concerns from nuclear accidents during this period. Subsequently, amidst the growing sense of urgency to act decisively to limit climate change, another oil price shock and broader economic policy responses to the great 2008–09 financial crisis expanded significantly the market-creating and industry-supporting policies for wind and solar PV generation technologies and in time automobiles with battery electric drivetrains. It is this sequence of energy shocks and policy measures that have advanced low-carbon alternatives in what are now seen as relatively easy-to-decarbonize sectors like much of electric power generation and light-duty road vehicles. Looking forward, the broader economic policy responses to the public health and economic crisis caused by the novel coronavirus pandemic could usher in the next wave of reforms that advance low-carbon technologies. But regardless of how these policies develop, the 2015 Paris Agreement commitment to net zero emissions in the second half of this century is potentially significant in orienting further investments in innovations and market creation for low-carbon technologies that are necessary to achieve this goal (Chapters 5 and 9). This commitment, though, has yet to attract the implementing reform strategies that would strengthen its credibility and economic impacts.

Available evidence suggest that there are material market imperfections confronting innovations related to low-carbon technologies. Some are similar to those for new technologies in general, such as knowledge spillovers from R&D. For example, evidence on knowledge spillovers from low-carbon-technologies patents in electricity and transport are significant and comparable to those from information technologies. Similarly, there is significant clustering of low-carbon-technology-related innovations among their manufacturing firms and locations. But perhaps the most distinguishing features of innovations related to low-carbon technologies and their early deployment are associated with external scale economies and network effects among complementary technologies within energy systems. Such positive cost spillovers from early deployment of low-carbon technologies has been a significant feature of the ongoing energy-system transformation.

The galvanizing policies so far in advancing low-carbon technologies and transforming energy systems have been heterodox industrial policies, departing from the economic orthodoxy of using government supports for R&D and emissions pricing alone to incentivize and guide change. In countries with capabilities and specializations in innovation, national innovation systems were increasingly oriented towards low-carbon technologies, as in China, Denmark, France, Germany, Japan, South Korea, the United Kingdom and the United States (Chapter 8). The main means of directing private R&D activities were market-creating and industry-supporting policies for new renewable generation technologies and alternative drivetrain vehicles, alongside government R&D spending directed towards low-carbon technologies. Government fiscal incentives for private R&D investments, which underpin these innovation systems, remained largely targeted at activities deemed to be R&D rather than specific technologies. Industrial policies also helped realize scale economies in low-carbon technologies, especially in manufacturing solar PV modules and lithium-ion battery packs which are amenable to mass production. A key departure from 'first-best' policies espoused by economic orthodoxy was this use of market-creating and industry-supporting policies for low-carbon technologies rather than emissions pricing to advance directly these alternatives. The next chapter turns to emissions pricing.

NOTES

1. The focus here is on incentives associated with fiscal policy instruments rather than patents. Evidence on the economic impacts of patents is that their existence imparts stronger incentives for R&D and promotes greater patenting. However, marginal variations in the strength of patent protections, such as lengthening coverage periods, elicits little incremental response. See, for example, Hall and Harhoff (2012) and Maskus (2012).
2. For a theoretical and empirical assessment of the potential for product differentiation in a competitive global automobile manufacturing industry to support innovation, see Hashmi and Van Biesebroeck (2016).
3. Lane (2020) provides a useful survey of evidence from microeconomic evaluations of industrial policies on which this section draws.

5.   Calibrating emissions pricing

Levying a price on carbon dioxide and other greenhouse gas emissions consistent with long-run climate goals would align the direction of innovation and investment choices with these goals. This policy would encourage firms and households to internalize the climate externality in their short- and long-run decisions, which is central to transforming energy systems and stabilizing the climate.

In principle, the policy appears straightforward to implement. Consider a socially optimal climate goal informed by the social benefits and costs of limiting climate change as best assessed. These benefits include the avoidance of net social and economic losses from climate changes that would not occur, and the costs are the additional real resource costs of cutting emissions while maintaining economic and household activities. The socially optimal pathway for reducing emissions is the one along which marginal social benefits and costs are balanced. The social cost of carbon dioxide emissions (SC CO_2) reflects this balancing of marginal benefits and costs. One way to implement this goal thus is to impose a tax (or equivalent) on carbon dioxide emissions at the SC CO_2. However, implementation of this approach is encumbered by highly uncertain and widely ranging estimates of SC CO_2, which can hold back the long-run investments necessary to achieve the Paris Agreement goals.

Often in economic planning, whether in a large organization or for an economy, quantities form the basis of planning goals and serve as instruments of control in place of or together with prices (Weitzman, 1974). For example, the Paris Agreement goals are broadly expressed quantities—a global warming limit and attainment of net zero emissions in the second half of this century. These expressions have associated with them a 'shadow' price of the long-run binding emissions constraint, which is the cost of meeting the constraint. It could be implemented by either quantitative regulation of emissions or emissions pricing. In the latter case, the IPCC finds that the range of estimated cost-effective economy-wide emissions prices for a given target emissions pathway is much narrower than those for the SC CO_2 (Rogelj et al., 2018a). This finding is unsurprising because many of the uncertainties buffeting the SC CO_2 are subsumed within an assumed given quantitative emissions target.

The challenge for formulation and implementation of climate goals, including through emissions pricing, is not to make the inherent uncertainties confronting the goals and implementing policies somehow disappear. That would be seeking the impossible. Rather, the challenge is to set out the goals and implementing policies in ways that manage relatively effectively the inherent uncertainties which they confront and to which they must respond over time (Jaeger and Jaeger, 2011). Moreover, in assessing their potential effectiveness, it is necessary to consider not only the uncertainties involved, but also other factors that can influence policy effectiveness. These factors include missing markets for future emissions prices, arbitrary changes in government policy implementation, distributional impacts of emissions pricing and potential time inconsistencies in its application (Newbery 2016b; Stiglitz, 2019; Stern, 2021; Stern and Stiglitz, 2021). These considerations point to the importance of using complementary policies to build policy credibility, including fostering domestic interests in low-carbon alternatives and emissions pricing (Chapters 8 and 9). For example, there are potentially significant complementarities between government policies that support innovations and market creation for low-carbon technologies and emissions pricing, especially if well sequenced. An important test of credibility of energy reform strategies is their effectiveness in accelerating the long-run investments in low-carbon alternatives, including upfront investments in innovation and market creation for low-carbon technologies (Chapter 9).

The focus here, however, is on the empirical bases for calibrating emissions pricing, which is assumed to take the form of an emissions tax rather than a cap-and-trade system. A tax manages better interactions with complementary innovation and industrial policies to advance low-carbon technologies and provides greater certainty for business planning and investment (Metcalf, 2019, pp. 78–85). Consider first the empirical basis for cost-effective emissions pricing consistent with a given quantitative emissions target, as this is the simpler approach and can be adapted to sector contexts. The main conceptual building block for this heterodox policy approach is the long-run marginal costs of cutting net emissions, allowing for the impacts of market-creating and industry-supporting policies to advance low-carbon alternatives. For example, these policies have contributed significantly to lowering the costs of cutting emissions in so-called 'easy-to-decarbonize' sectors, but others remain 'difficult'.

The evidence for emissions pricing using orthodox social cost–benefit analysis is much more complex and highly uncertain and applies to whole economies. In addition to long-run marginal costs of abating emissions, long-run marginal costs of adapting to climate changes are necessary too. Social cost–benefit analysis also requires projections of net social and economic losses from climate changes while allowing for cost-effective adaptation measures. These losses reach far into the future, and are inherently and highly uncertain and subject to value judgements in their equity weighting and discounting to present values.

These elements of the evidence base for calibrating emissions pricing are considered in turn. The chapter concludes with observations about how to effectively combine time-bound goals to achieve net zero emissions, government policies for supporting innovations and market creation for low-carbon technologies and emissions pricing consistent with a quantitative emissions constraint. These observations are also compared to emissions pricing so far.

EMISSION-ABATEMENT COSTS AND COST-EFFECTIVE EMISSIONS PRICING

Abatement costs in cutting and managing carbon dioxide emissions arise from substituting relatively more expensive low-carbon technologies for incumbent ones in energy systems to sustain production of useful energy services and materials. Some but not necessarily all incur additional operating and capital costs and hold back somewhat productivity of the economy. Such changes in relative costs tend to reduce output, along with consumption, saving and investment as firms and households adjust to relative price shifts. However, these abatement costs are not static. Rather, they change over time, including with government policies that support innovations and market creation for low-carbon technologies, and some give rise to only temporary adverse productivity shocks that are reversed through technology advances.

The Intergovernmental Panel on Climate Change (IPCC) projects that abatement costs increase with greater depth of emissions cuts and shorter time periods over which they are achieved (Table 5.1). In a scenario which limits the expected change in global average surface temperature in 2100 to 2.5°C, emissions cuts are relatively backloaded—at 7 and 92 per cent from the 2010 emissions level by 2050 and 2100, respectively. In this scenario, projected median consumption losses from mitigation actions are 1.7 per cent in 2050 and 2.3 per cent in 2100. These consumption losses are relative to a baseline without climate mitigation actions and impacts from inaction. In a scenario which limits expected global average temperature to well below 2°C (1.6°C), emissions cuts are around 56 and 98 per cent in 2050 and 2100, respectively. The projected foregone consumption is 3.4 per cent in 2050 and 4.8 per cent in 2100. An emissions-reduction scenario that falls between the two other scenarios sees intermediate consumption losses.

'Easy-to-Decarbonize' Sectors and Inframarginal Abatement

Beneath these projected aggregate abatement costs is much heterogeneity across sectors and over time (Chapters 3 and 4). Some near-term emission reductions have relatively low-cost and cost-saving abatement opportunities. Such opportunities arise primarily from energy and material efficiencies.

Table 5.1   *Global emission-abatement costs in cost-effective and delayed-action scenarios*

2100 temperature change in °C	Change in CO₂e emissions compared to 2010 in %		Consumption losses in cost-effective mitigation scenarios in % GDP			Cost of delayed action to 2030: peak emissions <55 GtCO₂e in % increase	
Scenario	2050	2100	2030	2050	2100	2030–2050	2050–2100
1.6	-56	-98	1.7	3.4	4.8	28	15
2.2	-33	-70	0.6	1.7	3.8	3	4
2.5	-7	-92	0.3	1.3	2.3	3	4

Notes: Change in CO₂e emissions compared to 2010 are mid-range values based on published studies that implement three emissions-reduction pathways for limiting global warming in 2100. The consumption loses in cost-effective mitigation scenarios are median values in per cent of gross domestic product (GDP). The costs of delayed action are increases in consumption loses due to backloaded reductions in emissions after they peak at above 55 GtCO₂e. Ranges are reported in IPCC (2014b) and IPCC (2019).
Sources: Data are from IPCC (2014b), Tables SPM1 and SPM2, and, for projected mean temperature change in 2100, IPCC (2019).

Other near-term opportunities are—or were—relatively expensive, such as early deployment of wind turbines, solar photovoltaic (PV) units and battery electric vehicles (BEVs). However, with favourable cost dynamics, their abatement costs fell significantly over time, allowing deeper emissions cuts at relatively low additional costs to the economy. Nevertheless, the very deep cuts in emissions necessary to stabilize the Earth's climate have associated with them relatively high and likely enduring costs in some sectors. For example, some activities that are otherwise very hard to decarbonize could require negative emissions from other activities to manage them. Such negative emissions could be produced by fitting carbon dioxide capture, use and storage technologies to thermal power generation plants burning sustainable biomass, or direct air capture of carbon dioxide and its safe storage. Both are inherently more expensive than current technologies.

Policy evaluations of interventions to cut emissions, especially those using experimental and quasi-experimental methods, provide useful insights into specific types of abatement costs. Consider first those policies expected to have relatively low abatement costs and thus potentially high in the merit order for policy interventions. For example, evidence shows that some energy-saving policies deliver an overall cost saving, even without a cost of carbon dioxide. Such savings can arise from well-designed policies that overcome biases in beliefs about energy costs, as well as inattention or limits to making sound economic judgements about them (Gerarden et al., 2017).

For example, behavioural nudges and information provision can encourage households to make better overall choices of equipment, appliances and energy use. An extensively studied policy is the OPOWER programme in the United States, which inserts in residential utility bills a comparison of customer energy use with the anonymized usage of neighbours. Such comparisons encourage significant energy savings, the value of which is significantly larger than program costs. While the savings diminish over time, some persist after several years, and this persistence points to enduring changes in consumer behaviour or investments in energy efficiency (Allcott and Rogers, 2014). A recent survey of such programme evaluations points to more potential for negative-cost abatement opportunities, though it cautions on potential un-costed welfare losses from policy nudges (Hahn and Metcalf, 2016).

Energy efficiency standards for new appliances can also help cut carbon dioxide emissions. If well designed, these policies address behavioural constraints in making lower- and low-carbon energy choices, like high personal discount rates and principal–agent problems in rented buildings. For example, there is evidence that households modestly underestimate the present value of future energy costs when purchasing a new appliance, implying a private discount rate somewhat above a market rate, with this high discounting being especially significant among low-income households (Cohen et al., 2017). There is also evidence that a principal–agent problem with respect to energy efficiency choices exists in the US residential rental market and could be economically significant. Owners of residences that also pay for energy bills are more likely to insulate their residences and invest in more energy-efficient appliances and boilers than those with tenants that have this responsibility (Gillingham et al., 2012; Myers, 2018).

But not all energy-saving policy interventions prove to be cost-effective. Another well-studied programme is the US Federal Weatherization Assistance Program, in which the government supported heating and thermal-efficiency measures for low-income households (Fowlie et al., 2018). This programme paid for improvements deemed cost-effective in engineering appraisals of dwellings. Typical measures were boiler replacements, loft and wall insulation, and sealing gaps in building fabrics to cut uncontrolled air flows. In a randomized controlled trial of the programme, energy savings were in the range of 10–20 per cent, but well short of the estimated technical potential. The shortfall in energy saving was attributed primarily to an over-estimate of the technical potential of savings (by 60 per cent), as there was no evidence of increased comfort-taking from higher average indoor temperatures after the measures were adopted. Because of the shortfall in savings, the upfront investment costs were about double the present value of realized energy savings (Allcott and Greenstone, 2017). However, a better-designed programme with more precise technical estimates of energy-saving and training of installers might have been able to achieve better cost-effectiveness.

Other negative- or low-cost options to cut carbon dioxide emission include use of ethanol as an alternative to petroleum-based octane boosters for gasoline and switching to natural gas from coal for generating electricity. Blending ethanol with gasoline, up to a 10 per cent blend, cost-effectively boosts the octane of gasoline while cutting fossil carbon dioxide emissions because this biofuel is less expensive than the benzene and toluene for which it substitutes (Gillingham and Stock, 2018, based on Irwin and Good, 2017). Beyond this level, however, the fuelling infrastructure and vehicle engines must be adapted to accommodate higher ethanol blends. Thus, at low blend levels the cost of conserved carbon dioxide emissions from ethanol blending is negative. But above the so-called blend wall of 10 per cent, the costs rise sharply. Similarly, switching to relatively low-cost natural gas from coal for electric power generation in the United States, amidst its relatively abundant supply from shale resources, can cut carbon dioxide emissions, although its displacement of new renewable generation must also be considered. Absent policies, these two effects are projected to be broadly offsetting in the United States (Gillingham and Huang, 2019).

These relatively low-cost and cost-saving measures that cut carbon dioxide emissions are, given the scale of the challenge of achieving net zero emissions, small and bounded in potential scale (Gillingham and Stock, 2018). While they make important contributions to lowering emissions, much more fundamental changes to energy-producing and energy-using technologies are necessary if the Paris Agreement goals are to be achieved. For such goals, low-carbon alternatives are necessary.

Many low-carbon technologies are significantly more expensive than the incumbent technologies for which they substitute, at least initially. Some low-carbon technologies, though, have experienced significant cost declines as their deployment and use have increased, especially those with potential for global deployment and mass production. If abatement opportunities bring added benefits such as knowledge accumulation and scale economies that lower future abatement costs, they should be prioritized by more than their static costs alone would indicate (Goulder and Mathai, 2000; Popp, 2004; Acemoglu et al. 2012). This is especially so if both the scale of future abatement opportunities they afford and their cost-reduction potentials are large (Bramoulléa and Olson, 2005). It is also important to consider the scale of investment required to achieve net zero emissions and the expected asset lives of incumbent technologies, given time-bound emissions-reduction goals (Vogt-Schilb et al., 2018). These considerations all point to inherent limitations of a narrow focus on static emission-abatement costs.

Table 5.2		*Levelized costs of electricity across countries and over time (in 2019$/MWh)*

	World			United States		
	2019			2020	2025	2040
	Low	Average	High	Capacity-weighted average		
Conventional thermal plants						
Advanced CCGT	41	56	71	42	37	38
Of which: variable costs	26	41	56	26	26	29
Variable renewables						
Onshore wind	30	59	85	50	40	36
Utility-scale solar PV	42	52	99	60	33	30

Notes: World LCOE figures for 2019 are as follows. The LCOE of advanced CCGTs is from Lazard (2019b), with the range based on global variation in natural gas costs, assuming a low capital cost, 70 per cent average load factor and mid-range efficiency. The world average is the mid-range value. The LCOE of renewables generation are from IRENA (2020). The average is capacity-weighted for new plants commissioned in 2019. The range is the 5–95 per cent interval. The US EIA LCOE figures are projected capacity-weighted averages for US plants commissioning in 2020, 2025 and 2040, with the 2020 figure updated to 2019 fuel costs.
Sources: Data are from Lazard (2019b), IRENA (2020) and US EIA (2018, 2020).

With large declines in their combined investment and operating costs, variable renewables are becoming increasingly competitive with conventional thermal generation technologies such as advanced combined-cycle gas turbines (CCGTs) and coal plants. Table 5.2 shows the extent of this technological progress using levelized costs of electricity (LCOE) of new generation plants deployed worldwide in 2019 and those projected for the United States in 2020, 2025 and 2040. These levelized costs are the present value of expected capital and operating costs per megawatt-hour (MWh) of electric power the plant is expected to generate over its operational life, with both costs and output expressed in present values. While levelized generation costs do not include most system-integration costs, they nevertheless provide a partial indication of how cost-competitive renewable technologies are becoming.

On average across the world in 2019, the LCOE of new onshore wind and solar PV generation capacity was at or below that of advanced CCGTs.[1] However, there is wide variation in levelized costs, with renewables being cost-advantaged where renewable resources are relatively abundant, and conventional thermal generation where they are not. Looking forward, US Department of Energy LCOE projections for 2025 and 2040 point to continued cost declines for onshore wind and solar PV, as well as some decline in natural gas generation costs with lower fuel prices.[2] The projected declines in solar PV would be increasingly sufficient to cover their additional system-integration costs, at least up to 30–40 per cent shares in total generation (Chapter 3). These costs would also depend on how other complementary technologies deploy in power systems to manage the profile costs of wind and solar PV generation. Across much of the world, the levelized costs of onshore wind and solar PV are projected to fall below those for new natural gas and coal plants operated at baseload capacities, supporting their cost-effective integration into electric power systems. Moreover, costs of complementary flexibility technologies are also projected to decline towards and, for thermal and battery energy storage, below those of current balancing technologies, such as natural gas peaking plants (Chapter 4).

There is similar potential for the cost of battery electric light-duty road vehicles to decline as their deployment and use increase. The cost of lithium-ion battery packs, a key cost component of these vehicles, declined substantially over the past decade—by 18 per cent for each doubling of the deployed battery capacity (Schmidt et al., 2017 and 2019). Their cost in 2019 was about $156/kWh of power capacity, so a 95 kWh battery pack providing a vehicle range of around 475 km (300 miles) costs $14,820 (BNEF, 2019; IEA, 2019b, Annex). If the recent rate of cost decline for each doubling of deployed battery capacity and rate of growth of BEV deployment are sustained, battery costs would decline by 37 per cent in 2025 and 54 per cent in 2030 to $98/kWh and $71/kWh, respectively, in 2019$. At the same time, the overall energy efficiency of these electric drivetrains is expected to improve, so the same vehicle range in 2030 could be served with an 87 kWh battery pack (IEA, 2019b, Annex). On this cost trajectory, battery pack costs would decline to $6180 in 2030, ignoring battery disposal costs.

With further declines in battery pack costs, the total cost of owning and operating a BEV would become competitive with an incumbent vehicle, even though their purchase costs would remain higher. Table 5.3 sets out the total costs of owning and operating a prototypical mid-sized BEV and conventional (internal-combustion engine, or ICE) vehicle. This includes the purchase costs less present value of any residual value as well as expected operating costs and vehicle kilometres travelled in normal operation over its asset life (assumed to be 10 years and 15,000 km per year). This operating cost advantage is particularly significant in Europe, where gasoline costs around $1.50/l compared with $0.80/l in the United States, although much of this cost difference is European fuel taxes (a transfer within these economies rather than a real resource cost aside from their environmental costs). The assumed cost of delivered electric power is $0.7/kWh in the United States and $1.0/kWh in Europe. BEVs also have somewhat lower ongoing maintenance costs than conventional vehicles.

The difference in present value of total ownership costs between this prototypical BEV and a conventional vehicle in 2019 was about $5350 in the United States and approximately nil in Europe, assuming a 10 per cent private real discount rate. At current rates of cost reduction in battery backs, the differences

*Table 5.3   Total cost of ownership differences: new BEV versus ICE
                vehicles (in 2019$/1000 vehicle-km)*

	United States			Europe		
	2019	2025	2030	2019	2025	2030
ICE vehicle prototype						
Present value of capital costs	298	298	298	298	298	298
Operating costs	156	156	156	222	222	222
Present value of total costs	454	454	454	521	521	521
BEV prototype						
Present value of capital costs	433	369	339	457	369	339
Operating costs	80	86	86	86	89	89
Present value of total costs	512	457	424	519	460	427
Memorandum item						
Battery pack cost ($/kWh)	156	98	71	156	98	71

Notes: Key prototype vehicle cost assumptions from IEA (2019b) are: vehicle glider $23,000;
ICE drivetrain $4500, electric motor $2100 plus cost of 95 kWh battery pack. Residual value
of ICE vehicle is 20 per cent and BEV is nil (assumes battery is fully depreciated). Vehicle
ownership is over 10 years operated at 15,000 km per year. The present value of vehicle
kilometres travelled is 92,169 km at a 10 per cent private discount rate. Energy efficiency of the
prototype ICE vehicle is 2.7 MJ/km and BEV is 0.75 MJ/km. Unit energy cost assumptions are
noted in the text. The 2019 battery cost is from BNEF (2019). The projections assume an 18 per
cent learning rate for lithium-ion battery packs conservatively based on Schmidt et al. (2017,
2019) and cumulative deployment of 100 million BEV in 2030, up from 5 million in 2019.
Annual sales grow from 2 million in 2020 to 20 million in 2030 (a 20–25 per cent market share).
Sources: Data are from BNEF (2019) and IEA (2019b). Author's projections as noted.

in total cost of ownership would fall to nil during the 2020s in the United
States. However, this simple comparison of real resource-cost differences
between the two vehicle prototypes does not allow for switching costs (loss of
utility) that some consumers may incur in changing technologies. Developing
extensive vehicle-recharging networks at dwellings and commercial buildings
and along road networks is necessary to minimize these switching costs, with
early adopters having less access to such facilities and facing higher switching
costs than later adopters. Over time, however, expansion of both BEV stocks
and recharging networks would become mutually reinforcing through complementary falls in costs.

   While some low-carbon technologies are becoming competitive with existing technologies that use fossil fuels, getting to this point was very costly
and not all low-carbon alternatives will necessarily become competitive with
incumbent ones. For those that do, the cost of transforming the sector or activity is the present value of the real resource costs incurred during the period of
change less those that would have been incurred using current technologies to

provide the same energy and energy services.[3] Consider variable renewable technologies, for example. A measure of their additional resource costs is the difference in system costs of generation between a variable renewable technology and conventional thermal generation for which it substitutes, times the present value of expected generation from normal operation of the renewable technology over its asset life.

The total cost difference is cumulated over the period of formative deployment of variable renewables until they reach broad competitiveness with existing technologies—from 2000 through the mid-2020s. A broad estimate of the present value of the real resource cost of the early deployment of solar PV and onshore wind technologies over this period is about $1.1 trillion and $800 billion (2019$), respectively. This estimate uses the deployment of new variable renewables capacity from 2001 until they are expected to become competitive with new conventional gas generation in the mid-2020s and their difference in cost with an advanced CCGT.[4] Variable renewables during their formative deployment are assumed to substitute for natural gas generation because it has a higher operating cost than coal generation. But because new variable renewable capacity is an imperfect substitute for new dispatchable thermal generation, the cost difference allows for only a partial capacity credit of variable renewables. The assumed capacity credit for onshore wind turbines is around 20 per cent and solar PV 30 per cent (Newbery, 2018). This in turn assumes that the balance between deployment of wind turbines and solar PV across countries is oriented towards the relative abundance of renewable resources—wind and solar irradiance. So, while deployment of solar PV and onshore wind save the short-run marginal costs of gas generation, they save only a fraction of the levelized fixed costs of natural gas generation reflected in their partial capacity credits. This non-displaced natural gas capacity is assumed to be used for managing profile- and system-balancing costs of variable renewables.

Similarly, the broadly estimated present value of the real resource cost of early deployment of BEVs is $0.25 trillion. This estimate uses the difference in the present value of the total cost of ownership between prototypical mid-sized BEV and ICE vehicle times the deployment of BEVs (Newbery and Strbac, 2016). It uses the deployment of new BEV capacity from 2005 until this vehicle is expected to become cost-competitive with its conventional alternative in the second half of the 2020s.[5] The present value of operating cost uses US fuel and electric power prices and a private discount rate of 10 per cent. This estimate accounts for neither the utility that customers may derive from premium vehicles with battery electric drivetrains nor the disutility that early adopters may face from sparse recharging networks. It also does not account for the environment benefit of improved local air quality from zero-emission vehicles.

On flexibility technologies for electric power systems, the levelized cost of energy storage of utility-scale lithium-ion battery storage is currently above \$300/MWh. But given learning rates for the technology, driven in large part by developments in the automobile sector, it is projected to become a cost-effective storage technology for many power-system-balancing services in the second half of the 2020s (Schmidt et al., 2019). The cost comparison is with the levelized costs of a natural gas peaking (combustion turbine) plant—\$185/MWh at a 10 per cent load factor and \$85/MWh at a 30 per cent load factor (Lazard, 2019b). There are also potential cost savings from integrating battery storage with variable renewables generation on site rather than at grid level (R. Fu et al., 2018). This is an example of low-carbon technology developments spilling over from one aspect of energy systems—road transport with high transport fuel costs—to another—electric power generation with relatively low thermal fuel costs—as low-carbon technology costs decline.

These past and expected technical performance improvements over the decade ahead in onshore wind turbines, solar PV and battery electric storage point to significant permanent emissions cuts at large yet temporary costs in terms of foregone output and consumption. If these technical performance improvements continue as expected through the 2020s, the negative productivity shocks that arose from their early deployment would be subsequently reversed. Moreover, if long-run investments in skills accumulation are largely invariant to these persistent but passing productivity shocks, the supply side of the world economy would return to its original output path, along with those for consumption, saving and investment. This essentially assumes that investments in skills and their accumulation over time are not affected by these temporary negative productivity shocks. These technological advances could thus contribute substantial emissions cuts in electric power systems and much of road passenger transport in the 2030s and beyond with no further loss in consumption, and indeed some recovery.

The estimated real resource costs incurred in their early deployment spread over 30 years centred on 2015 are equivalent to about 0.1 per cent of the present value of world consumption over this period. This scaling of the real resource costs treats these new technological capabilities as global public goods which are necessary for mitigating the risks of climate change and to which all would have contributed. However, they are quasi-public goods, and these costs were largely incurred in a few countries and the European Union, so their economic costs are more concentrated. So too are potential benefits such as economic growth from comparative advantages and new economic specializations (Chapter 8).

Experiences in transforming energy systems so far thus point to the importance of low- or negative-cost emissions-abatement opportunities from lower- and low-carbon alternatives, as well as those low-carbon technologies with potential for mass production, large-scale cost reductions and global scale. Looking forward, there appears to be further potential for such opportunities. As with energy efficiencies, especially in dwellings and automobiles, there could be similar potentials with aspects of material efficiencies, such as using buildings and vehicles more intensively as well as recycling materials more cost-effectively at the end of products' lives. Societal responses during the global novel coronavirus pandemic demonstrated some of these potentials. Moreover, as with mass production of solar PV, lithium-ion battery packs and, to a lesser extent, onshore wind turbines, there appears to be similar potential for knowledge accumulation and scale economies in electrolysers and fuels cells for hydrogen production and use, as well as offshore wind turbines. They exhibit, for example, similar experience curves to those for solar PV and onshore wind turbines, respectively, and share similar characteristics that would enable their deployment at global scale and scale economies in their production.

Hard-to-Decarbonize Sectors and Long-Run Marginal Abatement Costs

Not all low-carbon technologies have such cost characteristics, however, and some alternatives to incumbent technologies appear likely to remain relatively expensive. They include some flexibility and non-variable generation technologies for electric power, low-carbon fuels for commercial transport, and alternative process technologies for heavy industries, such as steel, cement and chemicals (Chapter 3).

Consider the virtual decarbonization of electric power. This would likely require a substantial share of variable renewables, because of their cost competitiveness, as well as complementary flexibility technologies and non-variable low-carbon power to minimize whole-system costs. For example, a diversified mix of variable renewables, system interconnections, dispatchable and non-variable low-carbon generation technologies would likely be part of a least-cost technology mix for achieving net zero emissions from electric power (Kriegler et al., 2014). In addition, demand response and energy storage would be significant sources of flexibility, including battery, thermal and kinetic energy storage systems. The current costs per tonne of avoided carbon dioxide emissions from use of alternative low-carbon technologies that could contribute to the virtual decarbonization of electric power are significant. They range from around \$75/tCO$_2$ for new pumped hydroelectric storage to \$330/tCO$_2$ and \$365/tCO$_2$2 for compressed air and lithium-ion battery storage, respectively (Schmidt et al., 2019).[6] The counterfactual assumption used for calculating the shadow cost of avoided carbon dioxide emissions is the levelized cost of a new natural gas peaking plant (30 per cent load factor). For non-variable generation, the cost of avoided carbon dioxide emissions from a new nuclear generation plant is $250/tCO$_2$ (Lazard, 2019b), while that of carbon dioxide capture and storage (CCS) fitted to a CCGT is in the range of $70–115/tCO$_2$ (Budinis et al., 2018). The counterfactual assumption is a new CCGT plant without CCS.

Consider also the costs of low-carbon fuel alternatives. With currently available technologies, the least expensive way of producing low-carbon hydrogen is steam reforming of natural gas with CCS. The cost of this 'blue' hydrogen is in the range of $1.5/kg (United States, Middle East and Russia) to $2.5/kg (Europe and China) (IEA, 2019b). On an energy basis, the cost is $13–21/GJ. The cost of hydrogen from electrolysis is about $3 per kg at 70 per cent capacity factor for the electrolyser and electricity at $40/MWh, which is equivalent to $25/GJ. At a wholesale natural gas price of $3 per metric million British thermal units (MMBtu), this energy costs $3.2/GJ. So as a low-carbon alternative to natural gas, low-carbon hydrogen is four to eight times more expensive per unit of energy depending on where and how it is produced. So as a thermal fuel, hydrogen remains far from competitive with natural gas with a shadow cost of avoided emissions of $160/tCO$_2$ for least-cost 'blue' hydrogen and $400/GtCO$_2$ for 'green' or electrolytic hydrogen.

Hydrogen, however, is potentially more competitive as a transport fuel. In addition to its production cost, that of distributing hydrogen as a transport fuel is approximately $2.4 per kg ($20 per GJ) once stations reach their normal operating capacity (IEA, 2019b, Annex). The cost of delivered hydrogen is thus in the range of $33–41/GJ. This compares with the delivered diesel price of $0.8 per litre in the United States or $21/GJ in energy terms. Moreover, if used in a fuel-cell electric truck, its overall energy efficiency is greater in delivering transport services, requiring about 30 per cent less energy to deliver the same tonne-kilometre freight (IEA 2019b, Annex). The cost of avoided carbon dioxide by using least-cost blue hydrogen in a fuel-cell electric heavy-duty truck in lieu of diesel in a conventional one is $25/tCO$_2$.[7] However, this calculation does not account for the payload sacrifice from hydrogen storage, which is currently significant (Hunter and Penev, 2019). The current cost of avoided emissions of advanced biofuels is in the range of $55–550/tCO$_2$ (Chapter 3).

While abatement costs in heavy industry vary significantly across sectors and activities, long-run marginal abatement costs of backstop technologies can help to inform a cost-effective emissions price that covers all hard-to-decarbonize activities. Such backstops include carbon dioxide removals through CCS applied to biomass emissions as well as its direct air capture and storage. Estimates of the marginal costs of managing carbon dioxide emissions at scale with these technologies is in the range of $100–300/tCO$_2$ (Fuss et al., 2018; Hepburn et al., 2019). Estimates of abatement opportunities and costs of other backstops such as afforestation, reforestation, enhanced weathering of rocks and other nature-based solutions extend below the lower end of this range. However, policy frameworks to ensure their sustainability and tradability remain underdeveloped, so they are omitted here (Chapter 9).

On estimates of long-run marginal abatement costs, the IPCC finds, for modelling results it surveyed, that estimates of the economy-wide, cost-effective emissions price in line with limiting peak global warming to 1.5°C vary widely across energy-system models and their underpinning assumptions. Key assumptions relate to energy and material efficiency policies (Chapter 7) and access to carbon dioxide removal technologies. The cost-effective emissions price estimates range upward from $250/tCO$_2$ in 2050 (Rogelj et al., 2018a), and reach into the many thousands of dollars. Such high cost-effective emissions prices could arise in a 1.5°C scenario, for example, if there is little recourse to technologies for removing carbon dioxide from the atmosphere and slow advances in other low-carbon technologies.

The general characterization of emission-abatement costs in Table 5.1 is thus broadly borne out by experiences in transforming energy systems so far. Deeper and more rapid emissions cuts would likely have higher costs in terms of real resource costs and foregone consumption, including from more rapidly deploying relatively immature low-carbon technologies. This experience points to the importance of early prioritization of not only those emission-abatement opportunities that have low and negative costs, but also those with significant potential for cost reductions and deployment scale. These cost dynamics are set to enable substantial (but not complete) decarbonization of electric power, surface transport (road and rail) and buildings with little long-term sacrifice in consumption, perhaps more so than suggested by the IPCC estimates in Table 5.1. However, the process of commercializing these alternative technologies is very expensive, with quasi-public-good characteristics, and allocation of these costs has significant distributional impacts. Most countries that incur them also specialize in innovation and manufacturing (Chapter 8).

In addition, some aspects of transforming energy systems appear inherently more expensive than incumbent technologies and traditional use of fossil fuels, even as these alternatives mature. This is especially the case for heavy industry (steel, cement and chemicals), heavy transport (commercial aviation and shipping) and some aspects of electricity generation (supply in tight markets). The costs of cutting these hard-to-decarbonize activities will determine the long-run marginal abatement costs, along with those for backstop technologies for carbon dioxide removals from the atmosphere. The development of backstop technologies and approaches, including policies to govern the sustainability and tradability of nature-based solutions, is important for managing cost-effective emissions prices to levels that can attract broad-based support across societies.

The potential for high economy-wide, cost-effective emissions prices, together with the wide variation in long-run marginal abatement costs across sectors, also points to significant scope for sector differentiation in cost-effective emissions pricing to support stretching emissions targets. An important aim of such differentiation would be to lessen adverse distributional impacts that would be associated with uniform, economy-wide emissions pricing, especially with stretching time-bound net zero emission targets. However, such sector differentiation could entail some economic inefficiencies in abatement efforts. In contrast, the SC CO_2 affords no basis for sector differentiation in emissions pricing, leaving distributional impacts to be managed through other means, such as progressive labour tax cuts or lump-sum transfers.

SOCIAL COST–BENEFIT ANALYSIS AND THE SC CO_2

While cost-effective emissions pricing takes as given a quantitative emissions target, social cost–benefit analysis attempts to project jointly the socially optimal emissions pathway and the marginal social benefits and costs of cutting emissions along this optimal pathway. The SC CO_2 is the emissions price that in principle balances these marginal costs and benefits.

So-called integrated assessment models aim to provide a foundation for such projections, although some have eschewed a capability for social cost–benefit analysis.[8] For those that aim to do so, they integrate several elements (National Academies of Sciences, Engineering and Medicine, 2017, pp. 39–44). They start with scenarios for global population and economic growth, typically to at least 2100, and translate them into energy use and carbon dioxide emissions trajectories using models of the global economy. Further scenario inputs to these projected economic activities include a changing mix of traditional use of fossil fuels and low-carbon alternatives, with the latter share increasing over time due to government policies and technological advances. Climate models then use these emission trajectories in an effort to project climate outcomes. These scenario-based outcomes include changes in surface temperatures, rainfall patterns, sea levels and ocean acidification. The modelled negative and positive social and economic impacts of climate changes are then monetized and discounted, with their net present value expressed as an equivalent loss of current consumption to compare with current marginal abatement costs. The monetization and discounting functions of integrated assessment models are the most uncertain and dependent on value judgements.

Long-Run Socioeconomic Scenarios and Cumulative Emissions

The social and economic losses associated with emissions of carbon dioxide and other greenhouse gases are scenario- and path-dependent. They are expressed as expected future consumption foregone associated with an additional tonne of greenhouse gases emitted from human activities at a point in time. This foregone consumption depends on both projected net social and economic losses associated with future climate changes and investments in adaptation to manage them. But the present value of future costs of an emission at any point in time depends on both cumulative emissions that preceded it and scenario-based projections of subsequent social and economic developments, including technology advances. They determine the size of populations and economies exposed to potential climate change impacts and risks. Therefore, assessment of the economic consequences of emissions needs to consider demographic and economic developments and cumulative emissions as well as their future trends that are inherently and highly uncertain. A tool for exploring these uncertainties are socioeconomic scenarios that consider ranges for potential future trends deemed plausible given preceding developments and future enabling social, political, economic and technological contexts (O'Neill et al., 2017a).

But projecting what the world would look like over the next 50 to 100 years—even without taking account of climate change—is very challenging. For example, the United Nations projection for global population in 2100 is 11 billion, with a 95 per cent confidence interval of 9–13 billion (United Nations, 2019). Alternative projections point to a wider confidence interval with a significantly lower mean estimate of 9 billion and range of 7–12 billion (Vollset et al., 2020). Much of this uncertainty involves population growth in developing countries in Asia and Africa, around which critical uncertainties are changes in fertility rates with female education and rising per capita incomes. There is even greater uncertainty around the pace of productivity gains and per capita income growth in both advanced industrialized and developing countries. A projection of annual average growth in world GDP per capita from 2010 to 2100 based on long-run time-series analysis of its past growth rates is 2.2 per cent (likely confidence interval of around 1 to 3 per cent) (Christiansen et al., 2018). With compounding, the expected level of world average GDP per capita in 2100 is six times its 2010 level, with a range of 3–12. There is thus very wide uncertainty around world population, average real income per capita and total world real GDP, with that around the per capita income trend being the widest.

The range of world real GDP growth carries with it a range of demands for useful energy services and materials that underpin these economic and household activities (Chapter 2). At the same time, there is a transforming energy system, and projections for energy use and emissions must take account of policy actions and progress in advancing low-carbon technologies. For example, scenarios and projections from 2020 should take account of government policies and low-carbon technology advances already in train (Grant et al., 2020; Morris et al., 2020). Recent business-as-usual scenarios from 2020, with no further policy measures or new technology trends apart from those already underway and deemed largely irreversible, are consistent with limiting the expected temperature change to around 3°C (Hausfather and Peters, 2020; Morris et al., 2020). These scenarios use central projections for world population and per capita GDP growth. This projected climate outcome is below the global warming of more than 4°C in 2100 (and rising) associated with a previous business-as-usual scenario that is widely used as a reference scenario for integrated assessment modelling (IPCC, 2013, p. 19–23).

Differences between the previous and updated business-as-usual scenarios in terms of projected climate change outcomes reflect impacts from policy measures and technology advances over recent decades (Hausfather and Peters, 2020; Morris et al., 2020). In other words, climate actions taken so far and reasonably expected to continue reduce significantly the likelihood of previous worst-case scenarios of little or no climate action. That said, recent business-as-usual scenarios from 2020 still fall well short of the Paris Agreement goals. But by learning lessons from and building on past climate actions, new policy measures and technology advances, as well as new choices by firms and households, could bring society closer to these goals.

SOCIAL AND ECONOMIC LOSSES FROM CUMULATIVE CARBON DIOXIDE EMISSIONS

The IPCC's reasons for concern regarding climate change focus on those climate change impacts that go beyond biophysical systems to possible consequences for society and ecosystems (Oppenheimer et al., 2014; O'Neill et al., 2017b; Hoegh-Guldberg et al., 2018). It groups these impacts into five categories and assesses their risks in relation to degrees of global warming. One reason for concern is risks to unique and threatened ecological and human systems, such as coral reefs and indigenous communities. A second reason is risks associated with extreme weather events, such as heatwaves and more intense and frequent cyclones, and their risks to human health and livelihoods as well as ecosystems. A third is the risks from potential large-scale singular events, such as accelerated loss of Greenland and Antarctic land ice, methane releases from thawing Arctic tundra and degradation of coral reefs. The two other reasons relate to risks of adverse distributional impacts of climate changes, especially on vulnerable populations and groups, and overall aggregate impacts on social and ecological systems, such as those to biodiversity, human lives and livelihoods.

Integrated assessment models with social cost–benefit functions focus on projections of aggregate social and economic losses from climate changes for a given degree of global warming. These scenario-based losses aim to reflect foregone consumption associated with potential climate change impacts and investments in adaptations to them. For example, a recent meta-analysis of such loss functions projected world aggregate losses at global warming of 3°C of approximately 2 per cent of world GDP (with a ±1 standard deviation range of nil to 4 per cent) (Nordhaus and Moffat, 2017). Another recent meta-analysis of modelled losses from climate change projects higher economic losses at warming of 3°C, with mean estimates in the range of 4 to 11 per cent of world GDP depending on the methodology used (Howard and Sterner, 2017). A difference in approach between the two meta-analyses is the weighting scheme applied to the underlying studies included in them. The former uses the authors' subjective quality weighting of each underlying study, which may reflect their modelling approach. The latter omits studies that use the same or similar loss projections to exclude overlaps and includes studies that have not been peer-reviewed, widening the range of projections used in the meta-analysis.

Such aggregate social and economic loss functions face significant methodological challenges and have attracted several criticisms. One is that they use arbitrary functional forms and lack directly relevant empirical evidence (Pindyck, 2013 and 2017). For example, some estimates of losses extrapolate from impacts of variations in weather over relatively short periods of time to changes in climate over longer periods of time. A specific concern with such studies is that they may not account well for long-run adaptations to climate change and potential technological advances related to them (Auffhammer, 2018). Another criticism is that these loss functions underestimate climate impacts, especially in scenarios with high emissions pathways and projected temperature change, but also over much of a range of future climate outcomes (Weitzman, 2009; Stern, 2013). A specific concern is that such functions neglect potential impacts that are less well measured or unmeasurable, including at both low and high degrees of global warming. Examples of the latter include impacts from potential large-scale singular events, such as potential acceleration of land-ice loss from Greenland and Antarctica and associated sea-level rise (IPCC, 2019).

Projections for specific potential impacts of climate change illustrate some of the challenges and issues involved. Consider, for example, those for human morbidity and mortality and agricultural yields across the world. The former is one of the main impacts in most social and economic loss functions, while the latter has particular significance for developing countries.

On morbidity and mortality impacts, a recent comprehensive study finds a significant statistical correlation across 41 countries between exposure to temperature extremes, adaptation costs and human mortality based on past weather variations. For example, exposure to temperatures above 35°C in a day increases the mortality rate across countries and age groups on average by 0.4 per 100,000 compared to a 20°C day, and below -5°C by

0.3 per 100,000 (Carleton et al., 2020). Adaptation costs, which cannot be observed, are inferred indirectly from observed variations in temperature sensitivities of mortality rates and per capita incomes, which are assumed to reflect adaptations to temperature extremes (for example, investment in air conditioning where incomes afford). In an emissions scenario with moderate global warming (1.6°C in 2100), the long-run projections based on these past statistical relationships point to increased mortality in developing countries and higher adaptation expenditures in advanced industrialized economies. These overall impacts, though, are sensitive to underpinning socioeconomic and emissions scenarios and range widely (from positive to negative in a scenario with moderate global warming to significantly negative in a worst-case scenario with a high degree of global warming). Translating these projections into net social and economic losses is made even more imprecise because they also rest on value judgements about and assumptions for the value of a statistical life, its variation across countries and social discount rates.

On agricultural yields, a recent comprehensive study projects the impacts on maize, rice, soy and wheat crop yields worldwide for average surface temperature changes in the range of 1–3°C (F.C. Moore et al., 2017). This assessment takes account of temperature and fertilizer effects of more atmospheric carbon dioxide, cost-effective adaptations of agriculture to climate change (more intensive and extensive irrigation) and terms-of-trade shifts for countries that export and import food. With an average global temperature change of 1°C, crop yields are projected to remain broadly stable, supported by the fertilizer effect, with increases in maize, rice and wheat yields though a decrease in soy yields with higher temperatures. With warming of 3°C, yields of maize and wheat would be projected to decline as well, with rice yields remaining broadly stable at a global level. However, the confidence intervals of the projected changes in crop yields are much wider than the mean estimates. The projected net economic gains and losses from agriculture arise from changes in crop yields, adaptation costs and terms-of-trade shifts that affect countries that gain or lose agricultural production and export or import more food. Regions projected to be more vulnerable to such loses with global warming of 3°C include South Asia, sub-Saharan Africa, Latin America and China.

Such potentially significant pressures on livelihoods in developing regions, moreover, could give rise to migration as a way of adapting to climate changes. For example, there are significant statistical associations between past episodes of extreme temperatures, drought and domestic conflict in developing countries on the one hand and increased asylum applications to potential host countries on the other (Missirian and Schlenke, 2017; Abel et al., 2019). Strengthened governance and investments in adaptation like irrigation could improve domestic resilience in developing countries, and well-managed migration could also have net economic benefits in host countries and for migrants. But on an unmanageably large scale, migration could have potentially disruptive impacts within countries and across borders. Projections of net social and economic losses from climate changes do not account for such potentially disruptive migrations for want of a robust estimation methodology.

Another potential impact that is difficult to assess, and for which it is hard to gauge the mitigation response, is sea-level rise from thermal expansion of oceans and land-ice losses from glaciers, Greenland and Antarctica. These climate changes are very slow-moving relative to changes in surface temperatures, with significant inertia over centuries (Mengel et al., 2018; IPCC, 2019). The juxtaposition in social cost–benefit analysis of current emission-abatement costs and social benefits of avoiding such slow-moving but potentially large-scale impacts well into the future points to an inherent difficulty in using this analysis to calibrate mitigation responses.

Discounting and the Present Value of Future Losses

While projections of net social and economic losses from climate changes are inherently and highly uncertain, there is little consensus in economics on how best to translate projected future losses to present values. In a growing economy with rising per capita incomes, as has been the case since the industrial revolutions but not before, investing for future benefits is equivalent to asking current consumers to forego consumption for the benefit of future consumers who are expected to be significantly wealthier. But in a utilitarian framework, current consumers would only do so if investment returns are large enough to compensate for lower current consumption and they valued sufficiently the utility of future consumers. The so-called Ramsey rule for approximating the social discount rate assesses this trade-off using a simplified model of economic growth in which consumers today maximize the present value of utility over time and across generations given expected production technologies (Ramsey, 1928; Gollier, 2013, pp. 36–8).[9] According to this stylized rule, the social discount rate consists of two components: the pure rate of time preference and rate of per capita income growth times a measure of aversion to income variation.

There is wide variation in implementation of the Ramsey rule, reflecting differences in value judgements, assessments of the rule's parameter values and relaxations of its strong simplifying assumptions (Dasgupta, 2008). On parameter values, one view is that the pure rate of time preference should reflect the value judgement that it should not matter when in time a good is consumed and by whom (Cline, 1992, pp. 247–55; Stern, 2007; pp. 49–59). Another view is that the pure rate of time preference should be informed by how this intertemporal trade-off is reflected in observed consumer choices, saving behaviour and market interest rates (Lind, 1982; Nordhaus, 1994, pp. 122–31; Nordhaus, 2008, pp. 59–62). The second component of the rule typically uses trend growth rates of real per capita incomes and a measure of aversion to income variation.[10] Assumptions about these parameters are informed by past economic growth rates and parameters estimated from models of aggregate consumption for advanced industrialized economies. In various combinations, assumptions about parameter values together with relaxation of some simplifying assumptions yield social discount rates that range widely (National Academies of Sciences, Engineering and Medicine, 2017, p. 165). Use of higher rates in this range would heavily discount climate change impacts, especially those that could arise well into the future.

ESTIMATES OF THE SC CO_2 AND PARIS AGREEMENT GOALS

Estimates of the SC CO_2 thus rest on foundations of wide-ranging value judgements and scenario-based projections of marginal social benefits from avoided net social and economic losses and marginal abatement costs. These estimates effectively span the full range of conceivable stringency of emissions pricing (Wang et al., 2019). However, similar uncertainties also pertain to quantitative goals such as those in the Paris Agreement to limit global warming and cumulative emissions if they were to be viewed through the lens of social cost–benefit analysis.

The choice between comparable emissions prices and quantitative emissions goals rests not with their removing uncertainty but rather how well they manage inherent uncertainties and contribute to policy design and implementation credibility. A particularly important litmus test for credibility is the ability of a policy goal and its implementation to galvanize government, firm and household actions to achieve the desired outcome. Central to achieving climate goals are long-run investments in low-carbon alternatives by firms and households, including upfront public and private investments in innovations and market creation for low-carbon technologies.

Assume that the socially desired climate outcome is in a range where the marginal social benefits and costs of additional climate actions are high, and substantial uncertainties exist around them. For example, the IPCC sets out substantial social benefits of limiting global warming to 1.5°C compared with 2°C (IPCC, 2018). The marginal social costs of implementing this goal are also high, as outlined in the section on cost-effective emissions pricing. Consider now two alternative approaches to framing climate goals and implementing policies. On the one hand is a social cost–benefit approach that frames and implements the climate goal as an emissions price pathway in line with the current and future SC CO_2. On the other is a comparable economy-wide goal to achieve net zero emissions by a given date informed by the same evidence and judgements as the SC CO_2.[11]

To see whether the SC CO_2 or emission quantity goal is more effective, consider which one leads to a more stable and informative focal point for coordinated actions among governments, firms and households.[12] If both expected marginal social benefits and costs are steeply sloping in the neighbourhood of the climate goal, as seems plausible, unexpected shifts in them as would lead to significantly greater variations in the SC CO_2 than in the quantitative emissions goal. Moreover, translating the quantitative emissions goal into a time-bound goal for achieving net zero emissions converts the latter uncertainty into time-bound changes in policies, technologies, choices and behaviours necessary to stabilize the climate at a given temperature. This creates a more stable and directly informative focal point for coordinated actions of governments, firms and households in transforming energy systems. In contrast, the SC CO_2 can direct the focus on shorter-run and inframarginal emission-abatement options because of its profound uncertainties.

Consider also implementation of a net zero emission goal through heterodox government policies that support advances in low-carbon alternatives in combination and sequence with emissions pricing (Chapter 4). Through such innovation and industrial policies, governments bear some emissions price risks and impart subsidies to internalize the external benefits of innovations and market creation for low-carbon technologies. They in effect address market imperfections that can hold back technology advances, as well as some missing markets for emissions pricing and potential time inconsistencies in its implementation, especially in the early stages of transforming energy systems. In this framing, cost-effective emissions pricing across sectors consolidates the transformations to low-carbon alternatives and supports their widespread diffusion as they become commercially available. There is in effect a sequencing of government support for innovations and market creation for low-carbon technologies and emissions pricing, with the latter creating the exit from market-creating supports. This heterodox policy approach adds resilience to the overall policy framework, with the potential to elicit directly and early technology advances that are mutually reinforcing of the climate goal, especially with stretching policy goals for limiting emissions.

In a policy framework founded on a time-bound goal for attaining net zero emissions, cost-effective emissions pricing brings forward in time the 'shadow' emissions price associated with the long-run emissions constraint and promotes economy-wide changes to it (Kaufman et al., 2020). A simple and transparent way to do so—yet consistent with cost-effectiveness—would be to use the cost of a backstop technology as the long-run economy-wide marginal cost of achieving net zero emissions and to discount it to the present using a private or social discount rate.[13] Such emissions pricing would do two things: it would encourage changes in firm and household activities towards lower-carbon fuels and greater energy efficiency, and it would raise prices of emitting goods and services, reducing demand and encouraging substitutions into lower-carbon ones (King et al., 2019). For example, such target-consistent emissions pricing in heavy industry (cement, steel, chemicals and plastics) would encourage material-intensive sectors, such as construction and manufacturing, to improve material efficiency and use lower-carbon materials over time—well in advance of the binding emissions constraint. This impact of emissions pricing is particularly important in hard-to-decarbonize activities such as heavy industry and commercial transport, where the long-run marginal cost of abatement is expected to be relatively high. In easy-to-decarbonize sectors, however, such as buildings, light-duty surface transport and electric power, the long-run structural adjustments in related sectors are likely to be less extensive apart from those related to power-system flexibility.

With substantial differences in long-run marginal abatement costs across sectors, overall efficiency of emissions reductions and structural adjustments to them could benefit from differentiated and targeted emissions pricing by sector rather than a single economy-wide emissions price (King et al., 2019). Such differentiated emissions pricing, moreover, would ease adverse distributional impacts of emissions pricing, especially on households. This could add further resilience to the overall policy framework.

EMISSIONS PRICING IN PRACTICE

Consider actual implementation of emissions pricing schemes, using a broad benchmark deemed as an adequate price of emissions, $40–80/tCO_2$ in 2020 (€30–60/tCO_2 in 2018) (High Level Commission on Carbon Prices, 2017; OECD, 2018). An Organisation for Economic Co-operation and Development (OECD) measure of the carbon-pricing gap is the proportion of emissions in energy-intensive sectors covered by market-based policies (emission taxes, permit prices and fuel taxes) with effective tax rates of at least €30/tCO_2. In OECD and G20 countries, the emissions gap is systematically widest in the electricity, industry and building sectors and narrowest in transport, especially in Europe with its long-standing high import dependency on crude oil and oil products, and high transport fuel taxes. Countries with the lowest economy-wide carbon-pricing gaps are those in Europe with relatively low-carbon electricity sectors—Denmark, Finland, France, Luxemburg, Norway, Slovenia, Sweden and Switzerland—owing primarily to high shares of hydroelectric and legacy nuclear generation in the power sector.[14] Among these countries, only Denmark has a relatively high share of new renewable generation. Progress across all countries in raising the level of carbon pricing and broadening its sectoral coverage over the last decade to close these carbon-pricing gaps was slow (OECD, 2018).[15]

It is important to recognize that causal relationships between effective emissions pricing and low-carbon technology advances can run in both directions. An economics lens sees causality running primarily from pricing to technological progress via incentives. Political economy considerations point to the importance of low-carbon technological advances in easing distributional impacts of emissions pricing. As low-carbon alternatives advance, these adverse impacts of emissions pricing ease as opportunities to substitute away from fossil fuels expand and new economic interests in change emerge. Experiences of transforming energy systems so far find that emissions pricing gaps are narrowest in countries that benefit from abundant traditional renewable resources like hydroelectric power and, in some cases, legacy nuclear power, as well as those with long-standing energy-security concerns. Moreover, these gaps are relatively wide in countries that specialize in innovations and manufacturing low-carbon technologies—China, Germany, Japan, South Korea and the United States (Chapter 8).

CONCLUSION

Given climate externalities, emissions pricing is perhaps the simplest policy to prescribe yet among the more challenging to implement in practice. The challenges in emissions pricing relate primarily to managing inherent uncertainties and value judgements around marginal social costs and benefits of mitigation actions, as well as distributional impacts of emissions pricing and potential time inconsistencies in its implementation. Emissions pricing raises the prices of energy services and materials costs faced by most firms and households and reduces the profits and real incomes of some. Such distributional impacts can and have contributed to time inconsistencies in implementation of emissions-pricing policies. The issues of missing markets for future emissions prices, abrupt shifts in government approaches, distributional impacts of emissions pricing and potential time inconsistencies in its implementation matter for long-run investments necessary for transforming energy systems.

This chapter examines how framing of climate goals and designing emissions pricing, together with complementary policies, enable it to become a more effective policy within a comprehensive energy reform strategy. It explores why time-bound commitments to net zero emissions can provide a more stable and informative focal point for coordination government, firm and household climate actions than emissions pricing based on inherently and highly uncertain estimates of the SC CO_2, especially in early stages of transforming energy systems. It illustrates that with plausibly steep marginal social benefits and cost curves in the neighbourhood of what is best assessed as the socially optimal climate goal, a quantitative emissions goal would be more stable in response to unexpected shifts in social benefits and costs than an emissions price. Moreover, expressing the quantitative emissions goal as a time-bound goal for achieving net zero emissions translates its uncertainty into one of timing for achieving the binding constraint. This creates not only a relatively stable but also directly informative focal point for coordinated government, firm and household actions, in particular for those to transform energy systems to low-carbon alternatives. Government supports for innovation and market creation for low-carbon alternatives directly target the advance of these alternatives. Such well-designed policies hold few regrets, even if the optimal timing of system decarbonization changes with shifting assessments of marginal social benefits and costs of climate actions. To the extent successful, these policies also strengthen economic interests in and credibility of the quantitative emissions goal.

A potentially coherent and credible reform strategy, therefore, has as its goal the time-bound attainment of net zero emissions. Complementary policies to support innovations and market creation for low-carbon technologies and cost-effective emissions pricing strengthen its credible implementation and overall resilience, especially if effectively sequenced. Moreover, some sector differentiation of emissions pricing—for example, between relatively easy- and hard-to-decarbonize sectors—would ease its adverse distributional impacts without significantly sacrificing economic efficiency. Further complementary policies within this unfolding energy-reform strategy are reforms to government-designed markets for electric power and energy infrastructures, along with policies to promote greater energy and material efficiencies. They are the focus of the next two chapters.

NOTES

1. An indicative LCOE range for new coal generation plants spans a low of $58/MWh in India to a high $98/MWh in Germany (in 2019$). On India, see the IEA: www.iea.org/data-and-statistics/charts/levelised-cost-of-electricity-lcoe-for-solar-pv-and-coal-fired-power-plants-in-india-in-the-new-policies-scenario-2020-2040. The German figure is the mid-range for new hard coal plants (Kost et al., 2018).
2. The US LCOE figures include grid integration but not profile- and system-balancing costs.
3. This measure of incremental real resource costs does not allow for any net social and economic losses associated with continued investments in current technologies.
4. The renewables deployment data from 2000 to 2018 and renewables capacity factors and renewables levelized cost data from 2010 to 2022 are from the International Renewable Energy Agency Data & Statistics Query Tool, www.irena.org/Statistics/View-Data-by-Topic/Capacity-and-Generation/Query-Tool,

and International Renewable Energy Agency (2020). The levelized cost data for 2000 to 2009 are back-cast using the change in LCOE over this period in the United States from the Open EI database, https://openei.org/apps/TCDB/transparent_cost_database.Variable renewables deployment projections use the IEA Current Policies Scenario (IEA, 2019a, Annex A). The assumed learning rates for solar PV and wind turbines are from Schmidt et al. (2018) and Lafond et al. (2018), respectively.

5. The BEV deployment data from 2005 to 2019 is from IEA (2020f). It is assumed that the accelerating deployment trend for BEVs continues through the 2020s. The assumed learning rate for lithium-ion battery packs for vehicles is from Schmidt et al. (2018).

6. Carbon dioxide emissions from a new advanced CCGT are assumed to be $0.4tCO_2$/MWh. See IPCC (2014c). The energy efficiency of a new combustion turbine is assumed to be 73 per cent of that of a CCGT (Lazard, 2019b). Emissions factors are well-to-plant.

7. The carbon dioxide emissions per GJ of gasoline, diesel and jet fuel is 72–4 g/GJ. See IPCC (2014c). The emissions factor is well-to-tank.

8. Integrated assessment models with a capacity for social cost–benefit analysis include DICE (W. Nordhaus), FUND (D. Anthoff and R. Tol) and PAGE (C. Hope). Examples of similar models but without an explicit social cost–benefit function are IGSM (Massachusetts Institute of Technology), IMAGE (Netherlands Environmental Assessment Agency), MESSAGE (International Institute for Applied Systems Analysis) and REMIND (Potsdam Institute for Climate Impact Research).

9. Gollier (2013, pp. 41–57) and Wagner and Weitzman (2015, pp. 68–78) examine implications of relaxing some simplifying assumptions of the Ramsey rule, especially with respect to uncertainty.

10. This single measure of aversion to income variation can reflect either risk aversion to future income uncertainty or aversion to income inequality, or both if deemed to be the same.

11. See Hänsel (2020) on the potential consistency between social cost–benefit analysis and the temperature goals of the Paris Agreement.

12. On the role of focal points in coordinating behaviours and choices, see Schelling (1960, pp. 54–8, and Mehta et al., 1984a, 1984b). If it is not feasible to establish a credible emissions price signal, especially in the early stages of transforming energy systems, a quantitative emissions goal could serve as a focal point for limiting emissions if its credibility can be more readily established.

13. Use of a social discount rate would aim at promoting inframarginal emissions cuts as well as long-run non-marginal changes in the capital stock. A private discount rate would aim at efficient economy-wide structural changes in line with the long-run emissions constraint.

14. The primary energy shares in generation are from IEA Data & Statistics. Retrieved from www.iea.org/data-and-statistics.

15. See also the World Bank Carbon Pricing Dashboard for recent policy developments in emissions pricing. Available at https://carbonpricingdashboard.worldbank.org/map_data.

6. Adapting energy-market designs and infrastructures

Energy markets are central to transforming energy, but some are atypical and inherently imperfect. In a low-carbon energy system, primary energy would likely flow primarily from renewable resources through electricity networks to customers, in contrast to the current system that relies heavily on fossil fuels and their processing and distribution and conversion to electricity. But electricity markets differ significantly from others because of electricity's physical characteristics. Electric power has the strict physical condition that supply and demand—generation and load—must continuously and almost instantaneously balance, otherwise the frequency of a network's alternating current deviates from its reference value. The equipment that generates and uses electric power must operate at this reference frequency or risk damage. If there is too much demand relative to supply, the frequency decreases, or, vice versa, the risk of equipment failures rises. To help avoid these risks and ensure overall reliability, electric power systems have system operators tasked with the responsibility of physically balancing supply and demand in real time.

The system operator's role is vital because market imbalances can create a cliff edge for the whole system. A systemic risk arises if the frequency falls too much and generators begin to disconnect from the network, worsening the imbalance and potentially creating a cascade of plant disconnections and system collapse—a power blackout. Falling over this edge is an economic catastrophe to be avoided at least cost. Much of the onus of system balancing currently falls on dispatchable thermal generation because most demand is unresponsive to electricity prices in the short run. Storing energy from electricity is also costly and limited. So as demand reaches available generation capacity, electricity prices rise sharply to curb price-responsive demands. In extremis, system operators can ration demands through rolling blackouts of some customers, in ways agreed with governments or regulators, to avoid complete system blackouts if other available balancing options are exhausted.

But nearing this cliff edge plays an important role in funding investments in the system. In normal circumstances, electricity markets work like most competitive markets, balancing supply and demand with a market price reflecting marginal (operating) costs of suppliers and marginal benefits of customers (at least among those that face real-time prices). But as markets tighten, electricity prices rise to attract marginal generators and, if necessary, above short-run operating costs to curb demands once available generation capacity is reached. This 'peak-load' pricing, including administrative-scarcity pricing during rolling blackouts, is central to creating an incentive for adequate investment in generation capacity and flexibility technologies. Even in systems with adequate capacity, peak electricity prices should be much greater than 'normal' electricity prices and such peak prices should prevail for only a few hours per year. This is an efficient way of allocating fixed investment costs in electric power generation (Ramsey, 1927; Boiteux, 1960), but one that carries some vulnerabilities and risks.

Such extreme peaks in electricity prices can be politically and economically problematic, largely because of their adverse distributional impacts, concentrated commercial risks and vulnerabilities to potential abuse of market power and government fiat. Many customers, such as hospitals, are largely unable to adjust demands to peak prices, and they would shoulder much of the system's capital costs. Moreover, to recover capital costs, generators need to be available during tight markets. Potentially tight electricity markets, however, could be prone to market manipulation by generators and traders (Borenstein, 2002; Léautier, 2019, pp. 85–93). This potential creates a concern that high prices could arise from strategic behaviour regarding plant availability and reflect unfair profits rather than economic fundamentals. Governments also have an incentive to intervene arbitrarily in tight markets to curb price spikes (Blackmon and Zeckhauser, 1992). Wholesale markets for electric power are thus inherently imperfect due to its physical characteristics. Their government designs must balance complex trade-offs between efficiency and market power, and do so credibly and predictably. Growing shares of variable renewables in generation make this balancing act more difficult.

At the same time, low-carbon fuels would likely be a key part of low-carbon systems to serve those energy demands that cannot be met with low-carbon electric power given foreseeable technologies. Natural gas markets have some similarities with those for electricity—but also important differences. Supply and demand for natural gas too must physically balance, but imbalances between them can last much longer than in an electricity system before the network fails. Because of its low cost and large scale, natural gas storage also plays a substantial role in system balancing, especially in serving peak seasonal demands. Supply and demand imbalances for natural gas are thus mostly corrected via market transactions rather than system-operator interventions.

The challenges in transforming natural gas networks to low-carbon fuels arise not from complex wholesale market design and system balancing, but rather from coordination of investments in low-carbon fuel supply and end-use technologies. In contrast to electricity-using technologies, for which the main adaptations would be to make their electricity use more responsive to power prices and grow their roles, those that use natural gas would require more fundamental transformations. The extensive deployment in buildings of space-and water-heating technologies that use fossil fuels is a particular challenge. They are largely locked in to their current roles by both energy network infrastructure as well as building fabrics and their poor thermal efficiencies. The current combination of relatively low-cost heating with fossil fuels and thermally inefficient building stocks would likely need to adapt to the performance characteristics and increase in the relative cost of low-carbon energy for space heating. This energy could be either low-carbon electric power or fuel such as biomethane or hydrogen. But transforming large building stocks to accommodate a new low-carbon energy source for heating would involve investments in better thermal efficiency of buildings, alternative heating equipment and low-carbon fuel supply. Coordinating these investment decisions is challenging, because of network externalities and because investments in buildings' thermal efficiency are most cost-effective when buildings undergo periodic extensive refurbishments.

The challenges in adapting energy markets and infrastructures are perhaps more pressing in electric power than natural gas systems, though both would need to change. Low-carbon energy systems would depend crucially on electric power markets and networks that are well adapted to variable renewables as their key energy resource. Designing electric power markets well is fundamental to cost-effective investments in variable renewable generation and complementary flexibility technologies necessary to grow these systems while maintaining their reliability. The main challenges for natural gas networks arise from investment-coordination challenges rather than complex market and regulatory reforms. Prioritizing investments in buildings' thermal efficiency measures holds few policy regrets and eases the lock-in of fossil-fuel boilers. Achieving credible and predictable government policies to transform both aspects of energy systems is proving challenging, but building on sound economic understandings of these markets and infrastructures is key to unlocking their transformations.

WHOLESALE ELECTRIC POWER MARKETS

Unlike most markets, which emerge from interactions among firms and customers, those for electric power are government-created and government-designed. They emerged from electricity system restructurings that began in the United Kingdom in the early 1990s and spread across the United States, Europe, Japan and beyond over subsequent decades. Before their restructuring, electricity systems were vertically integrated monopolies consisting of generation, transmission, distribution and supply and typically subject to rate-of-return regulation on their assets (Figure 6.1). The unbundling of these vertically integrated monopolies was aimed at introducing markets and competition, where feasible, especially competition in generation and wholesale electricity markets. Competition in electric power supply and retail markets is also feasible, but can be more difficult to achieve in practice, especially where customers have grown accustomed to monopoly provision. Only the transmission and distribution networks are natural monopolies, and in liberalized systems they are owned by separate transmission and distribution companies. There are, of course, intermediate system reforms between sector monopolies and fully liberalized wholesale and retail markets. For example, introduction of independent power producers in otherwise vertically integrated systems increases competition in generation. Moreover, not all liberalized systems have competitive retail markets. In such systems, regulated distribution companies provide retail supply services to customers.

Wholesale exchanges of electric power can begin years before its delivery and initially they take the form of long-run bilateral contracts negotiated between generators and wholesale customers (Cretì and Fontini, 2019, pp. 59–71; Kirschen and Strbac, 2019, pp. 34–41). These contracts usually exchange large amounts of power over long periods of time. For example, a generator planning to build a new power plant will often sell forward a substantial proportion of the expected output to an interested counterparty like a large electricity supplier (Figure 6.2). There are also standardized forward contracts exchanged in over-the-counter markets for smaller amounts of electric power. They are for well-established periods of days and weeks as well as seasons. Generators and suppliers use these contracts to refine their net physical positions as delivery draws near. Exchange-traded electricity futures contracts also enable market participants to manage future electricity price exposures ahead of delivery.

These physical and financial exchanges continue to the day ahead of actual delivery, when large-scale market participants aim to balance their expected physical positions in centralized wholesale markets. In the day-ahead market, participants exchange electric power for one-hour intervals over that day, alongside already-agreed bilateral commitments. But at this point in the time sequence of wholesale markets, paths diverge in different market designs, with the key difference being the system operator's role. In some systems, common in North America, the system operator optimizes centrally the scheduling and dispatch of generation plants based on day-ahead supply offers and demand bids. An alternative design found in Europe and Japan uses exchange-based trades between supply and demand in the day-ahead and intraday markets to balance generation and load, with self-dispatch of generation plants.[1]

For example, the day-ahead markets in North America take the form of generators offering supply schedules (quantities and prices) which reflect plant start-up costs, their minimum efficient scale and offers of supply above

Vertically integrate utility with independent generator and large off-taker

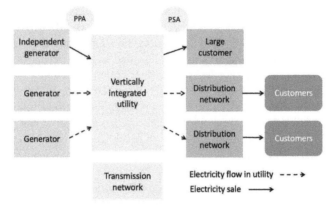

Competitive wholesale and retail markets

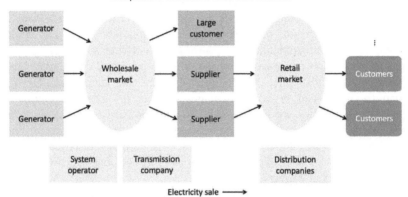

Notes: *PPA* is a power purchase agreement and *PSA* a power sales agreement. The two stylized representations of electric power system organization and market design reflect two boundary cases—one a traditional vertically integrated and publicly regulated utility, albeit with an independent generator, and the other a liberalized system with competitive wholesale and retail markets for electric power. There are many possible intermediate cases.
Sources: Hunt (2002, pp. 42–7) and Kirschen and Strbac (2019, pp. 2–7).

Figure 6.1   Electric power market structures: from vertically integrated utilities to competitive wholesale and retail markets

minimum scale shaped by marginal costs. Similarly, the demand bid schedules reflect the diminishing marginal value of electric power. The system operator uses these supply and demand offers to schedule the day-ahead dispatch of generation to meet expected demand, considering bilateral commitments submitted to the system operator and any network constraints. Market participants

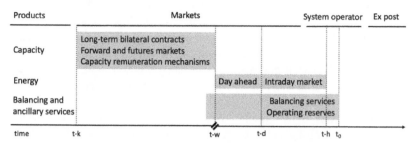

Products	Markets		System operator	Ex post
Capacity	Long-term bilateral contracts Forward and futures markets Capacity remuneration mechanisms			
Energy		Day ahead : Intraday market		
Balancing and ancillary services			Balancing services Operating reserves	

time   t-k   t-w   t-d   t-h t_0

Note: t_0 is a point in time when generators are scheduled to deliver energy to the market. Working backwards in time, *t-h* and *t-d* are hours and the day ahead of delivery. Going further back in time, *t-w* is the one week ahead. Initial contracting for delivery of electric power from a plant begins years ahead at *t-k*. *Ex post* settlement of the market takes place on a specified day after delivery.
Sources: J. Fasel (2012), Timeline of the power market, Wikimedia Commons (CC BY-SA 3.0), and Creti and Fontini (2019, p. 62).

Figure 6.2   *Time dimension of electricity markets*

must submit any changes to their offers and bids as well as bilateral commitments after the day-ahead market closes to the system operator for updating dispatch schedules.

Wholesale markets in Europe and Japan follow an alternative design. There, once the day-ahead market closes, the intraday market opens and continues until near delivery of electric power. The intraday market allows generators, large-scale customers and suppliers to continue to adjust their positions. The intraday market exchanges power for hourly and longer intervals, as well as shorter ones. When an intraday market first opens, its operation is similar to that of the day-ahead market in balancing expected supply and demand. But as delivery draws near, the intraday market increasingly operates as a balancing market among generators, large-scale customers and suppliers, enabling market participants to respond to unexpected developments in supply and demand. After the intraday market closes, typically one hour before delivery, the system operator assumes responsibility for running and balancing the system.

From years ahead to just hours before delivery, electricity markets exchange power volumes and establish prices that vary across market intervals— minutes, hours, days, weeks and seasons. They guide capacity investments and efficient dispatch of available capacity either through optimized dispatch schedules of system operators or self-dispatch of plants through market exchanges. However, it is often questioned whether wholesale electric power markets provide adequate incentives for investment in generation capacity and flexibility to meet expected demands and manage reasonably expected variations in supply and demand and system component failures. These investment incentives depend crucially on peak-load prices, which pose commercial risks and can be politically problematic. Moreover, as variable renewable shares in generation increase, challenges mount for market design and regulation of networks to reflect well the true value of electric power over time and grid location, while limiting the scope for potential abuse of market power and building policy credibility.

The functioning of wholesale electricity markets as the role of variable renewables rises must also be seen in the wider system context. This includes the real-time balancing market run by the system operator and its acquisition of ancillary services to ensure system reliability. Network regulation and spatial dimension of pricing as well as retail markets and time-of-use prices can also make significant contributions to system efficiency and reliability. But before turning to these other elements, it is important first to focus on wholesale markets, their adaptation to variable renewables and adequate incentives for investment in generation capacity and flexibility.

Wholesale Market Impacts of Variable Renewables

As variable renewables gain greater generation shares, electric power markets must reflect wider potential imbalances as variations in generation become more correlated and less predictable due to their dependency on renewable resource availability. When available, variable renewables tend to displace thermal generation in the merit order of plant dispatch and attract lower electricity prices. Efficient dispatch, whether through markets or by system operators, sees available capacity run in its merit order from low to high operating costs to meet demand while minimizing operating costs. Wind and solar photovoltaic (PV) generation have low operating costs because these energy resources are free once generation sites are procured, so they tend to dispatch if available. Moreover, because supply—but not necessarily demand—becomes increasingly correlated with wind and solar resource availability, variable renewables tend to displace thermal generation plants with high operating costs, such as those using natural gas, lowering wholesale electricity prices, especially in slack markets. In addition, when renewable generation is unavailable for want of renewable resources, wholesale markets can become tight, with prices potentially rising above short-run operating costs of generation to align inelastic short-run demand with available supply. These impacts of variable renewables on plant dispatch and wholesale electricity prices are increasingly apparent as their generation share rises (Clò et al., 2015, Kyritsis et al., 2017; Seel et al., 2018; Figueiredo and da Silva, 2019). Variable renewables thus contribute to greater variation in these prices, including more frequent periods of low and peak-load prices, at least until other system components adapt to their growing role.

At the same time, flexible assets that curb demand when markets are tight and use electric power to store energy when they are slack, as well as dispatchable low-carbon electricity generation like hydroelectric power, would become more valuable. They would increasingly provide the system balancing and reliability currently provided by dispatchable natural gas and coal generation plants in traditional electricity systems, which benefit from low-cost storage of thermal fuels. In addition, transmission and distribution networks must adapt to new generation locations where renewable resources are available, and more electricity demands in transport, buildings and industries. The value of extending the geographical reach of systems, including interconnections among them, also increases if this lessens correlation of variable renewable generation and diversifies demand within a system. The challenge is thus to design electricity markets that reflect well the variation in value of electric power by time and location—and costs of carbon dioxide emissions—to serve effectively demands with new low-carbon technologies, including a cost-effective mix of variable renewable and flexibility technologies.

Wholesale Electricity Markets and Investment in Capacity

In principle, a well-functioning electric power market should enable profitable investment in adequate generation capacity and flexibility, as well as efficient dispatch of generation that minimizes short-run operating costs. But a concern with such 'energy only' markets is that they lead in practice to under-investment in capacity because their revenue is insufficient to cover long-run costs, including fixed investment costs. Such 'missing money' problems are typically associated with government interventions in electricity markets and out-of-market actions by system operators to balance systems (Cramton and Stoft, 2006; Joskow and Tirole, 2007; Joskow, 2008; Cramton et al., 2013). For example, many wholesale electricity markets have price caps aimed at curbing the market power of generation plants in tight markets. Economic fundamentals suggest that if such caps are imposed, they should be set at the average value of lost load (VoLL)—the average opportunity costs incurred by electricity users from any involuntary loss of supply in rolling blackouts.[2] A price cap at this level would create an incentive for adequate investment in capacity to cost-effectively meet expected electricity demands (Léautier, 2019, pp. 31–9). But not all caps are set at an economically efficient (sufficiently high) level. Many reflect broader political aims like low electricity prices. A missing-money problem can also arise if the system operator inadequately or arbitrarily remunerates ancillary and balancing services through non-market-based actions.

'Missing markets' can contribute too to under-investment in generation capacity and flexibility. Such problems arise if investment risks cannot be efficiently allocated through bilateral long-term contracts and forward contracts, or if carbon dioxide emissions are not expected to be costed adequately (Newbery, 2016b). For example, the longest-dated liquid maturities of forward contracts in most electric power markets span only several years, far short of the 25-year life of a wind turbine or natural-gas-fired thermal plant and 40-year life of a coal-fired or nuclear plant. So long-run price and volume risks of generation cannot be allocated efficiently in this way. Missing markets for emissions pricing further compound this issue (Introduction and Chapter 5). There are no available hedges for such material policy risks (Léautier, 2019, pp. 4–6). These missing markets raise long-run revenue risks for project developers and their financiers, making it more difficult to risk-adjust cash flows and gauge expected profits of investments.

The missing-money and missing-market problems would likely become more significant as variable renewable generation shares rise, at least until other system components adapt to them. This means more frequent periods of both tight and slack markets. In tight markets, the value of electricity is high and its price should approach the VoLL as demand nears available capacity, including voluntary demand response. In slack markets, power prices can go to nil or even turn negative if it is costly for thermal plants to curb output or renewables benefit from feed-in tariffs that can offset negative prices. This higher variability in wholesale electricity prices makes missing forward markets more problematic for long-term investors and more frequent peak prices more problematic for governments and customers. Emissions pricing adds a further layer of complexity to this issue. Moreover, as systems transform, system operators can turn to administrative and regulatory means to obtain ancillary and balancing services rather than market-based approaches.

Capacity Remuneration Mechanisms

Capacity remuneration mechanisms are found in many systems to make up for missing money and missing markets, especially where capacity margins are tight and new investments are seen as necessary for system reliability. Whether such mechanisms are necessary or desirable in principle is the subject of a long-running debate. Nevertheless, many exist in practice, and they take a variety of forms. Two that remunerate future capacity through market-wide mechanisms are considered here: physical capacity auctions and financial reliability options.[3] Capacity is defined as the ability to generate electricity or shed load when wholesale markets are tight. Capacity so defined is electric power in tight markets. If a generation plant or demand load receives remuneration for capacity but fails to perform during a scarcity period, the recipient would typically face a substantial non-performance penalty or incur a significant financial loss. In addition, short sales of capacity are usually prohibited to limit gaming opportunities. An entity that offers capacity must control the amount of offered generation or load, or be reasonably expected to do so for new capacity.

Capacity auctions procure periodically (annually) an adequate amount of physical capacity for the year ahead and for several years ahead to allow for new market entrants to participate. The government or system operator determines the amount of capacity to procure, in theory at its socially optimal level.[4] The Pennsylvania–New Jersey–Maryland (PJM) system in the United States and those in Brazil, Great Britain and Western Australia use this approach. The remunerated capacity must supply electricity or shed load when the markets are tight, or face a financial penalty. This obligation follows demand, so if a generator has a 10 per cent share of auctioned capacity, it has an obligation to supply 10 per cent of demand during shortages. Capacity defined and auctioned in this way is comparable to a reliability option that hedges demand from scarcity prices that would otherwise arise, provided governments choose appropriately adequate capacity and market participants are bona fide (Cramton and Stoft, 2006 and 2008). In systems that are expected to have inadequate capacity in the future, the price for capacity ($/MW) is influenced by the investment costs of new entry. In systems that are expected to have adequate capacity, capacity auctions are likely to clear at a low price for capacity.

Reliability options are a financial alternative to a market-wide physical capacity auction. The system operator (or obligated electricity suppliers) buys these call options at periodic auctions from generation plants and load-controlling entities for the year ahead and several years ahead to allow for new entry. These options are used in the New England system in the United States and those in Columbia, Ireland and Italy. The seller of the option receives a fee for the contract, which gives the system operator the right to call the option when the electricity market price rises above the option strike price. When called, those entities that sold options must pay the system operator the difference between the prevailing electricity market price and the option strike price times the power volume associated with their share of the total option volume. In principle, the system operator should set the option volume at its socially optimal level and the option strike price at the short-run operating cost the marginal plant needed for adequate capacity. This creates an incentive for all entities that sold the call options to perform in tight markets when the system operator calls the options. Failure to perform would leave the entity exposed to the difference between the actual electricity market price and the option strike price, creating a strong financial incentive for those that sold the options to contribute to system reliability.

Auctions of physical capacity or reliability options are complex schemes to design and implement in practice. Their effectiveness depends on assessments of system parameters and future developments that are difficult to gauge and inherently uncertain. Both require the government or system operator to make judgements about future demand and generation, including by variable renewables. Availability of electricity supply from interconnections with other systems must also be considered. Based on these outlooks, the amount of physical capacity or volume of reliability options to procure at auction is set with a view to achieving the socially optimal level of capacity. To do so, the operator allows the actual amount procured to respond to the auction price for capacity, balancing at the margin social benefits and costs.

For example, consider first the traditional engineering reliability standard for electricity systems, which is typically three hours of expected unserved energy from rolling blackouts per year—a reliability standard of 99.97 per cent. In other words, adequate system capacity is technically defined as that level of nameplate capacity, adjusted for expected plant availability, at which expected peak demands will be lower 99.97 per cent of the time.[5] In a well-designed auction of physical capacity or financial reliability options, adequate capacity arises from a balance at the margin between the cost of expected hours of energy unserved per year valued at the average VoLL and the rental capital cost of additional capacity to meet potentially unserved demand. The expected reliability of the system reflects this balance of marginal costs and benefits, rather than a rule-of-thumb reliability standard.

A concern with capacity remuneration mechanisms is that their implementation can err towards over-procurement. Governments, regulators and system operators can be risk-averse in assessing adequate capacity. They are exposed politically to any system failures, but do not directly bear the costs of excess capacity (Newbery, 2016b). For example, projections of future expected demand peaks and VoLL assumptions can be inflation-prone, and electric power availability from interconnectors with other systems heavily discounted. Such 'over-procurement' of physical capacity puts downward pressure on wholesale prices, while raising capacity and overall system costs beyond those which the benefits of more system reliability would warrant. The mechanisms are also relatively costly to run because they require administrative oversight to enforce short-sale restrictions by ensuring only bona fide bidders with credible physical positions participate in auctions. Physical capacity auctions also require performance monitoring and levying penalties for any non-performance. Reliability options have the advantage of providing a strong financial incentive for performance, rather than requiring administrative enforcement. Nevertheless, the administrative vetting of the bona fides of auction participants in both mechanisms can create barriers to market entry by new technologies through heavy discounting of their untested contributions to system reliability.

Operating Reserve Demand Curves

Another way to address the missing-money problem in electricity markets, at least in part, is to integrate them with the demand for operating reserves. The system operator uses such reserves to meet expected load increases or supply failures in real time—they are an essential component of all electricity systems. They would be unreliable without such reserves, which take several forms in terms of availability and duration—spinning (available at short notice and for a short duration) to 30 minutes (available with some notice and for a long duration). Since many generation technologies can provide both electric power and operating reserve, the wholesale electricity market and the system operator's demand for operating reserves are linked (Hogan, 2005).

This link can be exploited to ensure that the wholesale electricity market price adequately reflects that scarcity value of electricity because generators must be indifferent between supplying electric power and providing reserves to the system operator. If, for example, a generator providing reserves is called upon to produce electric power to avoid a rolling blackout, it would earn the administrative electricity price of the average VoLL (assuming scarcity pricing is optimal) less its short-run operating costs. Its expected profit in providing reserves is thus approximately the short-run loss of load probability (sLoLP) times the average VoLL minus short-run operating costs. The generator's alternative is to provide electric power in the wholesale market. To be indifferent between the two options, the electric power price must reflect a premium of sLoLP times VoLL above the short-run operating costs of the marginal plant.

The integration of the wholesale market and system-operator demand for operating reserves is found in some US electricity systems. They also use average VoLL scarcity pricing. For example, the Electricity Reliability Council of Texas (ERCOT), the independent system operator for the Texas system, currently adds a scarcity premium of sLoLP times average VoLL to wholesale prices. This 'adder' is usually nil or low, when the sLoLP is negligible, but occasionally high when the market is tight (Potomac Economics, 2019). The PJM system operator is considering a similar approach (Hogan and Pope, 2019). This integration of wholesale markets and system-operator demand for operating reserves, combined with VoLL scarcity pricing, provides incentives for optimal investment in capacity and reserves. It adds transparency to the value of capacity in electricity prices, but does not necessarily lead to more investment in them than could be achieved under VoLL scarcity pricing (Léautier, 2016).

The scarcity 'adder' is suitable for energy-only markets as well as ones with capacity-remuneration mechanisms if a 'belts-and-braces (suspenders)' policy approach is the government's preference. In the latter case, a benefit of a capacity-remuneration mechanism is potential smoothing of investment cycles that might otherwise occur in energy-only markets, in effect a compromise between planning- and market-based approaches to system investment. It also remains untested whether an energy-only wholesale market together with a well-designed market for ancillary services can remunerate adequately investment in low-carbon generation and flexibility technologies.

The challenge of mobilizing investment in low-carbon generation of electric power, especially variable renewable and complementary flexibility technologies, is central to transforming energy systems. Moreover, incentives for investment in flexibility technologies like energy storage, electricity demand management and system interconnections arise from the variation in electric power prices over time and across location. Having wholesale electricity markets that reflect well the value of electricity and its variations is thus fundamental to successful transformations. Most fundamentally, this requires administrative scarcity pricing at the average VoLL and wholesale market price caps if imposed at or above the average VoLL. The integration of system-operator demands for operating reserves with wholesale markets can improve transparency of the value of capacity in wholesale markets. In addition, if governments prefer capacity remuneration mechanisms, they should design and implement the mechanism to allow bona fide flexibility technologies to enter the market and compete on a level playing field.

Any market design choice, nevertheless, involves balancing complex trade-offs between market efficiency and potential abuses of market power as well as credibility and predictability of the framework. To foster this credibility, several liberalized systems, such as those in the European Union (EU), ERCOT, Great Britain's and PJM, established independent market-monitoring agencies. They provide regular transparency on the functioning of wholesale and retail markets, to promote improvements over time and strengthen policy predictability and credibility.

BALANCING MARKETS AND ANCILLARY SERVICES

System operators have responsibility for continuously balancing supply and demand across networks in all electric power systems. An operator does this by procuring from and requiring of generators, large-scale customers and suppliers a range of actions to support network reliability. For example, in liberalized systems, operators run a final balancing market from closure of the wholesale market until an operational window begins. The balancing market is similar to the wholesale market, except that the system operator stands between generators and wholesale customers and transacts on both sides of the market to align expected supply and demand across the network. Generators and their customers, however, are typically discouraged from relying on the balancing market. Prices for unsold supplies are usually below prevailing wholesale prices, and those to meet unmet demands above. System operators also typically procure operating reserves, which are available at short notice to meet demand in case a generator fails at the last minute or demand surges unexpectedly. The balancing markets and system-operator demand for operating reserves relate to and can be integrated with wholesale electricity markets, to strengthen their functioning and ensure that wholesale prices reflect the value of capacity to network reliability.

In addition to balancing markets and operating reserves, system operators procure or require generators and large customers to provide ancillary services like frequency response and voltage support to maintain system operation within its technical parameters. Frequency responses are changes in supply and demand in response to real-time deviations of system frequency from its reference value. Frequency responses are not specific to network locations, so many system participants can provide them. Some systems have markets for frequency-response services. They operate in advance of service delivery (monthly for the month ahead) and typically reward this service with payments for both capacity ($/MW) and power ($/MWh). However, in many systems, operators require generators to provide frequency-response services and pay administratively set compensation for energy supplied. Voltage support within an electricity network helps manage the flow of power and maintain its quality for customers. This so-called reactive power must be provided locally across the network, so its provision is less amenable to competition. System operators and distribution-network operators typically use mandatory requirements to provide reactive power at fixed prices as well as network-owned reactive assets. The other main type of ancillary service is bootstrapping capabilities of power plants to restart in the event of a large-scale power outage—so-called black-start capabilities. Only specialized generators and some conventional thermal plants can provide this service.

The costs of all ancillary services are a small fraction of the wholesale market value of electric power (a few per cent), with frequency response typically the largest among them. As the share of variable renewables in generation rises, however, operator demands for ancillary services are expected to increase (Pollitt and Anaya, 2019). For example, more wind and solar PV generation, which are not synchronized to system frequency, leads to wider frequency variations. In contrast, thermal generators, which are synchronized, automatically adjust to frequency variations—their so-called inertial response. Their rotation speeds up or slows down to counter frequency variations and, in the case of a plant failure or demand surge, this automatic response gives the system operator valuable seconds to dispatch available alternatives. Such instantaneous responses have gone unremunerated because they were simply an inherent feature of thermal generation technology. But as the share of thermal plants in generation decline, this inertial response and other ancillary services they currently provide must be replaced with alternatives.

Just as thermal generation technologies provide not only electric power but also ancillary services like frequency control, so too can variable renewables. Wind turbines provide inertial response through their power electronic converters, which control the voltage and frequency of electric power provided to the grid. They also vary their generation of electric power by controlling wind turbine blade angles and rotational speed. Solar PV and batteries can provide synthetic inertial response if their inverters, which convert direct current into alternating current, are designed to do so. These resources can also provide voltage control at a range of locations across networks, as can distributed demand response. Therefore, as the need for ancillary services increases with an increasing share of variable renewables in generation, so too can the supply of these services. The challenge is for system operators to adapt their specifications of ancillary services to new technologies and to remunerate the services competitively or adequately. It may also be cost-effective for system operators, including those of distribution networks, to own and operate resources that provide ancillary services, especially for voltage control (Pollitt and Anaya, 2019).

LONG-TERM CONTRACTS FOR ELECTRIC POWER

While well-designed wholesale and balancing markets and those for ancillary services can help to ensure adequate revenues for new investments, power purchase agreements (PPAs) are widely used contracts for matching long-term electricity supply and demand and sharing investment risks. They feature in both liberalized and traditionally organized sectors. The power purchasers—or off-takers—typically provide a guarantee of the amount of electric power purchased and in many cases the price paid over the contract length. This sharing of long-run quantity and price risks helps to lock-in project revenues and unlock commercial finance for new investments. The power purchasers—companies as well as private and public utilities—also benefit from securing electric power supplies and managing power price risks.

For example, there is a small but growing demand for corporate renewable PPAs, primarily in Europe and the United States, driven by companies committing to 100 per cent renewable power and seeking to differentiate their products and services by the sustainability of their inputs (IRENA, 2018; Simonelli et al., 2019). Utilities and electricity suppliers are also entering into PPAs with renewable project developers, especially in the United States (Bartlett, 2019).

These PPAs help to match long-term low-carbon power demand with supply, increasingly so as the cost-competitiveness of variable renewables improves. In addition, there are competitively allocated, 'subsidy-free' long-term contracts for variable renewables in which governments obligate electricity customers to purchase the electricity generated at or below expected average wholesale prices (Chyong et al., 2019). These contracts shift price risks arising from unexpected policy, technology and demand developments to customers without conveying an expected electric power price subsidy to variable renewable generators. These government-supported contracts, however, potentially distort investments away from stronger system reliability and lower-cost grid locations by shifting risks to customers and enabling investors to ignore price variations over time and across grid location (Newbery et al., 2018).

In traditionally organized sectors in many developing countries, PPAs perform a different role. They aim primarily at managing risks of incomplete power markets, evolving sector policies and regulations, and often unreliable system operation and poor financial performance. These systems often lack cost-reflective tariffs and financial viability, with governments using them to transfer resources for social and political goals. As a result, they are fundamentally difficult contexts for new private investment. However, some are open to participation by independent power producers through PPAs and have basic wholesale power markets such as cost-based electric power pools (Rudnick and Velasquez, 2018).

In such financially and operationally challenging contexts, PPAs help unlock private investment in new generation capacity to meet growing demands. These agreements tend to be between creditworthy governments (or government-backed utilities) and independent power generators and project developers. The proportion of power-generation capacity developed through PPAs varies depending in part on the extent of system unbundling—the separation of generation from transmission, distribution and supply. Most PPAs are in partially liberalized systems in Asia, Eastern Europe and Latin America and support investment in new conventional thermal generation. To improve project economics, they are often structured for baseload generation, for example, with take-or-pay provisions calibrated to baseload generation to improve project economics, but potentially at the expense of total system costs.

With rudimental wholesale electricity price signals, these systems rely primarily on system planning to manage expansion and reliability, the traditional sector approach. Economic development goals tend to instil a planning focus on baseload generation, often supported by PPAs, and grid extensions to demand centres to meet growing demands. Network expansions are typically funded directly by governments. But often network costs are not fully recovered through prices and their maintenance and reliability can be found wanting. These practices can direct systems away from cost-effective and reliable operation. Moreover, as the costs of variable renewables become increasingly competitive with conventional technologies, planning approaches to investment and system expansion must adapt. A predictable tipping point in these systems is when the long-run costs of variable renewables fall below the short-run costs of conventional thermal generation. At this point, planning assumptions that new thermal plants will operate at baseload over their asset lives clearly become untenable. The allocation and terms of PPAs must thus adapt to expected changes in the economics of generation to efficiently guide investment in reliable and low-carbon power systems. Such considerations are particularly important for coal generation plants, for which plant economics are very sensitive to load-factor assumptions because of their high capital costs.

System liberalization and development of well-functioning wholesale electricity markets can help strengthen adaptations to these technological changes. However, introducing wholesale electricity markets is complex, and suitable only for systems that are already functioning well (Rudnick and Velasquez, 2018). There are several preconditions to fulfil before their introduction. Firstly, for there to be meaningful wholesale market competition, the overall system must be financially sound. Distribution utilities, which are the main buyers in the market, need to be creditworthy counterparties to wholesale exchanges; otherwise they will not take place. This means retail tariffs are cost-reflective and collected, and unofficial grid connections cut. Nonlinear tariffs can help address energy access and affordability, but direct government subsidies may be necessary to allocate the costs of widening energy access equitably (Briceño-Garmendia and Shkaratan, 2011). Secondly, the electricity systems must be of sufficient scale for there to be enough generators and suppliers for effective competition. Few developing countries with small power systems (under 5 GW of capacity) have competitive wholesale markets (Jamasb et al., 2017). Thirdly, institutional arrangements for private investment in general and the electricity sector in particular must instil investor confidence that sector policies and regulations are predictable and property rights respected (Jamasb et al., 2005 and 2017). Establishing these conditions remains a priority in many developing countries.

NETWORK CHARGES AND SPATIAL PRICING OF ELECTRIC POWER

Transmission and distribution networks move electric power from generation to load with the laws of physics, rather than of supply and demand, guiding its flow along wires. The electric current heats the wires and dissipates away from them, resulting in transmission and distribution losses of electric power. There are also power losses from the step-down transformers that link high-voltage transmission lines to low-voltage distribution networks. Generators thus need to produce about 8–15 per cent more electricity than customers demand (Cretì and Fontini, 2019, p. 191). This is the main operating cost of transmission and distribution. Electricity also moves along all lines between generation and load, not just along the shortest route, so the task of system operators in running and balancing the system relies on complex calculations of electric power flows through systems.

Transmission and distribution wires, moreover, have maximum operational capacities. Breaching them risks physical damage to the wires, and their failure. When the flow of electric power is high, and transmission at capacity in at least parts of the network, the merit order of plant dispatch must adapt to the transmission constraints, prioritizing grid location over operating costs to balance supply and demand. This spatial segmentation of the market caused by congestion gives rise to so-called locational marginal (nodal) prices for electric power that vary across points (or nodes) on the grid. Nodal prices reflect the marginal cost of generation or marginal value of electric power to customers (or both) within each node on the grid, much as marginal costs and benefits determine the system-wide wholesale price (Schweppe et al., 1988, pp. 151–76). In some systems, proximate nodes that tend not to be segmented by network congestion are grouped together into well-defined zones, balancing the benefits of more market competition against efficiency loss of less granular pricing.

A key contribution to the economics of electricity market design shows that differences in nodal or well-defined zonal prices across the grid reflect the contribution of generation and load within the node to transmission losses, indirect power flows and congestion (Hogan, 1992 and 1998). If the network is not congested, these differences simply reflect marginal costs of transmission losses (Léautier, 2019, pp. 190–201). If it is congested, these differences also reflect the 'shadow value' of additional capacity in the congested part of the network. System operators calculate locational nodal or well-defined zonal prices, where they are in use, for market settlement after the real-time operation of the system when financial settlements for last-minute market imbalances are made. However, network congestion can also increase the market power of generators that are protected by a network constraint. For example, a generator protected by an import constraint into a node would have market power with respect to any residual power demands therein.

These 'shadow' prices are knowable only after the fact. However, system operators auction financial transmission rights (FTRs), which pay the holders the difference in nodal (zonal) prices between two nodes (zones) on the grid (Hogan, 1992). They are similar in purpose to reliability options for system capacity. Generators and wholesale customers purchase FTRs to help manage their exposure to nodal or zonal prices. These market arrangements are the basis for pricing network use and remunerating its operators. They also reveal the value of additional network capacity and help inform whether investment in its reinforcement and extension is cost-effective. But FTR can also enable generators to increase their exposure to their nodal prices, strengthening an incentive to exercise market power from any congestion constraint (Léautier, 2019, pp. 239–41).

Nodal and well-defined zonal pricing are found primarily in liberalized electricity systems in the United States. Their initial introduction was born out of the necessity of easing network congestion in recently liberalized systems. In contrast, most European systems rely on broad, politically defined zones in which countries are 'zones', with the notable exceptions of well-defined zones in networks in Italy and the Nordic countries. European utilities traditionally designed their networks to eliminate congestion within their country (Léautier, 2019, pp. 183–4). The foundations for a common European electricity market and interconnections among systems were laid only in 2015 with the EU Regulation on Market Coupling. While there tends to be little congestion within European zones, there is significant congestion across them. Moreover, the rising share of variable renewables in European generation is adding congestion between systems and causing it within them, especially in Germany and Great Britain.[6] Network charges in most European systems are levied like postage stamps. They are fixed charges per unit of electric power regardless of its origin or destination, and aim to cover average network costs. Congestion is managed within zones through countertrading by system operators and across zones by wholesale exchanges in day-ahead markets. The introduction of nodal or more well-defined zonal pricing would improve the efficiency and cost-effectiveness of the pan-European network, if not within each zone or country (Newbery et al., 2018).

Transmission and distribution networks must also adapt physically to the rising role of variable renewable generation. The current design of electricity transmission and distribution networks aims primarily to move electric power from large, centralized generation plants connected to transmission networks to widely dispersed customers connected to distribution networks. They are sized to accommodate almost all expected power flows at peak energy demands. However, introduction of variable renewables changes the demands placed on networks. Compared to locations of conventional thermal generation plants, those of wind and solar PV generation are much more widely dispersed, and many sites are connected to distribution rather than transmission networks. Distributed generation by small-scale solar PV installed on commercial and residential buildings and manufacturing plants adds generation to distribution networks. Transmission and distribution networks must thus adapt to variable renewables and their locations, including being designed and sized to manage surplus generation on distribution networks and its movement to other locations on the grid. Locational marginal pricing can help guide cost-effective network investments as well as least-cost locations for generation. They are useful too in informing investment decisions on interconnections between networks. However, these approaches require robust market-monitoring by the system operator to detect and manage potential exercises of market power.

RETAIL ELECTRICITY MARKETS AND TIME-OF-USE PRICING

While the main focus is on wholesale pricing of electricity and creating sound incentives for investment in a sector destined to change and grow in transforming energy systems, it is also important to consider customers. Customer choice is important in its own right. But the nature of these choices, especially how responsive they are to variations in the price of electric power over time and across location, is also fundamental to wholesale price formation and transmission pricing. In current electricity systems, customers are largely unresponsive to short-run variations in wholesale electric power prices, which contributes to price volatility and market power in electricity systems. Many retail customers in many systems have meters that measure only cumulative usage, so they are not exposed to time-varying power prices and there is no price signal to elicit a demand response. Large-scale customers typically do have meters that measure use over time intervals (hourly) and face real-time prices linked to wholesale prices.

With the development of digital control and communication technologies—smart meters that measure energy use over short time intervals, and smart appliances that can automatically adjust to real-time prices—there is more potential for suppliers to engage with retail customers on demand management. For example, smart meters enable suppliers not only to automate their meter and billing systems, saving substantial operating costs, but also offer more granular real-time pricing for customers. Together with smart appliances, these digital technologies enable more price-responsive demands. In addition, an increase in electric power demands for battery electric vehicles and electric space heating of buildings would expand the scope and scale of these digitally enabled demand-management services. At the same time, an increased role for variable renewables in generation would make these services more valuable to the electricity system. The challenge is to design electricity markets and business models that realize the potential of these complementary technological developments.

NATURAL GAS NETWORKS AND THEIR TRANSFORMATION TO LOW-CARBON FUELS

Compared to electricity networks and their adaptation to variable renewables and flexibility technologies, the adaptation of natural gas networks to low-carbon fuels is both simpler in some ways and more complicated in others. A simpler aspect is that wholesale markets for a gaseous fuel are more straightforward than those for electric power, system-balancing and network services.[7] But with well-designed wholesale and retail markets for electric power, investments in system transformation involve mostly wholesale market participants. Well-designed and well-regulated wholesale markets can coordinate much of the necessary investment in low-carbon generation and flexible supply and demand resources. Retail engagement with change focuses primarily on demand response as well as distributed generation and storage, especially where it alleviates network congestion. The more challenging aspect of transforming gas networks is that many retail customers must change technologies for space and water heating together with those for better thermal efficiency of buildings. Appropriate choices for space-heating and space-cooling technologies depend crucially on the thermal efficiencies of buildings' fabrics, which are poor in much of the existing building stock. The investment coordination challenges with transforming natural gas grids must thus involve their customers from the outset.

The conversion of British homes to natural gas for space heating and cooking in the 1960s and 1970s, following major resource discoveries beneath the North Sea, provides an example. In this case, the government shaped the transition of a sector dominated by a state-owned utility—British Gas—from limited use of town gas (a synthetic fuel made from coal) to widespread natural gas use for central heating and cooking. The Gas Council, a government oversight body, facilitated and coordinated investment in both upstream production and expansion of transmission and distribution networks. The Council also took the decision to make a complete and direct transition to natural gas from town gas (Arapostathis et al., 2013; Pearson and Arapostathis, 2017). This required millions of dwellings and properties to change or modify heating equipment and cooking appliances. To facilitate this change, the Council undertook training of gas engineers and information campaigns about the benefits of natural gas as a cleaner fuel and central heating as a better technology for thermal comfort and health. In contrast, the initial electrification of buildings in the early twentieth century faced similar investment coordination challenges, but little lock-in of incumbent technologies, such as kerosene lamps for lighting. In this case, electricity utilities and manufacturers of electricity-generation and end-use technologies promoted a widely shared vision of a modern society, including through their marketing of household electrical appliances (Chapter 1).

The transition from natural gas to low-carbon fuels and electrification of space and water heating poses similar challenges to those in the transition to natural gas from town gas—and then some. The effective coordination of investment plans and decisions of producers of low-carbon gaseous fuels and their many customers is necessary to achieve a viable market outcome dominated by low-carbon technologies. Markets and prices have an important role to play in informing and coordinating decisions on low-carbon energy and end-use technologies. But in contrast to previous waves of disruptive new energy carriers for buildings and their end-use technologies, low-carbon gaseous fuels such as biomethane and hydrogen appear likely to be more expensive per unit of energy than natural gas, even once they mature and scale (Chapters 3 and 5). The lower-cost low-carbon alternatives for space heating and cooking could well be use of electric power, such as heat pumps and electric resistance heating, rather than a fuel. But in either case, these alternatives to natural gas would only likely dominate incumbent fossil-fuel-based technologies with the support of government policies, such as polices to support development and early deployment of low-carbon alternatives and emissions pricing (or low-carbon subsidies).

For buildings, there are three critically interdependent investment decisions that require coordination—thermal efficiency measures for existing and new buildings, low-carbon heating and cooling technologies for thermal comfort and provision of low-carbon electricity and fuels to buildings. Emissions pricing consistent with a quantitative emissions target, such as a time-bound commitment to achieving net zero emissions, is one way to align these investment decisions. However, substantial increases in natural gas prices and heating costs would have regressive distributional impacts, hitting poorer households harder in the short to medium term and giving rise to concerns over distributional fairness, including inadequate winter heating in vulnerable households. This approach would likely require redistribution mechanisms (progressive tax cuts on labour income or transfer payments) to mitigate its adverse distributional impacts, and policy nudges such as winter heating payments to encourage healthy levels of thermal comfort taking.

One way for building owners to manage their exposure to higher energy and emissions prices is to substitute away from natural gas by undertaking energy efficiency measures—more efficient boilers, better heating controls and more loft (attic) insulation. While such incremental and typically cost-effective measures are helpful, the transformation of buildings to low-carbon technologies would require more fundamental changes. Low-carbon energy for space heating—be it electric power or low-carbon fuels—would likely be more expensive than natural gas in the long run. For example, shifting significant space-heating loads onto electric power would likely expose at least some of these demands to peak-load prices, especially during winter cold snaps. In advanced industrialized countries with cooler temperate climates, energy demands for space heating are several times that of existing electric power systems, so electrifying heating demands would have a significant impact on demand profiles and power market prices (for example, DECC, 2012, p. 12). Such a shift in relative prices of energy would point to greater cost-effectiveness of thermal efficiency measures in new and existing buildings.

In addition, the costs of thermal efficiency measures are lowest when buildings are being built or substantially refurbished. At such junctures in building life cycles, the least-cost combination of thermally efficient building fabrics and low-carbon electric or fuel heating becomes feasible to implement. But given the vast heterogeneity of building stocks, there is unlikely to be a single solution, and it is unclear which low-carbon technology combinations will dominate and scale in the long run in cooler temperate climates (Imperial College, 2018, pp. 20–26).

Apart from costing emissions, an energy-reform strategy that could begin to resolve this uncertainty leads with forward-looking, cost-effective building standards for new and existing structures undergoing substantial refurbishment. These building standards could be complemented with emission-performance standards for new and refurbished buildings that require their heat loads to be serviced by either hybrid (heat pumps plus boilers) or electric heating technologies. They would encourage innovation in cost-effective combinations of thermal efficiency measures and heating technologies for existing buildings and in low-carbon new buildings. For example, the Netherlands government prevented new residential buildings from connecting to the natural gas grid from mid-2019 and set high thermal efficiency standards for new buildings from 2021. The United Kingdom government plans to ban natural gas central heating in new dwellings from 2025. The German government strategy for substantially eliminating most emissions from new and existing buildings focuses strongly on improving the thermal efficiency of building fabrics (Bürger et al., 2016, pp. 36–50).

Such reform strategies have two important implications. One is that they begin to shift heat loads on to electric power while reducing this sizeable energy demand. Second, it leaves open how peak heating demands are served, albeit flattened by better thermal efficiencies for buildings and met with alternative heating technologies. They could be served either by electric power, including some at peak-load prices, or like hydrogen and biomethane in hybrid heating technologies. This strategy focuses market selection in serving peak heating demands in the transformation of buildings. It creates a product— low-emission dwellings—that could be potentially differentiated in rental and property markets, to support upfront firm investments in innovation and market creation. However, such regulations would likely have adverse impacts on market values of existing buildings that are relatively difficult and expensive to transform to low-carbon alternatives.

CONCLUSION

Adapting energy markets and infrastructures to low-carbon technologies is fundamental to cost-effective and decisive energy-system transformations. In particular, governments must adapt wholesale electric power markets, where they exist, to the performance characteristics of variable renewable and complementary flexibility technologies. Such reforms could involve integration of system-operator demands for reserves with wholesale market pricing and capacity remuneration mechanisms. More granular spatial pricing across networks and time-of-use tariffs for retail customers could further promote system efficiency and greater flexibility. Independent market-monitoring agencies in liberalized systems, such as ERCOT, Great Britain's and PJM, provide transparency on the functioning of wholesale and retail markets, helping to build confidence and credibility in market arrangements among participants. Where well-functioning wholesale markets are not in place, it is vital that more planning-led approaches to power-system development adapt readily to the growing cost-competitiveness of variable renewables to put their development on sustainable paths.

The challenges in transforming natural gas networks arise not from complex wholesale market-design and system-balancing issues, but rather from coordination of investments in energy end-use and complementary technologies, especially in buildings. This issue pertains primarily to advanced industrialized countries, Russia and parts of China with relatively cooler climates and extensive domestic natural gas networks. Building reform strategies that prioritize forward-looking, cost-effective thermal efficiency standards for new and existing buildings can help ease the lock-in of existing heating technologies based on fossil fuels. Complementary reforms aimed at least at partial decarbonization of heating technologies can also help accelerate innovation and deployment of alternatives to incumbent technologies. This policy strategy can avoid some of the regressive distributional impacts of emissions pricing on heating for dwellings, while at the same time creating conditions to facilitate use of emissions pricing in the long run to optimize the eventual system configurations.

NOTES

1. There are also other wholesale power market arrangements, such as cost-based electricity pools and single-buyer markets, found primarily in emerging markets that are in the process of reforming their electricity markets.
2. See Cramton and Stoft (2006) and Cambridge Economic Research Associates (2018) for useful descriptions of the average VoLL, its estimation and application in the United States and Europe.
3. Other types of capacity-remuneration mechanisms include a decentralized capacity market and strategic reserve. The former obligates electricity suppliers to purchase capacity certificates to ensure capacity adequacy and is used in the California, Mid-Continent and New York systems in the United States and in France. These certificates cover only relatively short operating periods. In the latter, the system operator procures capacity deemed necessary for reliable operation of the system. This capacity can be segmented from the wholesale market for the system operator's sole use. Several European systems use this approach.
4. The socially optimal amount of capacity is such that the average hourly marginal operating profit of an additional MW of capacity is equal to the fixed capital costs for that generation. The marginal operating profit is approximated as average VoLL times the expected number of hours of unserved energy demand. The marginal fixed cost of capacity is the annual capital rental cost of the capacity. For example, if VoLL is $20,000/MWh and annual rental cost of capital is $60,000/MW, the socially optimal expected number of hours of unserved energy demand is three. The amount of physical capacity auctioned should aim to deliver such expected system reliability. This is similar to the traditional engineering approach to system reliability but with the economic overlay of the marginal costs and benefits.
5. Suppose peak demands for a system are expected to be below 100 GW 99.97 per cent of the time, and the unplanned outage rate of each generation plant is 7 per cent—so their capacity credit is 93 per cent. Adequate system-generation 'nameplate' capacity in this simple example is 107.5 GW, which has a risk-adjusted or 'de-rated' capacity of 100 GW.
6. The introduction of variable renewables contributed to increased transmission between northern and southern regions of Germany, and Scotland and England in Great Britain.
7. Useful discussions of natural gas markets and their functioning in Europe and the United States are ACER (2019b) and FERC (2020).

7. Making better use of energy and materials

While low-carbon electricity and fuels are necessary for net zero emissions of carbon dioxide from energy, making better use of energy and material holds important potential—to improve current energy-system performance and facilitate its transformation to low-carbon alternatives. There is, for example, scope for cost-effective gains in energy efficiency with available technologies owing to market imperfections. Effectively addressing market barriers and failures as well as behavioural anomalies that impede least-cost combinations of investment and energy-use decisions yields improvements to the current energy system. Such policy measures help energy customers make better overall choices, save energy and avoid emissions. There is also potential to integrate in new ways energy and material-efficiency investments with complementary ones in low-carbon technologies that facilitate more cost-effective system transformations. For example, highly thermally efficient buildings enable use of low-carbon space-heating technologies such as electric heat pumps. At prevailing relative prices, this approach for new buildings is a cost-effective alternative to less thermally efficient buildings heated by natural gas or fuel-oil boilers (Lucon et al., 2014). Similarly, material-efficiency gains from improving designs of buildings, equipment and appliances, using them more intensively during their asset lives and recycling materials in them at the end of their asset lives can facilitate industrial transformations to low-carbon technologies. For example, recycling high-value materials like steel, aluminium and copper cost-effectively cuts energy use and emissions from metals production by avoiding the smelting of metal ores.

The technical potential to use energy more efficiently is significant. There are two main types of energy losses in the current energy system. One arises from energy conversions from one form into another. Examples include using coal or natural gas for thermal generation of electricity and using electric power for rotating motors. Conversion losses through the energy system are multiplicative. If the total loss from generating electricity is 60 per cent and that from a rotating electric motor is 10 per cent, 36 per cent of the primary energy in the fossil fuel is in the rotating motion of the motor. The second type of energy losses arise from how effectively conversion devices work together with buildings and appliances, vehicles, and factories and equipment that contain them to produce useful energy services and materials. For example, most heat pumped inside buildings for thermal comfort using boilers and radiators steadily dissipates outside, unless buildings are designed to the Passivhaus standard.[1] Together, the energy conversion and passive energy losses in the current system are many times larger than the amount of energy in embodied useful services and materials (Cullen and Allwood, 2010a). The technical potential to reduce them with available technologies is significant, and transforming energy systems creates more economic opportunities to boost energy efficiency and cut carbon dioxide emissions.

Similar efficiency potential arises with materials. In producing new buildings, equipment or vehicles, there can be technical potential to use fewer materials and alternatives to carbon-intensive ones while maintaining the usefulness of the goods produced. For example, building designs typically specify more steel than is necessary for their structural integrity alone. The design trade-off is between material and labour costs in construction (Moynihan and Allwood, 2014). It is also feasible and cost-effective to substitute engineered timber beams for steel in some structural applications. This is a design choice that sequesters carbon in buildings, at least during their lifetimes, and avoids carbon dioxide emissions from steel production. There is technical potential to use existing buildings and vehicles more intensively. Examples include Airbnb (multi-purpose dwellings), multi-purpose or shared commercial offices, digitally enabled flexible working and potentially shared autonomous vehicles. Life extensions of existing buildings, equipment and vehicles, and recycling materials in them at the end of assets' lives, also hold potential for cost-effective gains in material efficiency. Such material efficiencies could counter higher expected material costs from cutting emissions in heavy industries (Chapter 3).

Making better use of energy and materials with existing technologies holds significant potential, especially in producing and using buildings and road vehicles for which efficiency gaps appear wide from a technical perspective. Such gaps may reflect in part some past neglect of these performance characteristics of buildings and vehicles by investors in them, and correspondingly modest innovation efforts by firms. This neglect has roots in market imperfections related to energy and material efficiencies. Evidence shows that well-designed policies to promote greater efficiencies can improve market outcomes. Looking forward, efficiency policies should consider expected shifts in relative prices arising from government climate goals and the energy system transformations they would require.

ENERGY-EFFICIENCY GAPS: A TECHNOLOGICAL PERSPECTIVE

'A modern industrial society can be viewed as a complex machine for degrading high-quality energy into waste heat while extracting the energy needed for creating an enormous catalogue of goods and services' (Cullen and Allwood, 2010a; based on Summers, 1971). This quote puts in sharp relief the energy-efficiency challenge. That is, producing the enormous catalogue of modern goods and services with less energy by reducing energy losses—the low-grade heat that dissipates away from the energy system into the surrounding environment.

Primary energy resources such as fossil fuels, flowing water and fissile materials are high-quality inputs into this modern 'machine'. They are high-quality energy resources that are progressively degraded by the energy system as it works to provide useful energy services and materials. These losses arise from the conversion of the primary energy resources into more useful energy carriers like electricity and fuels, in their distribution to customers and again in conversion into even more useful motion, heat, light and sound. The latter conversions are performed by engines, turbines, motors, fuel burners and boilers, electric heaters and furnaces, lighting devices and electrical appliances. They work within buildings and appliances, vehicles, and factories and equipment to produce the energy services and materials that societies demand.

Energy losses occur with every energy conversion. They are consequences of the unavoidable laws of physics and limited technological capabilities in tapping theoretical potentials in converting energy flows through devices. Energy losses arise from both the amount of energy that flows through devices and their technical performance. The largest energy losses are associated with the largest energy flows through the system, though the technical performance of system components matters too.

In the conversion of primary energy into electricity and fuels, large energy losses arise from thermal generation of electricity using coal, natural gas and nuclear fuel and, to a lesser extent, refining crude oil primarily into liquid fuels (Cullen and Allwood, 2010a; Figure 7.1). In their further conversion into motion, heat, light and sound, the largest losses arise from engines in road transport as well as fuel burners and electric furnaces in buildings and factories. In comparison, the flow of energy through lighting devices and electronic equipment and appliances and their associated conversion losses are small. The conversion devices, be they turbines for generating electric power or internal combustion engines, aim to tap cost-effectively—but not necessarily reach—their theoretical potential to convert fuels into electricity, heat and motion.

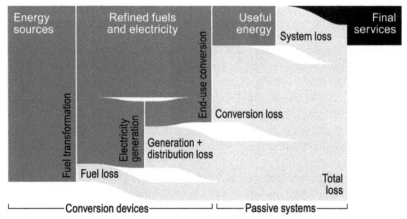

Notes: Approximate energy conversion losses are around 10 per cent from refining and processing fossil fuels and 75 per cent from thermal generation of electric power using fossil fuels, biomass and nuclear energy and its transmission and distribution to customers. The main conversion and passive losses from energy end-use arise from internal combustion engines for motion, their conversion losses and resistance to motion from air and road surfaces. The other main losses from energy conversions are boilers in buildings and factors and their associated passive losses, such as heat dissipating away from buildings and industrial processes. Final energy services are vehicle motion, useful heat in buildings and factories, light and material outputs. See Cullen and Allwood (2010a, 2010b) and Cullen et al. (2010) for technical estimates of these losses from existing energy technologies, although these efficiency factors do not consider technological progress over the past decade.
Source: Reprinted from J.M. Cullen and J.M. Allwood (2010), The efficient use of energy: Tracing the global flow of energy from fuel to service. *Energy Policy*, 38, 75–81, with permission of Elsevier. https://doi.org/10.1016/j.enpol.2009.08.054.

Figure 7.1   Global energy flows, conversion losses and useful services and materials

But there can be room for improvement. Several developments in conversion technologies in recent decades, for example, aim to capture heat that otherwise would have been wasted and convert it into useful motion and heat. Examples include combustion turbines combined with steam turbines (combined-cycle gas turbines) for electric power generation, internal combustion engines with turbochargers and superchargers for vehicles, and condensing boilers for buildings. These innovations in turbines, engines and boilers capture waste heat from fuel combustion that would have been dissipated into the environment and turn this energy into useful motion or heat, improving the efficiency of the conversion devices. There is technical scope for further improvement in each of these and other conversion devices, as with all technologies (Cullen and Allwood, 2010b). But while important, further innovations in conversion devices are perhaps not the main potential sources of cost-effective energy-efficiency gains in current or potential low-carbon-energy systems.

The least efficient parts of the energy system are its passive components: the buildings and appliances, road vehicles, and to a lesser extent factories and equipment that work with conversion devices to provide the energy services and materials that societies demand (Graus et al., 2009; Cullen et al., 2010). Buildings and appliances have the most technical potential for efficiency gains; factories and equipment the least. Light-duty passenger vehicles for roads have more potential; heavy-duty trucks, ships and planes less. The designs of factories and equipment as well as heavy-duty road vehicles, trains, ships and planes prioritize energy efficiency for their commercial use. The design priorities of buildings, appliances and light-duty passenger vehicles are elsewhere because the utility derived from them is not necessarily confined to their functional purposes. Social factors such as norms matter too for household and customer-facing business decisions.

The technical potential to save energy in buildings and appliances is substantial, mostly in space heating and cooling. New buildings designed to the Passivhaus standard and deep retrofits of existing buildings to a similar standard can cost-effectively and substantially cut the need for these energy services where construction capabilities and material supply chains afford (Bürger et al., 2016, pp. 36–50). Their added investments in thermally efficient building fabrics are largely offset by savings in heating and cooling systems and energy use. While deep retrofits of existing buildings appear to hold similar potential, investments in piecemeal improvements in their thermal efficiency often prove disappointing (Allcott and Greenstone, 2017; Fowlie et al., 2018; Davis et al., 2020). Appliances too have potential to use energy more efficiently, but they still have significant minimum energy requirements for hot water, cooling and heating food, lighting and information displays in televisions and computers. In transport, there are feasible design changes to passenger cars that would significantly boost their energy efficiency. They include using lighter materials in vehicle gliders,[2] reducing their size, improving their aerodynamics and reducing tyres' rolling resistance.

The transformation of an energy system towards one centred on electric power away from fossil-fuel combustion also holds significant potential to improve energy efficiency, although care is required in making this point. Electric motors are much more efficient in converting energy into motion than are internal combustion engines and combustion turbines. Electric heat pumps are much more efficient in converting energy into useful heat than a boiler because they move low-grade heat in the environment indoors rather than using high-quality energy to produce low-grade heat. Electric resistance heaters are almost 100 per cent efficient in converting electricity into heat, slightly better than a condensing boiler and significantly better than a conventional one. As energy systems become increasingly centred on renewables and electric power, there are significant potential efficiency gains in conversion technologies by reducing dependence on fossil-fuel combustion to produce motion. The large losses associated with thermal generation of electricity and internal combustion engines would be largely avoided.

New energy-efficiency losses would be introduced, however. Renewable technologies are inefficient in converting diffuse energy resources such as solar irradiance and wind into high-quality electricity. They are unable to capture much of the primary energy from these renewable resources. But this type of inefficiency is typically not scored as an energy system loss. It does not take the form of dissipated waste heat from energy captured by the system. Also, inefficiencies from the uncaptured primary energy do not give rise to higher operating costs because the solar irradiance and wind are free to use once generation sites are procured. Nevertheless, losses from transmission and distribution of electric power would increase with increasing electrification, as would conversion losses from batteries and fuel cells. The low energy density of batteries and their significant mass add to vehicles' inertial resistance. The largest new energy losses, though, would be from production of low-carbon fuels like hydrogen and advanced biofuels, which is a key reason why they are likely to be more expensive than the fossil fuels for which they would substitute. Materials made from bioresources would also require significantly more energy to produce than those using fossil-fuel feedstocks. Such changes in relative costs would require re-optimization of how various technologies are integrated in low-carbon energy systems.

ENERGY-EFFICIENCY GAPS: A DEMAND PERSPECTIVE

Much of the focus on potential barriers to energy efficiency in the current system is on household decisions related to their dwellings and automobiles. They have large energy losses with wide gaps between technically achievable energy efficiencies using available technologies and those realized. But this does not mean that achieving technically feasible minimum energy losses is privately or socially optimal. It is not. Rather, it is potentially important to examine these sectors for market imperfections that can impede cost-effective investment and energy-use decisions because of the scale of technically avoidable energy losses. In fact, available evidence suggests that households face several potential barriers to making cost-effective choices, and well-designed policies can help overcome them. The available evidence on whether firms face such barriers is more limited.

Conditions under which households' and firms' choices deviate systematically from those of the notional ideal of fully rational economic actors acting in perfect markets are both well-known and relevant to energy-efficiency gaps (Allcott and Greenstone, 2012; Gerarden et al., 2017). Relevant market imperfections arise from costly information about product and building characteristics, consequential information asymmetries among market participants and split incentives associated with information asymmetries. Split incentives refer to contexts in which investment and energy-use decisions are made by different actors with opposing interests, such as property owners and their tenants, and there is asymmetric information about these choices (principal–agent problems).[3] Capital market failures can in principle affect investment choices, but evidence on their salience to observed energy-efficiency gaps is weak and, to the extent they exist, should be corrected through capital market interventions (Gerarden et al., 2017). There are also departures from fully rational choices identified by psychology and behavioural economic studies. These departures include rational inattention to some information, bounded rationality, biases towards the present and status quo, and loss aversion (Gerarden et al., 2015; Laibson and List, 2015).

Evidence on split-incentive problems in dwellings is compelling. This effect is seen clearly in energy-efficiency comparisons between owner-occupied and rental properties. The latter are significantly less likely to have energy-efficient appliances, and more likely to have poorly insulated exterior walls and ceilings (Davis, 2012; Gillingham et al., 2012; Krishnamurthy and Kriström, 2015). The probability of having access to more energy-efficient appliances and better insulation is two to five times greater in owner-occupied properties than in rental properties. This difference is robust in household survey results for 11 Organisation for Economic Co-operation and Development countries, and remains so even after controlling for other differences in household characteristics between owner-occupied and rental properties. The overall size of impacts on energy use, though, may be small. That said, similar differences between property types are seen in conversions from relatively expensive fuel-oil heating to less expensive natural gas heating, where the potential investment and energy-cost savings are much greater than for more marginal gains from better appliances and insulation (Myers, 2018). Owner-occupied properties are much more likely than rental properties to convert to the lower-cost heating technology. This split-incentive problem is thus a potentially significant barrier to achieving transformative investments that integrate low-carbon heating technologies with building thermal-efficiency measures.

Information asymmetries, moreover, can create barriers to energy efficiency even in the absence of such principal–agent problems. Evidence shows that mandated and voluntary energy-efficiency labelling programmes provide useful information to consumers on energy usage, leading to more investment in energy efficiency and less energy use. Stated-choice experiments find that information well-tailored to the product choice at hand, such as the value of saving energy, helps consumers make better choices (Newell and Siikamäk, 2014; Davis and Metcalf, 2016). However, information that provides only a broad signal of product quality, such as the voluntary US Energy Star programme, can induce distortions in both product demand and supply. For example, there is evidence that consumers interpret energy certifications as a broader signal of product quality, and suppliers differentiate their products consistent with this broad interpretation, including pricing (Salle, 2014; Houde, 2018a and 2018b). Moreover, not all information provision adds value. For example, trials of costly energy provision by sales staff in stores had little effect on demand (Allcott and Sweeney, 2017). Evidence also points to significant heterogeneity in consumer preferences over product characteristics—some people more than others value energy efficiency more highly than other product characteristics. This heterogeneity in preferences is a potentially significant explanation for at least part of empirically estimated energy-efficiency gaps (Bento et al., 2012).

Energy-information provision can also affect residential property rental and sale values. Hedonic studies of buildings and their energy certification programmes find that, after allowing for other property characteristics, buildings certified as more energy-efficient attract higher market valuations. For example, in US city and German housing markets where such programmes are in effect, the variation in market values associated with energy-information provision is in line with variations in the present discounted value of energy costs (Walls et al., 2017; Frondel et al., 2020). However, in Austin, Texas, a US city with a voluntary certified energy-audit scheme, there is no evidence of higher uptake of audits by owners of more efficient houses, even though they have an economic incentive to do so (Myers et al., 2019). This finding suggests that property owners either do not know the energy-efficiency characteristics of their properties or have an imperfect understanding of the relationship between energy efficiency and property values. It points to the value of mandatory rather than voluntary energy-information disclosures for residences. However, there is also evidence from a US housing market study that variations in fuel costs among properties are well capitalized in variations in property values even without mandated energy-information disclosures (Myers, 2019). A similar effect is found for new and used automobile prices (Busse et al., 2013).

The main issue around provision of energy information is not about whether it is valuable to investment decisions involving buildings, appliances and vehicles. It is necessary for economically rational choices, given observed heterogeneity of products and costs incurred to produce, convey and assess information about them. Rather, key issues are the better ways to provide the information (markets versus mandates) and communicate it (details on projected energy use and costs versus broad product certifications). The value of detailed energy information communicated in ways that directly inform investment decisions appears clear. Whether it is better provided through market interactions or government mandates remains subject to debate. For business-as-usual choices market interactions may well suffice. But the value of information to households and firms can increase when new decisions must be made as customers move away from status quo choices (Allcott and Rogers, 2014). Well-designed information can help inform this change, and government mandates ensure that it is widely available when most valuable.

In addition to the market failures and barriers, there are several potential behavioural issues that could cloud investment and energy-use decisions. There is, for example, substantial evidence that consumers tend to ignore costs arising from purchase decisions apart from purchase prices. Notable examples are that consumers largely ignore the costs of replacement ink when purchasing printers, and out-of-pocket medical costs when choosing health care plans (Hall, 1997; Abaluck and Gruber, 2011). With respect to energy, almost half of surveyed automobile buyers in a US study reported making their choices without considering fuel costs (Allcott, 2011). However, the extent to which this inattention to energy leads to undervaluation of energy efficiency is unclear. There is some evidence that consumers undervalue changes in expected future fuel costs arising from variations in fuel prices when making automobile purchases, but the extent is not large (Busse et al., 2013; Allcott and Wozny, 2014). The value of used automobiles sold in wholesale auctions appears to reflect fully variations in fuel efficiency and fuel costs (Salle et al., 2016). Similarly, an experimental study on consumer lightbulb choices provides only mixed evidence of significant inattention to energy efficiency (Allcott and Taubinsky, 2015).

Experimental evidence, in addition, reveals that consumers systematically misinterpret information contained in energy-efficiency labels, providing some support for bounded rationality as an explanation for private energy-efficiency gaps. For example, the nonlinear inverse relationship between kilometres per litre (miles per gallon) and energy use is not well understood (Allcott, 2013). While fuel costs scale linearly with litres per kilometre, rather than its inverse, consumers often judge the relationship between fuel costs and kilometres per litre as linear too. This leads to underestimation of higher actual fuel costs for less efficient automobiles and lower actual fuel costs for more efficient ones; however, the effect is not large. In addition, there is evidence of simple heuristics guiding choices among automobiles, with stated preferences influenced by several yet highly correlated metrics of fuel efficiency (Ungemach et al., 2018). Proving multiple translations of energy-efficiency measures can help inform better choices.

Present bias and myopia are two types of human behaviours that can also affect investment and energy-use decisions. Present bias refers to situations in which decision makers discount the long-term future at a significantly higher rate than the near term, while myopia is simply a lack of foresight. These two types of human behaviours can be difficult to distinguish empirically because they lead to similar predictions about reversals of preferences and choices depending on how close in time are the costs and benefits of choices. Two recent surveys of energy customers elicited information about the preferences and choices to identify those that are present-biased and not. One study finds that present-biased customers tend to consume more electricity than those that are not (Harding and Hsiaw, 2014). A second finds that present-biased customers are less likely to have automobiles with high fuel economy and well-insulated houses (Bradford et al., 2017).

ENERGY-EFFICIENCY GAPS: A SUPPLY PERSPECTIVE

Just as customers could make better choices, firms could do more in supplying energy-efficient products, especially in the context of better energy-efficiency policies and customer choices. Firm decisions on investing in innovations, differentiating products, timing their introductions and pricing are all areas in which firms engage in strategic competition and earn market rents, including returns on their innovation activities. But this innovation engine may be sputtering when it comes to energy efficiency. There is evidence that tightening of energy-efficiency standards induces product innovations that both improve product quality and accelerate cost declines for household appliances beyond those for appliances not subject to the regulations. These effects are seen in both the Netherlands and United States (Van Buskirk et al., 2014; Houde and Spurlock, 2016). A similar effect on induced innovations is also observed with changes in Japanese fuel-efficiency regulations of automobiles on those domestically produced but not those made in the United States and Europe (Kiso, 2019).

Two explanations are possible of the observed product-offering improvements in response to changes in energy-efficiency standards that appear to be welfare-improving with no increase in price and costs. One is that regulations induced welfare-enhancing innovations that were not otherwise encouraged by the market. For example, this could be due to customer inattention to this particular product characteristic and learning-by-doing from producing more energy-efficient products. On the former, there is evidence of some customer inattention to energy efficiency. On the latter, there is evidence of unexpected product innovations and price declines associated with changes in product regulations beyond that anticipated by policy makers and manufactures at the time of their implementation (Taylor et al., 2105). But this evidence is only suggestive of a causal link from tighter regulation to innovation and dynamic efficiency gains. That said, energy-efficiency innovations in household appliances appear unresponsive to energy-price changes that could have induced such innovations (Cohen et al., 2017).

A second explanation is that firms may under-supply product quality in general and energy efficiency in particular because of market structures. For example, firms may be unable to engage in sufficient price discrimination for higher-quality products, which regulations could address (Gerarden et al., 2017). But there is significant evidence from US and UK appliance markets that manufacturers do engage in product differentiation and price discrimination for products with higher energy-efficiency certifications, and charge relatively high prices for them (Houde, 2018a and 2018b). Also, the more efficient products apparently induced by regulations are introduced by existing producers, rather than new market entrants, accelerating their model turnover and cannibalizing existing products (Brucal and Roberts, 2019).

Another potentially relevant market failure affecting the supply of energy efficiency could arise from firms supplying complementary products that are unable to internalize impacts of their product innovations on other firms' profits. These spillovers can lead to too little innovation and product differentiation (Chapter 4). One sector with potential for new technology combinations and cost synergies is buildings—both new and existing. For example, new buildings designed to the Passivhaus standard are comparable in cost—in terms of the net present value of investment and energy costs—to conventional buildings that are less thermally efficient and heated with fossil fuels. Similar potential exists when buildings undergo periodic deep refurbishments to renew structures, internal and external surfaces, and mechanical systems, including heating, ventilation and air-conditioning (HVAC). Key complementarities in building components are their air tightness, fabric thermal efficiency and HVAC systems, with the two former components complementing heat-pump and air-conditioning technologies in providing thermal comfort (National Academies of Sciences, Engineering and Medicine, 2009, pp. 94–6; Lovins, 2018). Highly efficient refurbished buildings appear to not raise overall costs until energy savings from current practices reach at least 70 per cent. However, construction firms and their supply-chain relationships have yet to tap such potential in many countries.

There are similarly significant complementarities among vehicle components, which change with alternative drivetrains. About two-thirds of energy use in existing light-duty passenger vehicles works to overcome inertial resistance from the vehicle's mass, with the remainder being used to overcome aerodynamic and rolling resistance (Chapter 2). Complementarities among lighter and more aerodynamic vehicle gliders, better battery management and more efficient motors are especially important in new battery electric vehicles, because of the large mass of batteries and their high cost. Battery electric vehicle gliders are increasingly being optimized to reduce battery-storage requirements for a given level of performance. In this example, automobile manufacturers and their suppliers appear able to internalize benefits arising from new technology complementarities.

POLICIES TO BRIDGE ENERGY-EFFICIENCY GAPS

Policies to address the market imperfections that impede energy-efficiency investments aim primarily at the demand for such investments. They fall into three broad categories: information programmes, building and product standards, and incentives (Gillingham et al., 2009; Cattaneo, 2019). For example, information programmes shape consumer choices and behaviours through building and product energy labelling, energy audits and feedback on energy use. Information programmes can also help overcome split-incentive problems if they improve the valuation of energy-efficiency investments in rental- and property-market values. Energy-efficiency standards for new buildings, appliances and vehicles remove those that are least efficient from the market. This policy can help address not only the split-incentive problem, but also some behavioural anomalies that lead to inadequate attention to or undervaluation of energy-efficiency characteristics of technologies. Financial incentives such as grants or subsidies, tax deductions or credits and rebates lower the relative costs of energy-efficiency investments. Each of these policy types, either on their own or in combination, can also induce energy-efficiency innovations.

The policy combination of information-provision programmes and efficiency standards for buildings, appliances and automobiles sees widespread application in Europe, North America and East Asia. But these demand-side instruments are only second-best interventions to address potential supply-side barriers to firms making more energy-efficient products. An example of a first-best policy would be a subsidy to internalize spillovers from learning-by-doing and complementarities in product innovations, but they are not usually used. Business models such as vertical integration and long-term business relationships in supply chains can also internalize at least some of such spillovers.

Assessing the effectiveness of energy-efficiency policy interventions is challenging, but evaluation methodologies are improving. Increasingly, empirical studies use experimental designs or control groups, and randomly assign policy treatments or control for selection bias using adequate instruments such as sample-matching techniques. The empirical findings on policy effectiveness surveyed here are from studies that use such methods to help ensure their results are robust. They pertain primarily to markets for household appliances and passenger vehicles, because these product choices are most amenable to these policy-evaluation methods.

Information Programmes

Energy labelling of buildings, vehicles and appliances discloses information about energy use typically associated with use of these assets. 'Field' experiments that investigate the usefulness of such disclosures test them in either a controlled or natural (in-store) environment. In controlled experiments, provision of information about electrical appliances and lightbulbs improves the cost-effectiveness of energy-efficiency investment choices (Newell and Siikamäk, 2014; Allcott and Taubinsky, 2015; Houde, 2018a). However, in natural experiments, provision of prescribed information by sales staff appears ineffective in increasing demand for energy-efficient investment (Allcott and Taubinsky, 2015; Allcott and Sweeney, 2017). A key difference between the two types of experiments is that control groups in in-store experiments have access to other sources of information available in the stores, such as energy labels, which may reduce the effectiveness of the information treatment. In addition, providing multiple translations of energy-use as well as energy-cost information improves household appliance choices (Davis and Metcalf, 2016; Ungemach et al., 2018). But an experiment in providing tailored information treatments on fuel costs finds no significant effect on energy efficiency of automobile purchases (Allcott and Knittel, 2019). These results point to variations in effectiveness of various policy interventions to provide projected energy-use information in the context of appliance and vehicle purchase decisions.

Energy audits of existing dwellings and commercial buildings provide detailed assessments of building fabrics, HVAC systems and appliances and make recommendations for improvement. They can improve energy efficiency if building owners are unaware of inefficiencies and scope for cost-effective investments, and if owners are willing to act on the recommendations. But studies need to control for selection bias arising from owners' choosing to have energy audits because of a strong selection effect associated with this choice. Among those that do control for this effect, a study of a German household energy-audit programme finds that on average audits increase investments in energy efficiency in refurbishments (Frondel and Vance, 2013). In a programme in Maryland, USA, household energy audits are associated with a reduction in energy use, but the investment decisions and behaviour changes that contribute to the outcome are not identified (Alberini and Towe, 2015). However, a randomized field experiment in Wisconsin, USA, found no evidence of behavioural anomalies that could justify government subsidies for household energy audits (Allcott and Greenstone, 2017). Also, the subsidies encouraged households to undertake energy audits that were less likely to implement recommended measures than those who had energy audits without a subsidy.

Energy-information programmes that provide ongoing feedback on energy use are widespread, and they use interventions such as in-home real-time energy-use displays and provision of information on comparative energy use in energy bills. Such feedback aims to inform better household choices on day-to-day energy use and energy-efficiency investments. Meta-studies of the large number of rigorous studies that examine the effectiveness of these programmes, mainly for European and North American households, find that they have a significant effect on energy use, with average programme savings in the range of 4–11 per cent (Karlin et al., 2015; Zangheri et al., 2019). Energy savings tend to be larger with direct feedback on energy use, such as real-time displays, and with more frequent information provision. However, there is evidence that the impact of feedback interventions diminishes over time in two ways. Firstly, energy savings tend to decay once an intervention stops, but some savings persist, perhaps reflecting energy-efficiency investments (Allcott and Rogers, 2014). Secondly, feedback interventions in the most recent decade have less of an impact on energy savings than those in the two previous ones, perhaps reflecting growing general awareness of energy efficiency over time and more energy-efficiency investments.

Regulations

Regulations that promote greater energy efficiency aim primarily at influencing investment choices. For example, efficiency standards for appliances and vehicles remove the least efficient product offerings from the market. Similarly, building regulations specify minimum standards for new buildings, and in some cases refurbishment of existing ones. Energy-performance requirements also exclude from the rental market properties that fall short of minimum standards and thus encourage energy-efficiency investments in this market segment. In general, these types of regulations impose a 'shadow cost' on relatively inefficient products and buildings. Available evidence shows that introducing or tightening of such regulations accelerates energy-efficiency investments and reduces energy use, especially for heating and thermal comfort (Aroonruengsawat et al., 2012; Jacobsen and Kotchen, 2013; Kotchen, 2017). They bring forward in time energy savings, but do not always reach the technical potential estimated when the regulations were adopted (Levinson, 2016).

Two types of issues arise from regulations related to their impacts on economic efficiency and distributional fairness. Because they are not adapted to differences among households, they can generate welfare losses because there are fewer available products, and can force changes on lower-income households that gain little from energy efficiency (Allcott and Taubinsky, 2015). Similar effects are found for fuel-economy standards for automobiles, once their effects on the market for both new and used vehicles are considered

(Jacobsen, 2013). In addition, new regulations applied to new buildings, appliances and vehicles do not encourage ongoing energy saving from existing ones.

Both regulations and emission pricing have potentially significant regressive distributional impacts. The impacts of pricing externalities are immediate and affect both ongoing energy use and investment decisions. Those associated with regulations arise mainly from periodic investment decisions, though they are capitalized in asset values such as those of buildings and vehicles. Policy choices in the energy transition so far have leant more heavily on regulations than emissions pricing in guiding household decisions, suggesting a greater political acceptance of the former's distributional impacts. This policy choice comes at the cost of static efficiency losses from not costing emissions from existing energy-using assets. But the dynamic efficiencies are perhaps more significant in the context of long-run climate-policy goals. Evidence shows that both regulations and prices induce dynamic efficiencies and innovations that are central to deep emission cuts, and that for some products such as appliances regulations are more effective in spurring energy-efficiency innovations.

Taxes, Subsidies and Price Incentives

Subsidies for more energy-efficient products or increased taxes on less efficient ones raise the relative cost of inefficient products. By lowering the relative cost of more efficient products, they facilitate energy-efficiency investments. Of these financial instruments, taxes on relatively energy-inefficient products would align more closely with the first-best policy of pricing the environmental externality. But neither taxes nor subsidies attached to investment decisions would affect energy use from existing buildings, appliances and vehicles—as with energy-efficiency regulations.

Of these financial instruments, subsidies for energy efficiency and other low-carbon technology investments are more widely used and thus researched. But their use raises several concerns (Cattaneo, 2019). Firstly, financial incentives can be associated with 'rebound' effects in appliance and energy choices from income effects of subsidies, dulling some of the potential energy saving (Alberini et al., 2016; Houde and Aldy, 2017). Secondly, subsidies tend to be poorly targeted and benefit substantially customers who would have purchased the products without the subsidy benefit. This effect is estimated to be large for subsidy programmes for both energy-efficient household appliances and hybrid and battery electric vehicles.[4] Thirdly, as with all forms of government expenditures, there is a deadweight economic loss from the taxation necessary to pay for the subsidies, including those poorly targeted with few policy benefits.

MATERIAL-EFFICIENCY POTENTIAL

Just as energy efficiency is the amount of useful energy services and materials produced from energy inputs, material efficiency is the amount of physical service provided per unit of material used in both its production and application (Allwood et al., 2011). Examples of physical services are cubic metres of sheltered and protected spaces, be they buildings or vehicles, paved roads and railways, or containments of liquids and gases. The most material-intensive sectors of a modern economy are its buildings and factories, physical infrastructure, transport vehicles and equipment. Improving material efficiency means creating and maintaining a similarly useful built environment, vehicle fleet and equipment stock with fewer materials and waste than at present. With the aim of achieving net zero emissions from energy, a particular priority is improving efficiency in production and use of energy- and emission-intensive materials such as steel, cement, chemicals and plastics, wood and paper, and aluminium (Worrell et al., 2016).

There are several strategies through which product designers, producers and users could improve material efficiency (Figure 7.2). More intensive use of buildings, vehicles and equipment would lead to greater utilization of the asset. Extending asset lives through repair, refurbishment or remanufacture and resale could extend the useful lives of existing assets. Reuse of components and recycling materials in buildings, vehicles and equipment would preserve at least some of the value and embodied energy in materials at the end of assets' lives. In the production of new buildings, vehicles and equipment, there can be scope for designers to light-weight designs so that the asset's purpose can be fulfilled with fewer new materials. In addition, processes for producing materials can be improved to increase yields and reduce waste. The technical potential for applying these material-efficiency strategies appears greater for buildings and vehicles; less so for electrical and electronic equipment (Hertwich et al., 2019). As with energy-efficiency gaps, there are market imperfections that impede a market economy in realizing cost-effective material-efficiency potential.

Buildings

There is significant evidence that demand for residential and commercial floor space increases with per capita incomes, but at higher income levels this rate diminishes significantly and per capita floor space tends to plateau. Moreover, for countries with high levels of per capita income (above around $50,000), there is wide variation across them in floor space per capita (Chapter 2). These differences reflect not only variation in urban densities and land values,

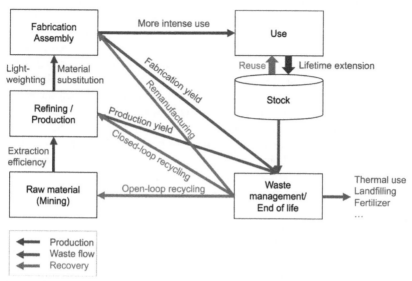

Notes: Examples of more intensive use are more space-efficient designs and multi-functionality of buildings, and higher utilization rates, such as ride-sharing in vehicles. Lifetime extension can be achieved through repair, refurbishment, resale and remanufacture of buildings, vehicles and appliances. Lightweight designs and material substitutions use fewer materials or less-emission-intensive materials in production of a product. Reuse of product components can be achieved through modularity and remanufacturing. Improved yields in the production of primary materials and use in manufacturing can reduce material wastes, such as through additive manufacturing.
Source: Reprinted from E.G. Hertwich et al. (2019), Material efficiency strategies to reducing greenhouse gas emissions associated with buildings, vehicles and electronics: A review, *Environmental Research Letters*, 14(4), 043004 (CC BY 3.0). https://doi.org/10.1088/1748 -9326/ab0fe3.

Figure 7.2   *Material cycles and material-efficiency strategies*

but also in the design and furnishing of dwellings. This variation points to potential for more compact designs with less material use in new buildings. There is also wide variation across countries in the average lifetime of existing residential buildings, ranging from more than 100 years in Europe to around 50 years in the United States and 25 years in China. Again, this wide variation in asset lifetimes suggests potential to extend building lifetimes through better design, construction and refurbishment practices. For example, feasible lifetime extensions of existing Chinese buildings could cut substantially material and energy use, as well as emissions, over the next decades (Cai et al., 2015).

At the same time, development in digital technologies creates economic opportunities to use existing buildings more intensively and thus increase material efficiency. For example, Airbnb is a digital platform that enables the use of dwellings to provide hospitality services at scale. One effect of this flexible use of dwellings is to serve peak demand for travel accommodation; for example, during large conferences and events (Zervas et al., 2017). Use of dwellings in this way lessens the need for hotel room capacity scaled to serve peak demands, leading to higher occupancy rates for both hotels and residential dwellings. In the retail sector, integration of digital ecommerce platforms with traditional 'bricks and mortar' retail stores enables significant increases in intensity of use of the retail floor space (sales per m^2) (Pozzi, 2013). The increased use of electronic equipment has material requirements too, though they are less materially intensive than those of buildings. Similarly, the sharp increase in flexible and home working for some jobs in response to the novel coronavirus pandemic, largely facilitated by digital technologies, points to new potential for optimizing jointly the use of dwelling and commercial building space.

Material-efficiency gains and emission reductions from construction of new buildings appear feasible through two means. One is to use materials more effectively in their construction. For example, load-bearing materials in buildings, such as structural steel beams, use significantly less than their full load-bearing capacities in many applications (Carruth et al., 2011; Moynihan and Allwood, 2014). One estimate puts the potential saving from closer specifications for steel use in buildings at around 20 per cent (Milford et al., 2013). The second is substituting wood for steel and concrete in buildings, the emission-reduction benefits of which are well established (Smith et al., 2014). They arise from the storage of carbon in the wooden biomass in buildings and displaced steel and cement in construction, which have high emissions from their production. However, to realize the full potential of these emission-reduction benefits, the harvesting of wood from forests must be sustainable so that land-use changes do not lead to carbon dioxide emissions or other environmental damage. This imposes a land-use constraint on the extent to which wood can substitute for steel and cement in new buildings.

There is significant potential to reuse and recycle building components at the end of buildings' lives. Waste from their construction and demolition are the largest waste streams to landfill materials in Europe and North America.[5] The potential to reuse building components like bricks, structural steel, steel cladding and potentially aluminium frames and cladding is significant (Cooper and Allwood, 2012). But tapping much of this potential requires better up-front building designs that enable material reuse, well-maintained records of building components and business models for buildings and their materials that capture the value of material reuse (Allwood and Cullen, 2015, pp. 215–30). It is also common practice to recycle relatively high-value metals like steel, aluminium and copper from construction and demolition wastes. Concrete and other mineral building materials are often downcycled into coarse aggregates, an alternative to quarrying for new materials.

Vehicles

Private automobiles see use on average only about 5 per cent of the time and at only one-third of their passenger capacity when used.[6] This largely idle fleet of vehicles provides much convenience and saves considerable time for households that own them. Measures that shift transport demand to public services away from private vehicles thus have significant potential to boost material efficiency by increasing the intensity with which the vehicle fleet is used. In urban areas with high population densities, public transport, cycling and walking are convenient alternatives. Car-pooling arrangements for daily commutes as well as car- and ride-sharing schemes increase vehicles' capacity use. In future, trip-chaining, autonomous taxis could substantially reduce the size of the vehicle fleet necessary to provide convenient passenger transport services. However, this technology remains at a pilot stage and its potential impact on vehicle fleet size remains uncertain. Lifetime extensions of the existing vehicle fleet would also increase material efficiency through more intensive usage, but lock-in for longer internal combustion engines and fossil-fuel use.

Light-weighting and right-sizing new vehicles can boost both material and energy efficiency. Vehicle designs seek to balance many objectives. They range from providing safe travel spaces, thermal comfort, convenience and entertainment for their occupants to using no more energy than necessary to provide the transport services demanded. The former considerations tend to increase vehicle size and materials use, and the latter focus on fewer materials, lighter weight and smaller vehicle sizes. Light-weighting often uses fibre composites and aluminium in place of steel in the vehicle glider. This reduction in mass of the vehicle enables further saving from drivetrain reductions—smaller engines or battery packs and motors. Advances in additive manufacturing also enable additional material savings from more precise component designs and less material waste in their production.

Recycling end-of-life vehicles is common, with recovery of about 85 per cent by mass of their materials in Europe and North America.[7] But metals from scrapped vehicles often undergo down-cycling because they contain many alloys and metal mixtures. For example, much high-quality steel in automobiles becomes construction steel, with alloying benefits lost (Ciacci et al., 2016). Moreover, shredding scrapped vehicles mixes materials and contaminates metal; for example, the quality of recycled steel is compromised by copper and tin contamination (Daehn et al., 2017). Similar issues arise in recycling aluminium from automobiles. More careful parts-based sorting and handling of materials from scrapped vehicles can feasibly preserve more alloying and avoid mixing metals (Ohno et al., 2017). But the economic benefits of doing so depend on the relative costs of the recycled and virgin metals.

With electrification of passenger vehicles, the demand for lithium-ion batteries and charging infrastructure will change substantially the sector's material requirements. In particular, battery manufacturing is an energy- and emissions-intensive activity and accounts for a significant share of the life-cycle energy use and emissions of battery electric vehicles (Peters et al., 2017). Recycling of the metals and other materials in lithium-ion batteries could significantly improve their environmental performance and reduce their material intensity. However, because of their complex mixture of materials, currently available recycling methods face difficult trade-offs, with the most promising route for their improvement being better material-separation processes (Gaines, 2018).

Avoiding and Managing Waste from Consumables

Municipal waste accounts for about 25 per cent of total waste in Europe (excluding mineral wastes), with the plastics, paper, food and garden wastes accounting for most by mass.[8] Business and consumer initiatives to reduce single-use plastics cut waste from consumables by increasing material efficiency. For wastes from many consumables, better product designs, including their packaging, and processes for sorting and collecting waste can cost-effectively improve their material efficiency. For example, for plastics, feasible design changes can make it easier to separate different types of plastics and recycle them mechanically (Material Economics, 2018, p. 80). With such 'feedstock' improvements, there is significant potential to increase scale and efficiency of waste recycling, including automatic sorting, processing and secondary plastics production. Also, new chemical recycling processes for plastics are being developed that avoid material degradation of plastics associated with their mechanical recycling.

Food waste too can be reduced through prevention interventions—better supply management by retailers and schemes to discount soon-to-expire products or their redistribution to food-insecure households (Ellen MacArthur Foundation, 2019, pp. 37–44). Biogenic waste also holds significant potential for energy conversion through anaerobic digestion and gasification. But biogas from these processes looks set to remain more expensive than natural gas, even allowing for the fact that much of the waste-handling and waste-sorting cost would be incurred anyway (Chapter 3). They would thus need emissions pricing or subsidies to advance.

MATERIAL-EFFICIENCY POLICIES

As with energy efficiency, there are market imperfections that impede greater material efficiency. Looking ahead, the market failure that is perhaps most material to greater material efficiency is un-costed environmental externalities. Materials such as steel, cement, plastics and other metals and minerals are highly energy- and emissions-intensive and their emissions are not priced at levels deemed adequate (Chapter 5). Decisions on their production, product designs and use therefore do not yet face emissions prices consistent with long-run climate goals. Nevertheless, emissions pricing is more extensive in heavy industries (and electric power) than in other relatively energy-intensive sectors, even if such prices have yet to reach levels deemed adequate (Chapter 5). Policies that establish and raise the explicit price of emissions or its shadow price in heavy industries are fundamental to better material efficiency.

But there are other potential market barriers and failures that can hold back material efficiency. An important market failure can arise if manufacturers have an insufficient incentive to design their products in ways that consider costs of material reuse, recycling and waste management (Söderholm and Tilton, 2012). There are many examples to suggest that this is the case (Material Economics, 2018, pp. 43–9). For relatively long-lived assets, such as dwellings and automobiles, discounting and present bias among households and firms can contribute to neglect of end-of-product-life disposal costs in initial investment decisions. The policy challenge is to create institutional and market arrangements that encourage firms to design products with a view to end-of-product-life costs that are largely borne—but perhaps neglected—by their customers. Constraints in designing such policies are information barriers faced by governments about specific products, heterogeneity in their material characteristics and recyclability.

Second-best policy approaches include product taxes, recycling subsidies, waste-disposal fees and extended producer-responsibility schemes (Calcott and Walls, 2000 and 2005; Walls, 2006). The latter are applied widely in Europe, North America and Japan for automobiles, appliances, electrical goods and batteries. Landfill fees are also widely used. Both instruments can induce innovations that reduce end-of-product-life costs, increase the value of reused or recycled materials and extend product lives. However, there is little evidence on their impacts and cost-effectiveness. Such policies remain a priority for policy development and rigorous evaluation.

COMPLEMENTARITIES AND TRADE-OFFS BETWEEN ENERGY AND MATERIAL EFFICIENCIES

There is significant potential to improve energy and material efficiencies, especially as energy systems transform to low-carbon technologies with expected shifts in relative prices of energy and materials given foreseeable technology developments. In considering ways forward, it is important to understand how changes in material and energy efficiencies interact—where there are complementarities to be exploited and trade-offs to be managed (Hertwich et al., 2019). Material-efficiency strategies that boost intensity of building use and cut their material requirements in new construction tend to reduce both material and energy use, along with carbon dioxide emissions. However, more thermally efficient new building fabrics require more materials, as do ground-source heat pumps. Nevertheless, case studies point to overall life-cycle energy and emissions decreases with these greater energy-efficiency investments (Chastas et al., 2016). The life-cycle estimates include the embodied energy and emissions in the additional materials. However, case studies of refurbishments indicate that they have lower life-cycle impacts only if refurbished to ambitious energy-efficiency standards (Pauliuk et al., 2013). Piecemeal approaches appear less effective. Lastly, when substituting wood for steel and cement, there is some loss of thermal mass in buildings, which can give rise to somewhat higher heating and cooling demands in spring and autumn (Heeren et al., 2015).

In vehicles, material-efficiency strategies such as downsizing, additive manufacturing and more intensive use also improve energy efficiency. Other strategies, such as light-weighting, lifetime extensions and electrification can involve trade-offs that need to be effectively managed. For example, light-weight composite materials like carbon fibre can be relatively energy-intensive to produce and difficult to recycle compared to aluminium or steel. It is therefore important to consider life-cycle impacts of alternative approaches.

Given the complex trade-offs that can arise in transforming buildings and appliances, vehicles and factories and equipment to low-carbon technologies, correcting price distortions is a clear policy priority. The price distortions that matter most in the long run are un-costed environmental externalities associated with energy and materials use. But there are also other market imperfections that impede energy and material efficiencies, which can be effectively addressed through information programmes, regulatory standards, waste-disposal fees and extended producer obligations. They address other non-price market distortions and can be useful early steps in initiating greater energy and material efficiencies, including eliciting innovative responses from firms.

CONCLUSION

Technical energy-efficiency gaps are widest and material-efficiency potential greatest with buildings, vehicles and appliances, perhaps reflecting performance characteristics that have been at least partly neglected by households and firms. The market imperfections associated with this neglect are several. Significant market failures arise from principal–agent problems and information asymmetries associated with rental properties and significant under-investment in their energy efficiency. But other barriers and behavioural anomalies matter too. There is significant evidence of information barriers, bounded rationality and biases towards the present and status quo. Moreover, their salience could gain in significance as energy systems transform through decisions about new low-carbon technologies and their energy supplies. But most fundamental are under-priced environmental externalities.

Available evidence from robust empirical studies of energy-efficiency programmes and policies points to their effectiveness when well-designed in helping households and firms make better energy-related choices. Tailored energy-efficiency information related to new investment decisions can contribute to more cost-effective investment decisions, and ongoing energy-use information in dwellings leads to energy savings. Energy-efficiency standards for buildings, vehicles and appliances remove the least efficient products from markets and can induce welfare-enhancing energy-efficiency innovations, at least for products for which this performance characteristic had been somewhat neglected by customers. These benefits of energy-efficiency policies are evident even in business-as-usual choices among existing technologies. These policy benefits may be larger, at least temporarily, during energy-system transformations, and complement policies aimed at fostering innovation and early-stage deployment of low-carbon technologies. Such information programmes and regulatory standards, especially in the early stages of transforming energy systems, can also help create conditions that facilitate implementation of effective emissions pricing in the long run to shape its eventual system optimization around alternative low-carbon technologies.

NOTES

1. Passivhaus buildings provide a high level of occupant comfort while using very little energy for heating and cooling.
2. A vehicle glider is the automobile body without the drivetrain.
3. There are two potential types of principal–agent problems. One arises when a tenant pays for utilities under the lease agreement and a property owner under-invests in energy efficiency, which is the main focus here. The other occurs when a property owner pays utilities under the lease agreement and a tenant over-consumes (wastes) energy.

4. For the evidence on household appliances, see Broomhower and Davis (2014), Alberini et al. (2016) and Houde and Aldy (2017). On electric vehicle subsidies, see DeShazo et al. (2017) and Sheldon and Dua (2019). Similarly, cash-for-clunkers schemes for new automobile purchases mostly move new automobile purchases forward in time. See Hoekstra et al. (2017).
5. See European Environment Agency (2020) and US EPA (2020).
6. Hertwich et al. (2019), based on Oakridge National Laboratory and US Department of Transportation, Federal Highway Administration, *National Household Travel Survey 2017*. See https://nhts.ornl.gov.
7. Hertwich et al. (2019), based on Eurostat, End-of-vehicle-life Statistics, and Automotive Recyclers Association, Industry Statistics. See https://ec.europa.eu/eurostat/statistics-explained/index.php?title=End-of-life_vehicle_statistics and www.a-r-a.org/industry-statistics.html.
8. Eurostat, Waste Statistics, retrieved from https://ec.europa.eu/eurostat/statistics-explained/index.php?title=Waste_statistics.

PART III

Energy-reform interests and strategies

8. Interests in low-carbon technologies and renewable resources

Enabled and encouraged by government policies as market imperfections warrant, firms and households must make alternative investment decisions and energy choices to transform energy systems. This transformation has been underway for some time, and it is important to learn from and build on these advances. They focused initially on the electric power, automobile and building sectors, primarily in advanced industrialized countries and China. These alternative technologies—solar photovoltaic (PV) units, wind turbines, lithium-ion battery packs, electric vehicles and thermally efficient buildings—are likely core components of low-carbon energy systems, but more low-carbon technologies beyond these are necessary to achieve net zero emissions from energy (Chapter 3). Moreover, almost all countries must transform their energy systems if net zero emissions from energy are to be achieved, whether by initiating disruptive technological changes to low-carbon alternatives or following in adopting them as they become successfully commercialized by others.

While energy transformations take root in economic sectors and activities, they occur in broader country contexts. These transformations foster new economic interests and comparative advantages among countries and pose challenges to the status quo. Several economic interests come to the fore in transforming energy through low-carbon technologies and renewable resources. One arises from specializations in innovation activities and domestic innovation systems that involve governments, firms, research institutions and venture capitalists in developing, demonstrating and deploying in their early stages new technologies. For example, innovation systems in the United States, Germany, Japan and Denmark, and increasingly those in South Korea and China, played leading roles in developing wind turbines, solar PV, lithium-ion battery packs and electric vehicle technologies.

A second comparative advantage that advances low-carbon alternatives is manufacturing capabilities. For example, China garnered relatively large export market shares in solar PV, a sign of revealed comparative advantage, more through its manufacturing capabilities than those in research and development (R&D), at least initially. In addition, as low-carbon technologies become increasingly global in scale, renewable resource endowments such as solar irradiance, wind, water cycles and sustainable bioresources help consolidate change. New sources of economic growth and comparative advantages from low-carbon technologies and renewable resources shape how countries and their sectors engage with transforming energy systems at the different stages in advancing new technologies.

Looking forward, the challenges are to expand the frontiers of energy transformations across countries and sectors. At country level, the challenge is to consolidate the transformations underway in the electric power, automobile and building sectors in countries that initiated these technological disruptions, and promote their widespread diffusion across all countries. In initiating countries, this would mean sustaining market-creating and industry-supporting policies to points at which low-carbon-technologies alternatives become commercially viable through cost reductions and/or politically sustainable levels of emissions pricing. Some low-carbon alternatives appear sustainable in the long run without significant emissions pricing. In following countries, the challenges are to remove non-price and, for some technologies, price barriers to their widespread diffusion. Removing non-price barriers includes investing in R&D to facilitate diffusion of technologies by adapting them to country contexts as well as reforming and adapting electric power systems, energy markets and network infrastructures to low-carbon technologies.

At sectoral level, the challenge is to extend the scope of low-carbon technology disruptions to hard-to-decarbonize sectors and activities. Decarbonization of all energy sectors would benefit from emissions pricing consistent with government climate goals, when and where governments can implement emissions pricing credibly. But in contexts where this policy falls short for want of political support and as other market imperfections warrant, market-creating policies can advance low-carbon-technology alternatives. These policies consist of a mix of investment and operating supports, potentially in combination with some emissions pricing, so that in combination they are sufficient to support investment in the early deployment and operation of low-carbon alternatives. Examples of such supports are investment tax credits and feed-in tariffs for renewable electricity generation. Competitive allocation of such supports where feasible helps to minimize policy costs and ensure timely exits from them. It is also important to ensure that policy costs are allocated fairly to avoid regressive policy impacts, such as those with feed-in tariffs and their cost recovery through electricity bills.

While economic interests in low-carbon technologies and politically sustainable energy-reform strategies help advance system transformations, interests in the status quo matter too to energy transformations. As with any deep structural change in economies, energy-system transformations give rise to significant distributional impacts, potential political resistance to change and concerns for fair and just transformations. Two such distributional impacts for which there is significant evidence are examined here, though this focus is not intended to diminish the potential importance of others. The impacts considered here relate to coal-mining communities and low-income households living in thermally inefficient and poor-quality dwellings. For these communities and households, targeted policy supports appear necessary because of their limited capacities to adapt to the structural changes that net zero emissions would require.

The challenge ahead is thus to forge energy-reform strategies that strengthen interests in change and policy credibility over time and foster upfront investments in innovation and market creation for low-carbon technologies. The constituent elements of such domestic policy agendas are well rehearsed—government supports for public and private R&D, emissions pricing, adaptations of energy markets and infrastructures to low-carbon-technologies, and energy and material-efficiency policies as market imperfections warrant. A policy grey area is the extent to which market-creating and industry-supporting policies are warranted by market imperfections and effectively addressed by government policies, although such policies have contributed substantially to energy transformations (Chapter 4). Comprehensive, coherent and credible strategies use these instruments to develop policy sequencings that minimize policy costs, manage distributional impacts, exploit technology complementarities and spillovers across sectors, and build interests that sustain and accelerate change (Chapter 9). Starting points for developing such strategies are the political economy of energy reform and country policies that advance low-carbon alternatives.

THE POLITICAL ECONOMY OF ENERGY REFORM: INTERESTS, IDEAS AND INSTITUTIONS

A majority of people around the world and in most countries see climate change as a major threat to their country, with a few exceptions, such as Russia, Nigeria and Israel (Pew Research Center, 2019). But the journey from broad public concerns to government, firms' and households' actions to limit climate change is complex. So far, countries are pursing widely ranging climate ambitions and actions—much more wide-ranging than if policies were simply determined by median voter interests in this single issue. Comparative political analyses of domestic policies, including climate change and environmental protection, typically focus on country factors in three broad areas to explain why they might differ so much. These factors are: *interests* such as the economic gains and losses from reforms and structural change; *ideas* like empirical evidence, political ideologies and societal values; and *institutions* such as those that guide political decision making (Lichbach and Zuckerman, 2009; Hughes and Urpelainen, 2015; Purdon, 2015).

Cross-country assessments of climate ambitions and actions need consistent measures of them, and there are several systematic approaches to gauging them. Two are highlighted here. One measure of country ambitions is the global temperature impact of each country's Nationally Determined Contributions (NDCs) set forward under the Paris Agreement. This measure assumes that such efforts should be distributed among countries based on an effort-sharing principle, but recognizes the lack of consensus on what this principle should be. Among several plausible burden-sharing principles, the measure uses the one most favourable to a country and places its climate ambition on a scale of 1.2–5.1°C based on the timing and extent of its committed greenhouse gas reductions (Robiou du Pont and Meinshausen, 2018). The embedded value judgements about a country's contribution to climate change mitigation, though, leave this measure open to challenge.

A second measure focuses on adopted country climate policies and emissions outcomes (Burck et al., 2019). It focuses on energy-efficiency policies, renewables deployment and emissions reductions. A composite index across these dimensions serves to rank countries, which are then grouped according to level of climate action, from very high to very low. However, this measure places no weight on innovations and early-stage deployment of low-carbon technologies beyond their direct emission reductions. In other words, their contributions to dynamic efficiency gains are unscored.

Although both measures are open to challenge, they paint broadly consistent pictures of cross-country variation in climate ambition and action among advanced industrialized countries and China. Many European countries are deemed more ambitious and active, and the United States, Canada, Australia and China less so. The scoring of developing countries is more sensitive to what is being measured—NDCs or actions—and how.

While the specific focus here is on energy-system transformations, available evidence on country factors that influence overall domestic climate ambitions and actions holds important insights. On economic interests that might advance or oppose climate initiatives, much analysis focuses on three. One is domestic fossil-fuel resources and production, the value of which is under long-run threat from climate action. There is strong evidence that domestic coal production is negatively associated with climate ambition and action, and weak evidence of such an association with domestic crude oil production (Bailer, 2012; Tørstad et al., 2020). These studies find no such association between natural gas resources and measured climate actions after allowing for other factors.

A second factor is a country's extent of industrialization and urbanization, often proxied by the level of per capita GDP, which can be interpreted as reflecting the costs of transforming accumulated capital stocks. The evidence finds that higher country average per capita income is negatively associated with climate ambition and action (Tørstad et al., 2020).

A third factor is country vulnerability to climate changes, such as small island states and least-developed countries. This vulnerability is significantly associated with higher measured country climate ambitions and actions (Tørstad et al., 2020).

There is thus significant evidence that some domestic economic interests shape country approaches to climate change. However, these analyses do not consider economic interests in low-carbon technologies and renewable resources. The implicit assumption appears to be that these low-carbon alternatives hold more economic costs than benefits, emissions reduction aside.

Before taking a closer look at economic interests in advancing low-carbon alternatives, it is important to consider other potential country factors that can influence climate action: ideas and institutions. On ideas, there is significant evidence that national and economic populism, which prioritizes domestic self-interest over international cooperation, is associated with antipathy to climate action (Lockwood, 2018; Huber, 2020). Cosmopolitanism—country populations self-identifying as citizens of the world—is weakly associated with stronger climate ambition and actions (Tørstad et al., 2020).

On political institutions, there are several potential hypotheses about their influence on climate action. One hypothesis, based on collective action theory, is that political leaders in democracies have strong incentives to provide public goods like environmental protection (Bättig and Bernauer, 2009). But democratic accountability and distributional impacts of short-run costs of climate actions can pose to barriers to governments' addressing the long-run challenges (Jacobs, 2011, pp. 10–17; Finnegan, 2019). So political institutions, such as voting rules (proportional representation versus plurality rules) and governance models (concentrated versus diffused power structures), can also influence country actions by facilitating or impeding the handling of distributional issues (Tobin, 2017; Finnegan, 2019). Evidence shows that countries with more democratic institutions have significantly higher climate ambitions and actions, other factors being equal (Tørstad et al., 2020). In addition, domestic political institutions that increase electoral safety (proportional representation) and diffuse organized opposition to reforms (corporatist states) are associated with higher climate ambitions and actions (Tobin, 2017; Finnegan, 2019).

SPECIALIZATIONS IN INNOVATION AND MANUFACTURING

In the context of economic interests, political ideas and institutions that influence climate actions, interests in advancing low-carbon technologies and renewable resources warrant close consideration. Energy-system transformations bring to the fore new fields of country specialization and comparative advantage (Fankhauser et al., 2013). One is innovation, in particular research, development and demonstration of new low-carbon technologies. Innovative new technologies and adaptations of existing ones are necessary to provide alternatives to incumbent technologies and fossil fuels. A second is specialization in manufacturing. Deploying low-carbon technologies at scale requires substantial new manufacturing capacities to produce equipment that capture renewable energy resources, produce low-carbon fuels and adapt energy end-use technologies to them. But not all countries will necessarily engage in this innovation and manufacturing production. As with much tradable-goods production, country specialization and comparative advantage hold sway over innovation activities and production of new low-carbon technologies.

It is important, moreover, to recognize that for any particular new technology, specializations in innovation and production are not necessarily bound together in the same place. With openness to international trade and multinational production, countries can specialize in either innovation or manufacturing of a technology. For example, since countries differ in their relative costs of innovation and production, multinational firms can separate geographically innovation and production through foreign direct investments and joint ventures (Markusen, 2002, pp. 17–22). Commercial transfer of intellectual property rights across countries also permits specializations in innovation and production in line with comparative advantages.

Many productive factors shape country comparative advantage and specialization. Some can be readily observed and measured, such as the relative abundance of labour, education and general skills, and natural resource endowments. But others, like scientific and engineering knowledge and its application to technologies and production processes, can be more difficult to gauge. Such technological sources of comparative advantage are typically inferred indirectly from revealed specializations in producing and exporting goods (Ballasa, 1965).[1] These revealed comparative advantages across countries and regions tend to persist over time, building on existing capabilities and expanding incrementally into related industries (Hidalgo et al., 2007). Moreover, capabilities as revealed by past patterns of specialization help predict the future areas of growth and specialization of countries (Hidalgo and Hausmann, 2009; Hausmann et al., 2019).

Measuring specializations of countries in innovation is difficult because innovation processes are complex and their outputs multifaceted. One attempt to reduce the 'noise' in any one measure takes a composite index approach, comprising both inputs and outputs of innovation, and compares them as a cross-check (Dutta et al., 2019). Measured country inputs include human capital and research, quality of governance and infrastructure, financial market depth and market scale, and business organization knowledge. Measured innovation outputs include patent applications, research article publications and citations, product quality certifications and intangible property. These composite indices suggest that among high- and middle-income countries there is relative specialization in innovation activities. Among high-income countries, Switzerland, Sweden, the United States, the Netherlands, the United Kingdom, Finland, Denmark, Singapore, Germany and Israel stand out for their relative innovation capabilities and outputs. Among middle-income countries, China far outperforms its income-level peer group, although India too has significant innovation capabilities. In terms of innovation scale, as proxied by public and private R&D spending, the big spenders are the United States and China, followed by Japan, Germany, South Korea, France, India and the United Kingdom.

A measure of country specialization in manufacturing goods that are relatively intensive in their use of technological innovations is a widely used index of 'economic complexity' (Hidalgo and Hausmann, 2009; Hausmann et al., 2014, pp. 20–31). Countries with higher economic complexity tend to specialize in producing and exporting products that are technologically more sophisticated, and vice versa for countries with lower economic complexity (Mealy et al., 2019). One interpretation of the economic complexity index is that it gauges the extent to which production and exports activities in countries cluster around goods that are relatively technologically advanced, such as machinery, chemicals, pharmaceuticals and information technology products. Countries with currently high measured economic complexity include Japan, Switzerland, Germany, the United States, Finland, Singapore, South Korea and the United Kingdom.[2] While most top-ranked countries have been relatively stable over recent decades, Singapore and South Korea significantly increased their economic complexity over the last two decades through their economic development strategies, as did China but from a lower starting point. Some previously top-ranked countries have seen significant declines in their level of economic complexity, such as Italy.

In addition to these productive factors that shape specialization across countries and regions, large domestic markets can also enable domestic producers to achieve scale economies and specializations in production (Krugman, 1980). Such home-market effects from product demands are perhaps most readily seen in a sector unrelated to energy: pharmaceuticals. In this sector, there is significant evidence that countries specialize in pharmaceuticals that are related to domestic health care needs (Costinot et al., 2019). By looking at the relationship between illness and economic activity, it is possible to control for potential reverse causation because specialization in domestic supply would not cause the domestic demands for the pharmaceuticals considered. A similar demand effect is seen in Finland's specialization in relatively complex manufactured goods, which traces back to having to make war reparations to the Soviet Union in the form of industrial goods after their 1944 armistice (Mitrunen, 2019).

OBSERVED SPECIALIZATIONS IN LOW-CARBON TECHNOLOGIES

Low-carbon energy equipment and products tend to have relatively high product complexities, requiring technologies more sophisticated than average for goods (Mealy and Teytelboym, 2018). For example, renewables equipment with relatively high product complexities include concentrating solar, wind turbine assembly and manufacture of several of their components, combustion turbines and environmental monitoring equipment. But some have low product complexity, such as several related to bioresource conversions. Solar PV panels and power inverters are of moderate technological complexity. The extent of technological sophistication of low-carbon technology products is thus likely to influence which countries specialize in these areas of innovation and manufacturing, at least initially, building on existing capabilities and activities. These emerging patterns of specialization in innovation and production of low-carbon technologies are increasingly apparent in specializations in low-carbon technology patents, which is a measure of innovation activity, as well as in manufacturing output.

Patents for solar PV, wind turbines, batteries and electric vehicles account for most of those for low-carbon technologies over the past two decades. Countries specializing in these innovations have tended to do so over time, though this persistence is less than in production activities. Five countries account for 80 per cent of low-carbon-technology patents since 1970: the United States, Japan, Germany and increasingly China and South Korea (Figure 8.1). France, the United Kingdom, Canada, Italy and Denmark also engaged in significant low-carbon-technology innovation activities. All of these countries also specialize in technological innovations in general—not just in those related to low-carbon technologies. It is also worthwhile to note that, among countries that specialize in low-carbon technology innovations, the United States, China and Germany have significant domestic fossil-fuel resources and per capita production, while Japan and South Korea specialize in fossil-fuel-intensive industries. These economic interests in the energy status quo have not necessarily posed impediments to these innovation activities in low-carbon technologies, though the counterfactual could be even more innovations in these countries in the absence of status quo interests.

In addition, countries tend to innovate in low-carbon technologies related to existing capabilities and technologies.[3] For example, specialization in solar PV tends to be associated with that in semiconductors (Zachmann and

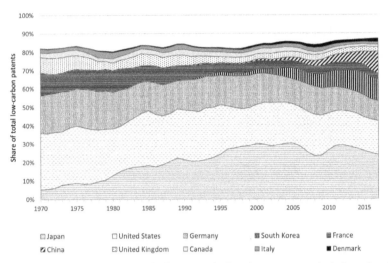

Notes: Countries appear in the chart in the same order from bottom to top as in the legend read from left to right. Patent data are for high-value patents defined as those filed in more than one jurisdiction. Lower and low-carbon technologies include: decarbonization technologies related to buildings, including housing and appliances; capture, storage, sequestration or disposal of greenhouse gas emissions; reduction of greenhouse gas emissions related to energy generation; decarbonization technologies related to transportation; and system integration technologies related to power network operation.
Sources: Data from the European Patent Office, PATSTAT worldwide patent database, retrieved from www.epo.org/searching-for-patents/business/patstat. Lower- and low-carbon-technology patent classification is based on European Patent Office and UN Environment Program (2014). Vivid Economics provided the data analysis.

Figure 8.1   *Shares of lower-carbon-technology patents by country (in percentage shares of worldwide lower-carbon-technology patents, 1970–2017)*

Kalcik, 2018). Countries specialized in solar PV innovations include Japan, South Korea, Germany, China, the United Kingdom and the United States, while those in wind include Denmark, Spain, Germany and the United States. Countries specialized in battery packs and electric vehicle innovations include Japan, South Korea and China in East Asia; Germany, France and Italy in Europe; and the United States, all of which have large domestic automobile industries and home markets.

Countries that specialize in innovations related to low-carbon technologies also tend to specialize in their production and export, though there are significant differences across sectors (Zachmann and Kalcik, 2018). For example, Denmark, Spain and Germany are relatively specialized in the production and export of wind technologies, as well as innovations. In batteries, Japan, South Korea, Germany and France are relatively specialized in both production and innovations, while in electric vehicles the United States is relatively specialized in production and export, but less so in innovation as measured by patenting. China is also relatively specialized in production and export of batteries and solar PV, relative to its patenting activities in these areas.

In solar PV, much of the original patenting activity for this technology was concentrated in the United States, Germany and Japan. The technology transferred to China through its import of production equipment for making silicon-based PV panels from these countries and their firms specialized in making semiconductors—a related silicon-based technology (de la Tour et al., 2011; Carvalho et al., 2017). As Chinese manufacturing capacity expanded rapidly, innovations increasingly focused on manufacturing costs reductions, especially in China. At the same time, Vietnam built production capabilities in low-carbon technologies, including through participation in East Asian manufacturing supply chains (Mealy and Teytelboym, 2018; ClimateWorks Australia and Vivid Economics, 2019).

While much of the focus is on technological innovations and new manufacturing capabilities for relatively complex products, energy transformations also increase demand for some mature technologies and products. Examples include ethanol produced from relatively sustainable feedstocks like Brazilian sugar cane and, to a lesser extent, American corn (Mekonnen et al., 2018). Brazil's Programa Nacional do Álcool and the US Renewable Fuel Standard, launched in 1976 and 2010, respectively, introduced mandatory blending of ethanol in gasoline. These programmes tapped into each country's substantial bioresources and induced scaling up of their supply chains and process innovations for ethanol production.

The examples of biofuel production in Brazil and low-carbon-technology manufacturing in Vietnam point to developmental pathways through which developing countries can participate in advancing low-carbon technologies, without necessarily specializing in innovation like China does. Developing such capabilities is, of course, important in its own right and a key to economic development and industrialization, whether through low-carbon technologies or other specializations (Lall, 1992; Bell and Pavitt, 1995). Investments in these capabilities also help advance the diffusion of new technologies, including low-carbon alternatives, by facilitating their adaptations to country contexts (Chapter 4). Supporting the development of these capabilities in developing countries is an important focus for bilateral and multilateral technical cooperation among countries.

MARKET DISRUPTIONS INITIATED THROUGH INNOVATION AND INDUSTRIAL POLICIES

While comparative advantages in innovation and manufacturing shape low-carbon-technology specializations across countries and sectors, government policies also play a role. In most countries specializing in innovation, their governments have innovation policies that invest in scientific skills (science, technology, engineering and medicine) and research institutions as well as providing public funding mechanisms for R&D activities (Chapter 4). The granting of patent protections also enables successful innovators to garner returns on investments in innovations from future market rents. The focus here, though, is on public funding of and tax credits for R&D activities because they achieve impacts over shorter time horizons than innovation policies aimed at building skills and institutions. In addition, evidence shows that tax incentives for private and publicly funded R&D activities are effective in spurring innovations and low-carbon technologies (Chapter 4).

Five countries account for much of the publicly funded low-carbon-technology R&D. Since the 1970s, the United States, Japan and Germany, and increasingly South Korea and China, have accounted for 80 per cent of this public funding (IEA, 2020g).[4] The composition of this funding has shifted progressively over time, away from nuclear energy in the 1970s towards energy efficiency, renewable energy (primarily solar PV and wind turbines), energy storage, hydrogen and fuel cells, and basic energy science by the 2010s. Renewable generation technologies, other power-related technologies and storage as well as energy efficiency account for about half of this public funding in the past decade. Governments in these five countries have funded public and private R&D activities across a range of low-carbon technologies in recent decades. Other countries, such as Denmark, France, Italy, Spain and the United Kingdom, specialized more narrowly in certain low-carbon technologies and committed fewer total public resources. The countries that in recent years have increased public funding for low-carbon technology R&D are the United States and China. The total global level of this funding over the past decade is in the range of $20–25 billion (2018$; IEA, 2020g).

While significant public R&D funding is directed to low-carbon technologies, especially renewables, most government incentives for private R&D activities are technology-neutral. In the main, they support firms' activities deemed to involve R&D rather than particular technologies. Government policies that boost demand for low-carbon technologies, along with broader market developments, do though play a significant role in directing private R&D activities towards low-carbon technologies (Chapter 4). Product differentiation and product-market competition also create and manage market rents that provide returns to upfront investment in innovation and market creation. To the extent that successful innovations create at least temporary market power or cost advantages, they generate returns on such investments.

The bulk of private R&D spending on energy technologies worldwide is the automobile and oil and gas sectors, with the former growing significantly over the last decade and the latter declining (IEA, 2020g). The global automobile industry engages in significant product differentiation and produces vehicles that range widely in terms of product quality and price, and this market structure supports relatively high levels of private R&D spending (Hashmi and Van Biesebroeck, 2016). This innovation is increasingly directed towards low-carbon technologies by government regulation of vehicle emissions and declining relative costs of alternative technologies. In contrast, government-designed wholesale markets for electric power appear less supportive of private innovation in electric power technologies (Jamasb and Pollitt, 2011 and 2015; Newbery, 2018). Globally, current private R&D spending on electric power generation technologies, both thermal and renewable, and transmission technologies are less than those on fossil fuels and automobiles, and have been relatively stable over the last decade. This stability prevailed despite government market-creating policies for renewable generation technologies (Chapter 4). In contrast, significant public R&D funding is directed towards renewable generation technologies, as well as those for power networks and energy efficiency (IEA, 2020g).

In addition to public funding for low-carbon technology R&D and tax credits for private R&D activities, the early-stage deployment of solar PV, wind turbines and battery electric vehicles attracted substantial resources over the past decade. The real resource costs incurred in early-stage deployment of solar PV and onshore wind from 2000 to 2018 were approximately $1.0 trillion and $0.8 trillion, respectively, in present value terms (2018$) (Chapter 5). These real resource commitments were an approximate order of magnitude larger than publicly funded R&D—almost $2 trillion versus $0.3 trillion (2018$; IEA, 2020g). Much of these market-creation costs appear to have been borne by incumbent electricity generators (lower load factors in existing thermal plants), electricity customers (feed-in tariffs) and taxpayers (investment and operating tax credits).

With some simplifying assumptions it is possible to give a broad indication of how early-stage deployment costs were allocated across countries. Consider renewable generation equipment and assume for simplicity that their manufacturers expected to earn competitive returns on their investments, and any expected future market rents were heavily discounted because of high uncertainties around technology developments, government electricity-market reforms and emissions pricing. If so, the incremental real resource costs incurred in early deployment of these then relatively expensive generation technologies were largely borne in countries where they were deployed.

European Union (EU) countries and China invested in much of the early-stage deployment by capacity, with EU countries clubbing together to share these costs through the 2009 EU Renewables Directive (Newbery, 2016a). The EU accounted for 43 per cent of installed solar PV capacity (primarily Germany and Italy), 34 per cent of onshore wind (Germany and Italy) and 82 per cent of offshore wind (the United Kingdom, Germany and Denmark) from 2000–18 (Figure 8.2). China accounted for 22 per cent of solar PV, 29 per cent of onshore wind and 16 per cent of offshore wind. Among other countries, the United States accounted for 8 per cent of solar PV and 19 per cent of onshore wind, Japan for 10 per cent of solar PV and India 7 per cent of wind and 3 per cent of solar PV. Around the end of this period (2017–19), manufacturing of solar PV panels and wind turbines was also dominated by these countries. China (30 per cent), the United States (15 per cent), India (7 per cent), Japan (6 per cent), Vietnam (3 per cent) and European Union (3 per cent) accounted for about two-thirds of the world's solar PV panel sales by capacity. China (40 per cent), the EU (38 per cent), the United States (12 per cent) and India (2 per cent) accounted for almost all of onshore wind turbine sales by capacity. These countries also accounted for much of offshore wind turbine production with significant market shares: the EU 68 per cent, China 27 per cent and the United States 4 per cent.

Figure 8.2 shows a strong association between cumulative early deployment of variable renewable-generation capacity and current market shares in manufacturing this equipment, consistent with home-market effects from market-creating and industry-supporting policies (Chapter 4). Similarly, there is evidence that the intensification of domestic policies to support renewables deployments during the economic recession and recovery from 2008–11 was associated with a significant subsequent increase in the equipment exports (Mealy and Teytelboym, 2018). The main exception to this potential home-market effect is EU support for solar PV deployment, the production of which shifted increasingly towards relatively low-cost manufacturers in Asia, such as those in China, India and Vietnam, for these mass-manufactured goods of moderate technological complexity.

For battery electric automobiles, a similar but perhaps less precise picture emerges. The real resource costs of early deployment of these vehicles from 2005 to 2018 were approximately $0.25 trillion in present-value terms (2018$) (Chapter 5). In principle, these costs could be borne by existing and future customers who are willing to pay a premium above unit costs for automobiles seen as high-quality, including a low-carbon drivetrain.[5] In addition, governments provided incentives to purchase battery electric vehicles, thereby allocating some early-deployment costs to taxpayers. Assuming for simplicity that these

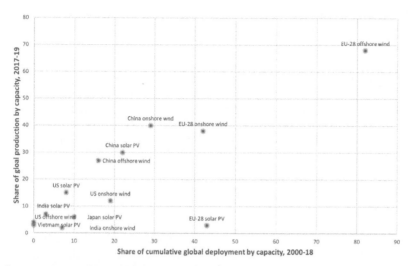

Sources: Data are from the International Renewable Energy Agency, Renewable Capacity Statistics, retrieved from www.irena.org/publications/2020/Mar/Renewable-Capacity-Statistics-2020, and Bloomberg New Energy Finance, Solar PV Module Providers, 2017–19, and Wind Turbine Providers, 2017–19, retrieved from https://bnef.com.

Figure 8.2   *Global new renewables deployment (2010–17) and manufacturing shares (2017–19) (in percentage shares of global capacities)*

costs were borne by current customers and taxpayers where battery electric vehicles were purchased, China (33 per cent), the United States (26 per cent), Europe (13 per cent—primarily France, the United Kingdom and Germany) and Japan (11 per cent) incurred much of the early-stage deployment costs. These countries also have large domestic automobile industries and major companies headquartered in them, which account for much of the world's automobile production.[6] Norway and the Netherlands also shouldered significant burdens through government electric vehicle purchase supports and early adopters, but they do not have significant domestic industries. These climate actions perhaps reflect societal values that lead governments to prioritize climate actions.

The countries that incurred the bulk of real resource costs for innovations and early-stage deployment of solar PV, onshore and offshore wind, and battery electric vehicles also specialized in manufacturing these low-carbon technologies. To the extent that government policies supported early-stage deployment of low-carbon technologies, which incur the bulk of real resource costs in commercializing low-carbon alternatives, they appear more politically sustainable if there are domestic economic co-benefits. These benefits could

be reasonably expected to include growth in high-value-added manufacturing and employment, along with global environmental benefits provided
by low-carbon technologies. However, such commercial successes are not
ensured by these policies, as the EU experience with solar PV shows. These
governments could also reasonably expect to garner domestic support for these
and other climate-policy actions through growing economic interests in structural change alongside environmental interests in climate action.

THE VALUE OF VARIABLE RENEWABLE RESOURCE ENDOWMENTS

While specialization, comparative advantage and political economy influences
on innovation and industrial policies shape low-carbon-technology developments and early deployments, their diffusion across countries reflects other
economic interests. In particular, these technology developments advantage
renewable resource endowments and could over time unlock substantial
economic value from them. Potential widespread diffusion across countries of
wind turbines and solar PV panels reflects the ubiquitous yet varied endowments of wind and solar irradiance. The world is well endowed with these
resources—the scale of energy that could be feasibly captured from them is
several-fold greater than projected total world energy demand through 2100
given energy-conversion technologies (Chapter 3). Bioresources are also
widely available, but their use must weigh carefully their relative value in
production of food, fibres, materials and bioenergy, as well as the value of
ecosystem services such as biodiversity and natural carbon sinks. Recycling
bioresources can ease some trade-offs among these alternative bioresource
uses (Chapter 7).

Given wind and solar resource endowments and increasingly cost-effective
technologies to capture them, their relative abundance across countries has the
potential to become a new source of growth and increasing prosperity as these
technologies advance. Typically, these resource endowments are assessed separately, but given their variability the correlations of their availability matter
too. The potential complementarities between wind, solar irradiance and other
renewable resources have both temporal and spatial dimensions. For example,
spatially distributed wind generation sites can help smooth their combined
output as distances among generation sites increase. But this benefit of spatial
diversification must take account of electricity transmission costs. An example
of temporal complementary is the annual pattern of wind and solar availability
in Europe. The former is relatively abundant in autumn and winter and the
former in spring and summer. Managing well these imperfect and seasonally
changing correlations among variable renewables helps smooth generation of
low-carbon electricity from them, lowering their overall system costs. Their

overall correlation with electric power demand—its variation within days and across days, weeks and seasons—must be considered as well.

A simple measure of complementarities between wind and solar resources for electric power generation covering both spatial and temporal dimensions is their critical overlap hours by geographical location. They are the expected hours of modelled generation from comparably scaled solar PV (single-axis tracking) and wind turbines (3 MW with 150 m hub height) in the same location (within 50 km²) that would need to be curtailed over a full year to meet a given electric power demand. In most parts of the world, critical overlap hours are less than 500 hours per year. That is, resource availability for wind and solar PV tends to be inversely correlated, with significant complementarity between these technologies in generating electric power (Bogdanov et al., 2019; Fasihi and Breyer, 2020). But there are some regions with high critical-hour overlaps. They include Patagonia and the Atacama Desert in Latin America, Sudan and the Horn of Africa, and Tibet in Asia, where daytime wind resources are extremely abundant.

Excluding overlaps, the total potential hours of combined wind and solar PV generation in much of the world are greater than 4000 hours per year, except in tropical forests and sub-arctic regions (Figure 8.3). This is almost half of the 8760 hours per year. Based on hourly solar irradiance at a spatial resolution of around 50 km², the full-load hours of electric power generation from single-axis tracking solar PV are greater than 1600 hours per years in most regions within ±45° latitude (Fasihi and Breyer, 2020). This area ranges from the northern United States to Patagonia, Southern Europe to South Africa and Northern China to Australia. That of wind is greater in most regions based on wind speed data at the same spatial resolution. The exceptions are equatorial regions in Latin America, Africa and Asia where wind speeds are typically low and winds inaccessible, as in tropical forests. In addition to Patagonia, the Atacama Desert, Sudan, the Horn of Africa, Tibet and Western Australia, the best wind sites with full-load hours of 4500 to 6500 per year are Northeast China, the Central United States, Newfoundland and Labrador in Canada, Ireland, the United Kingdom and coastlines of Western Europe and Australia.

While Figure 8.3 shows potential combinations of wind and solar resources that could be converted into electric power, their economic value depends on how they are combined with each other and flexibility technologies that firm-up supply and balance demand. Where wind resources are relatively abundant, cost-effective technology combinations would use wind turbines more intensively than solar. Solar-oriented systems are more cost-effective where solar resources are relatively abundant and correlate with intraday variation in electric power demand. Regarding flexibility technologies, battery technologies complement solar PV well, where intraday variability and battery cycling is high. Electrolytic hydrogen, hydrogen storage and combustion turbines

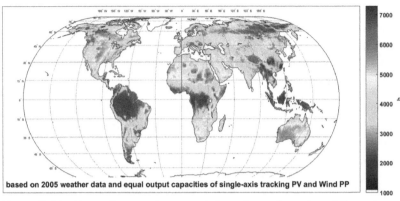

based on 2005 weather data and equal output capacities of single-axis tracking PV and Wind PP

Notes: Full-load hours of a power plant are equivalent to hours of operation at full capacity to provide the same annual output of electric power. Full-load hours of hypothetical hybrid solar PV-onshore wind generation plants exclude critical overlap hours. The hypothetical hybrid plants consist of comparably sized solar PV and wind generation capacities in terms of their maximum electric power output. When modelled outputs of the hypothetical hybrid plants exceed the maximum single capacity of either component, given expected solar irradiance and wind patterns, the excess generation is assumed to be curtailed. This curtailment excludes critical overlap hours from the estimated full-load hours of the hybrid plants. This measure indicates the physical feasibility of generating electric power from these variable renewable resources. A cost-optimized plant would be based on different shares of solar PV and wind energy together with energy-storage options to balance electric power supply and demand.
Source: Reprinted from M. Fasihi and C. Breyer (2020), Baseload electricity and hydrogen supply based on hybrid PV-wind power plants, *Journal of Cleaner Production*, 243, 118466 (CC BY 4.0). https://doi.org/10.1016/j.jclepro.2019.118466.

Figure 8.3   Full-load hours of hybrid solar PV–onshore wind generation less critical overlap hours (in hours per year)

complement wind technologies for which resource availability varies significantly at frequencies over days, weeks and seasons. Examining least-cost combinations of these technologies in different locations scaled to locational electricity demands indicates the potential economic benefits from these widely available yet variable renewable resources.

Given current technology-cost trends for wind turbines, solar PV, lithium-ion battery storage and electrolysers, Figure 8.4 shows how their sustained technological progress could affect projected electric power supplies from these resources across the world (Fasihi and Breyer, 2020). It shows projected electricity supply curves for dispatchable electric power (a supply that can be readily controlled to match demand) with sufficient flexibility to meet typical demand profiles using least-cost combinations of these four technologies. The projected supplies assume that cost trends are sustained over the next three decades and allow for transmission costs from generation sites to load centres.

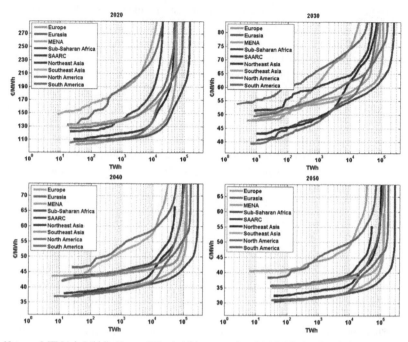

Notes: MENA is Middle East and North Africa countries. SAARC is South Asian Association for Regional Cooperation countries. Electric power supply curves are based on cost-optimized combinations of solar PV, onshore wind, battery storage and electrolytic hydrogen production, storage and conversion into electric power with a combustion turbine. These hybrid-plant configurations are optimized to renewable resource availability by geographic location—primarily solar PV-based, wind-based or mixed plants. To aid interpretation of the supply curves, note that the assumed exchange rate is $1.35/€, so a levelized cost of electricity of €40/MWh equals $55/MWh, about the levelized cost of electricity from an advanced coal plant in India. Terawatt-hour (TWh) x 10^4 equals 36 exajoules (EJ) and TWh x 10^5 equals 360 EJ. In comparison, the IEA (2019a) projects China's total final energy demand in 2050 to be about 120 EJ under current policies. These projected electric power supply curves based on this limited set of technologies serve to illustrate the potential value of renewable resources and comparative advantages they may present. However, the feasibility of such 100 per cent variable renewable electric power systems is the subject of much debate, with more diversified systems, including hydroelectric, biomass generation and nuclear, potentially affording lower overall system costs.
Source: Reprinted from M. Fasihi and C. Breyer (2020), Baseload electricity and hydrogen supply based on hybrid PV-wind power plants, *Journal of Cleaner Production*, 243, 118466 (CC BY 4.0). https://doi.org/10.1016/j.jclepro.2019.118466.

Figure 8.4   *Supply curves for dispatchable electric power from cost-optimized hybrid plants: nine regions, 2020–50 (levelized cost of electricity in €/MWh)*

The projections also assume hydrogen is stored in widely available salt caverns and is used to generate electric power with readily modified combustion turbines. If these technology trends are sustained and energy systems increasingly rely on variable renewables for their primary energy, the projections show that electric power supplies would be relatively low-cost in Latin America, Sub-Saharan Africa and the Middle East and North Africa. North America, South Asia and Southeast Asia (including Australia) would also benefit from relatively low-cost electricity. The regions less advantaged by these technology developments and primary energy resources are Europe and Eurasia. The future mix of low-carbon generation technologies that minimize overall system costs, however, is not necessarily confined to these technologies for generation and flexibility (Chapter 3).

The support for low-carbon technologies by EU member states and the United Kingdom, China, the United States, Japan, South Korea and other countries through development and early-stage deployment of solar PV, wind, battery storage and electric vehicles could unlock substantial value from variable renewables resources in developing economies. These technologies create potential for new sources of sustainable growth and comparative advantages for developing countries based on their relative abundant renewable resources, which would otherwise go largely untapped. Relatively low-cost and abundant electric power would be a direct benefit to households and firms in developing countries. There would also be potential to specialize in economic activities that are relatively intensive in their use of electric power, such as low-carbon fuels made from electric power—such as hydrogen and synthetic hydrocarbon fuels—which could be exported. Other potential comparative advantages could arise from electricity-intensive materials production.

EXPANDING FRONTIERS OF ENERGY SYSTEM TRANSFORMATIONS

Looking forward, the energy-system transformation challenges unfold along two dimensions. One focuses on countries and the other on sectors, although the two interact.

At country level, the challenge is to consolidate the transformations underway in the electric power, automobile and building sectors in countries that initiated these disruptions and promote their widespread diffusion across all countries. Consolidating these changes in initiating countries, for example, requires sustaining market-creating policies until exits emerge through technology-cost reductions and/or emissions pricing to support commercial viability. Mechanisms for competitive allocation of investment supports, such as auctions for long-run contracts with feed-in tariffs for electric power from renewable generation, serve to minimize policy costs and ensure timely policy exits. Energy markets and infrastructures must also adapt to variable renewable generation and battery electric vehicles. This involves reforms to wholesale electric power markets and pricing of network- and system-balancing services, as well as charging-network extensions for battery electric vehicles to lower switching costs for potential customers (Chapters 4 and 6). This could involve early-deployment support for flexibility technologies for electric power systems. But it is important to recognize positive technology spillovers from electrification of road passenger and freight transport using battery electric and potentially hydrogen and fuel-cell-electric drivetrains where unit energy costs of transport fuels are high.

Building-sector transformations in cooler climates with extensive domestic natural gas networks are also underway, especially for new buildings, with high thermal-efficiency and low-carbon heating-technology standards, as in Germany, the Netherlands and the United Kingdom. The main challenge is in transforming existing buildings. A reform strategy that prioritizes forward-looking, cost-effective thermal-efficiency standards for existing buildings when extensively refurbished also complements both potential electric and low-carbon-fuel heating alternatives. This policy approach holds few regrets while investments in innovation and demonstration projects de-risk low-carbon heating technologies. This approach would recognize that eventual low-carbon heating alternatives are expected to increase the long-run benefits of more thermally efficient, long-lived buildings. Complementary policies for at least partial decarbonization of heating technologies, such as emission performance standards for heating equipment, would also help accelerate innovations and deployment of low-carbon alternatives.

Promoting diffusion of these low-carbon technologies across countries from those initiating the technology disruptions to all, including developing countries, requires lowering non-price barriers to their deployment as they become commercially viable. As their cost-effectiveness improves, these technologies are increasingly attractive in countries with abundant solar and wind resources, albeit amidst domestic fossil-fuel interests and their embodiment in long-lived capital assets and employment (Grubert, 2020). In most countries, the state owns the title to subsurface resources and licences firms to develop them in return for licence fees (Flomenhoft, 2018).[7] Priorities to facilitate deployment of variable renewables and flexibility technologies are investments in capabilities to adapt these technologies to local contexts, including through bilateral and multilateral cooperation. These capacity-building efforts extend to planning-led processes for electricity-system development and long-term contracting for generation and network extensions that are adapted to changing technology costs. Creating the enabling conditions for and implementing market-oriented reforms of electricity systems is also important in the long run to guide efficient investment in them. This requires strengthening the cost recovery of systems while safeguarding energy access for all (for example, two-part tariffs which subsidize supplies for basic needs). It also requires support for communities and households that lack capacities to adjust to these deep structural changes in energy systems. Developing these institutional and societal capacities in developing countries is a focus for bilateral and multilateral technical cooperation among countries, and one that should be prioritized.

There are in addition significant complementarities between expansion of vehicle charging and electric power network extensions by increasing network loads, supporting their reaching cost-effective scales more quickly than from stationary power demands alone. Electric two- and three-wheeled vehicles are a rapidly growing market entry point for battery electric vehicles in developing countries, and they currently have lower total ownership costs than conventional motorcycles. Electric two-wheelers in advanced industrialized countries are also a growing alternative for urban commutes and short journeys.

At sectoral level, the energy-system transformation challenges are to expand the sectoral frontiers of disruptive technological changes to so-called hard-to-decarbonize sectors (Chapter 3). On low-carbon fuels, there is a growing cluster of innovation activity around fuel cells and electrolysers, which are closely related technologies (Chapter 4). This innovation activity tends to occur first in fuel-cell applications and then spill over to electrolysers, with automobile, automobile component and electronics manufactures in Japan, South Korea and the United States specializing in this area of innovation (Ogawa et al., 2018a and 2018b). Since hydrogen has a low energy density by volume, innovations also focus on storage-related technologies, including hydrogen compression and high-strength, lightweight composite fibre storage tanks. Other storage options are its conversion into hydrogen-rich compounds such as low-carbon methanol, ammonia and liquid organic hydrogen compounds that can be stored in conventional tanks as more energy-dense liquids.

Countries relatively specialized in biofuel innovations include the Unites States, Germany, Japan, France and the United Kingdom, which account for most high-value patents in this area (O'Connell et al., 2019). Among advanced biofuel technologies, those producing biodiesel and bio-jet fuel from waste vegetable oils and fats are commercially mature technologies (trans-esterification and hydro-processing). A few European companies specialize in these technologies, but face barriers related to feedstock availability and quality as well as production scale. Other advanced biofuel technologies that are relatively mature but not yet commercially viable included processes for producing ethanol from lignocellulosic materials and catalytic conversion of ethanol to biodiesel and bio-jet fuel (O'Connell et al., 2019). These processes still face significant technological barriers, but are potentially less constrained by feedstocks.

On low-carbon industrial processes, there are a number of demonstration projects using low-carbon technologies underway, and they could show viable ways forward. For example, a public–private partnership in Sweden is developing the first pilot project for low-carbon steel production using electrolysis to produce hydrogen and hydrogen for the direct reduction of iron ore.[8] Sweden has high-quality iron ore deposits and specializes in high-strength steel products, and the main Swedish steel producer and electricity generator are driving this project with government funding support. A similar public–private partnership in Canada is developing the first pilot project for aluminium smelting that eliminates carbon dioxide emissions and emits only oxygen.[9] This consortium includes Alcoa and Rio Tinto, as well as Apple, for which aluminium is a signature material in its premium consumer electronics products. In the United States, a public–private–university collaboration is demonstrating molten-oxide electrolysis technologies that can produce a broad platform of metals.[10] China, which is the world's largest producer and user of cement, is also an innovation hub for low-clinker and novel cements, spending the most public funds of any country on research, development and demonstration in this area and dominating its high-value patents (Lehne and Preston, 2018). Cement technologies with lower net carbon dioxide emissions are also being developed through public–private–university partnerships in the United States and Europe, such as carbon-cured, low-clinker cements.[11] These examples highlight the range of innovations in industrial process that are being advanced to reduce or eliminate emissions, especially in countries that specialize in innovations or seek to do so and the firms that participate in these efforts.

Innovation systems in a number of countries are thus working to advance technologies that could make inroads to achieving net zero emissions in hard-to-decarbonize sectors. Nevertheless, there is much more to do to commercialize and scale these innovations and to develop alternatives that are potentially better. But the substantial innovation responses to previous energy price shocks and security concerns, and ongoing ones spurred more directly by energy-reform policies, suggest that effective innovation capacities and policies are in place in countries that specialize or are developing specializations in innovation. The policy challenge is to direct these capacities decisively towards low-carbon technologies in a timescale consistent with government climate goals. At the same time, commercial approaches of firms that offer customers low-carbon products and services can help attract those willing to pay for this characteristic among early adopters.

There are three types of demand-pull policies for low-carbon technologies that could direct investments in innovation and market creation more strongly towards them. One is emissions pricing. However, progress in pricing has been incremental in most countries, with those having legacy renewable resources (hydroelectric) and low-carbon technologies (nuclear) as well as long-standing energy security achieving more progress than others (Chapter 5). This experience suggest that higher levels and a wider scope of emissions pricing follow from greater deployment and use of low-carbon technologies developed through mechanisms other than emissions pricing, at least in early stages of system transformations.

A second demand-pull approach involves market-creating and industry-supporting policies for low-carbon technologies that directly strengthen demand for these technologies, especially in countries that specialize in related innovation and manufacturing production. Such policies can be warranted by missing risk markets and potential time-inconsistencies in government policies such as electricity-market reforms and emissions pricing. Investment subsidies for investments in low-carbon alternatives during their early deployments would counter investment disincentives from any potential time-inconsistencies in government policies (Ulph and Ulph, 2013; Stiglitz, 2019). They would also target external scale economies and network effects from early deployment of low-carbon alternatives as they integrate into energy systems. Competitive government procurements of low-carbon technologies are also potentially effective demand-pull policies. Designing such market-creating and industry-supporting policies to work with market competition is key to their cost-effectiveness and timely exits. They should also be designed to ensure that policy costs are allocated fairly, avoiding regressive impacts such as the electricity bill charges from feed-in tariffs for renewable power generation.

A third is time-bound commitments to achieve net zero emissions at sector or country level, which impose high long-run 'shadow prices' on emissions, well above any short-run emissions pricing especially in hard-to-decarbonize sectors. Such long-run policy signals—to the extent credible—indicate future market sizes for low-carbon technologies and thus strengthen incentives for private investment in innovation and market creation for them. A challenge, especially for the deep decarbonization of heavy industry and commercial transport, is to develop energy-reform strategies that build the credibility of long-run commitments to achieve net zero emissions of fossil carbon dioxide from energy (Chapter 9).

The international automobile sector provides an example of the latter two policies sequenced at sector level and tacitly coordinated across countries to increase credibility and strengthen incentives for private investment in innovation and market creation. Countries with significant domestic automobile industries have long had policies to support customer purchases of alternative-drivetrain automobiles. But ten countries announced plans in 2016–17 in quick succession to phase out of conventional drivetrains in new automobiles (Meckling and Nahm, 2019). Countries announcing such intended bans included those with established automobile producers and exporters, such as France, Germany, Sweden and the United Kingdom, as well as emerging exporters such as China and India. Some automobile-importing countries also announced planned phase-outs, including Ireland, the Netherlands and Norway, perhaps motivated by broader societal interests in environmental protection rather than domestic economic interests. Together, these policy announcements signalled both the potential timing and size of future markets for alternative-drivetrain vehicles and thus potential returns to private investment in innovation and market creation. Their signalling credibility rests on both underlying technology trends as well as the credibility of the broader energy-reform strategies of which they form part (Chapter 9).

Similar policy strategies could be used for heavy-duty, long-distance trucks to expand the frontiers of low-carbon technology disruptions, especially in countries that specialize in their manufacture. They include China, France, Germany, Japan, India, the Netherlands, Spain and the United States. For example, European transport fuel taxes close much but not all of the total-cost-of-ownership gap between a diesel and fuel-cell electric heavy-duty truck and the fuel-price gap between diesel and biodiesel. This assumes that low-carbon fuels are exempt from fuel taxes (treats them as emissions taxes) and that biofuels are made from low-cost and sustainable technologies and feedstocks. As necessary, higher fossil-fuel taxes, low-carbon fuel subsidies and/or investment supports for alternative-drivetrain trucks could be used to strengthen market-creating policies for these alternative technologies and fuels. Given marginal cost pricing of transport services, policy designs would need to ensure that a combination of fossil-fuel taxes and low-carbon-fuel subsidies made sure that market prices covered operating costs of low-carbon alternatives. Well-designed investment supports could take the form of investment subsidies for alternative-drivetrain vehicles; for example, investment tax credits as used for R&D activities. They would counter investment disincentives from any potential time-inconsistencies in government policies and target positive spillovers from early deployments. This policy approach would also enable road freight transport service providers to offer differentiated services to customers with a willingness to pay for low-carbon services.

This policy approach could be adapted to other country and sector contexts, such as countries that are able to sustain only low fuel or emissions taxes, and the shipping, aviation and heavy industry sectors. Depending on the context, the policy mix could lean more on market-creating investment and operating supports than emissions pricing during early deployment of low-carbon alternatives. For example, emissions pricing on international shipping and aviation would require cooperation among countries on its determination. Absent this, reliance on other policy instruments such as investment and operating subsidies would be necessary to advance low-carbon alternatives. In heavy industry sectors, emissions pricing within countries is significant in its level and scope in Europe, and increasing in some other countries. But absent international coordination of emissions pricing in heavy industry, carbon border adjustments would also be necessary for tradable, emission-intensive materials and goods to ensure a level competitive playing field in home markets with costed emissions. Again, the mix of emissions pricing and low-carbon-fuel subsidies would need to ensure that marginal cost pricing covers the operating costs of low-carbon alternatives. This combined policy approach of emissions pricing for heavy industry (low-carbon-fuel subsidies), investment supports for early deployment if necessary, and carbon border adjustments would help advance low-carbon industrial processes. Complementary material-efficiency polices could help counter increases in unit material costs to help keep down any rise in material service costs.

The exit from these policies to support early deployment of low-carbon alternatives could arise from declining costs of low-carbon alternatives and increases in the level and scope of emissions pricing as low-carbon alternatives become more widely commercialized. Mechanisms for the competitive allocation of supports for low-carbon alternatives can help ensure that timely exits are achieved along with minimization of policy costs. Funding any investment and operating supports through government subsidies or tax credits (expenditures) would help ensure a fair allocation of policy costs.

DISTRIBUTIONAL IMPACTS AND FAIR ENERGY TRANSFORMATIONS

Looking forward, there are two impacts that would arise from the fundamental changes to energy necessary to achieve net zero emissions and affect vulnerable communities and households, and that warrant a particular policy focus. This attention is necessary because of limited capacities of some households, firms and communities to adapt to these structural changes. They relate in particular to coal-mining communities and low-income households living in poor-quality housing. Chapter 9 examines how policies such as emissions pricing consistent with a quantitative emissions goal can be designed to manage their potentially adverse distributional impacts.

The global coal-mining industry and many coal-mining communities would be highly exposed to energy-system transformations with significant potential for 'stranded' employees and regions (Spencer et al., 2018). Past examples of coal industry declines show that they can be associated with large and persistent economic and social dislocations in terms of employment, regional value-added and local government revenues. Typically, these community and regional impacts were poorly anticipated and managed. To the extent they were addressed, policy responses tended to be focused on short-run compensation rather than long-run regional adjustment. In current energy transformations, there is limited evidence of governments planning for large-scale adjustments in coal-mining regions.

Other potentially vulnerable communities exposed to energy transformations are low-income households living in poor-quality housing—especially poorly insulated houses in cooler climates that depend on fossil fuels for winter heating (Green, 2018). In transforming buildings and cutting emissions from them, the costs of providing thermal comfort in such dwellings would be expected to rise and tend to be capitalized in lower market values for such properties (Chapter 7). There is a risk that poorer households in poor-quality housing become locked into fossil-fuel heating as it becomes increasingly expensive from costing emissions either directly or indirectly, with little capacity to adapt. Public investments in adapting social housing and targeted supports for low-income owner-occupied and private-rented dwellings would likely be needed to address these adverse distribution impacts and health risks from under-heated dwellings. However, experience with such targeted programmes for low-income households to adapt privately owned dwellings is mixed (Chapter 7).

CONCLUSION

The experiences in transforming energy systems reveal that countries and firms initiating low-carbon technology disruptions to current systems are those that specialize in innovations, especially in related technology fields, as well as in manufacturing. These activities and economic specializations are central to the technological changes necessary for net zero emissions from energy systems. Government policies that support these upfront investments in low-carbon alternatives are domestic innovation systems, including public R&D funding and tax credits for private R&D investments, and early-deployment supports for low-carbon alternatives, primarily in the form of market-creating and industry-supporting policies. These policy measures incurred large real resources costs—more than an estimated $2 trillion (2018$), mostly for early-deployment supports (Chapter 5). While the emission cuts they afforded were not least-cost in static terms, they unlocked substantial dynamic scale economies that substantially lowered the costs of future investments in these low-carbon alternatives. The countries supporting these upfront investments were primarily EU member states and the United Kingdom, China, the United States, Japan and South Korea. These investments have quasi-public-good characteristics that hold potential benefits for all countries. They were largely incurred in countries that captured some economic benefits by specializing in innovations and production of low-carbon alternatives, advancing not only these technologies but also domestic interests in change.

Looking forward, the challenges are to learn from experiences in transforming energy systems to expand their country and sectoral coverage. Energy-reform strategies adapted to country and sector contexts are key to unlocking low-carbon technology and renewable resource potentials. These strategies depend on whether countries initiate low-carbon technologies or follow in adopting them as they become commercially viable. Reform strategies in countries that have initiated technological disruptions show how they could be extended to hard-to-decarbonize sectors. The mix of early-deployment supports include emissions pricing to the extent politically sustainable, as well as investment and operating supports for low-carbon alternatives. The appropriate mix of instruments and approaches depends on sectors contexts. On the widespread diffusion of low-carbon technologies across countries, a significant incentive for developing countries to adopt policies to facilitate their deployment as they become cost-competitive is to tap the value of their abundant renewable resources. Low-carbon technologies create the potential to derive significant economic value from their wind and solar irradiance resources. Developing capabilities and capacities in these countries to adapt and adopt these technologies is key, including through bilateral and multilateral technical cooperation. But regardless of the context, country energy-reform strategies must be comprehensive, coherent and credible if they are to create the market conditions that enable firms and households to make the time-bound investments necessary to achieve net zero emissions. Such strategies are the focus of the concluding chapter.

NOTES

1. A country's specialization in exports relative to the overall composition of global exports is a standard measure of revealed comparative advantage in production.
2. The Observatory of Economic Complexity, Country Rankings. Retrieved from https://oec.world/en/rankings/country/eci.
3. A country's specialization in patenting relative to the overall global composition of patenting is a measure of revealed comparative advantage in innovation. See Zachmann and Kalcik (2018).
4. IEA, Energy Technology RD&D Statistics. Retrieved from https://doi.org/10 .1787/enetech-data-en.
5. At least ex post, investors now expect a successful premium electric automobile innovator and disruptive market entrant—Tesla—to be able to sell its vehicles at a market premium in the future, as reflected in its stock market valuation. At end-2019, its Tobin's Q-ratio (average) was 13 versus those for the US stock market average and incumbent US automobile producers of less than 2 and 1. Tobin's Q is the ratio of the market value to book value, and the Q-ratios are retrieved from Yahoo Finance (https://finance.yahoo.com).

6. International Organization of Motor Vehicle Manufacturers, World Motor Vehicle Production Statistics 2019, www.oica.net/category/production-statistics/2019 -statistics.
7. A notable exception is the United States, where private landowners also hold the title to subsurface resources.
8. www.hybritdevelopment.com/steel-making-today-and-tomorrow.
9. www.elysis.com/en.
10. www.bostonmetal.com/moe-technology.
11. www.solidiatech.com.

9. Accelerating change

To transform energy systems to low-carbon alternatives, governments must address the multiple market imperfections that direct decisions of firms and households towards incumbent technologies and traditional use of fossil fuels. This transformation requires most firms and households in advanced industrialized and many in developing countries to make alternative investment decisions and energy-use choices to their current ones. The scale of investments is immense, and their informational requirements are demanding. Both considerations point to the fundamentally important role of markets in transforming energy systems, with their imperfections effectively addressed. These policies include those to support foundational investments by governments and firms in innovations and market creation for low-carbon alternatives. They create the market opportunities for all firms and households to make different decisions and choices about the technologies in which they invest and the energy, energy services and materials they consume. Over the next decades, most firms and households in most countries would need to invest in transforming energy systems if net zero emissions from energy are to be achieved.

A narrow government energy-reform strategy focuses on correcting two key market failures: knowledge spillovers from investments in research and development (R&D) and environmental externalities from emissions of carbon dioxide and other greenhouse gas emissions. The two corresponding policy correctives are government supports for R&D and emissions pricing based on the social cost of carbon dioxide emissions (SC CO_2). This policy paradigm holds important—but incomplete—insights. The two correctives alone appear vulnerable to becoming stuck at low levels of policy ambition, especially emissions pricing to internalize environmental externalities. The barriers that these policies confront are not epistemic—their rationales are readily grasped and widely understood by policy makers and the public. Rather, the barriers more likely reside in incompleteness of these policy prescriptions, distributional impacts of systemic transformations, potential policy time-inconsistencies and the political economy of deep structural reforms. But to be effective, government reform strategies for transforming energy systems must consider not only important market imperfections but also institutions and infrastructures that support energy systems and their transformation. Reform strategies must be sufficiently comprehensive to address all material market imperfections, and sufficiently coherent and credible to foster interests in sustained reforms and long-run investments in pursuit of clear climate goals. These considerations of comprehensive, coherent and credible policy strategies point to more hetero-dox policies than R&D supports and emissions pricing based on the SC CO_2.

For example, time-bound goals for achieving net zero emissions provide a relatively stable and informative focal point for coordinating the policies and choices of governments, firms and households (Chapter 5).[1] Market-creating and industry-supporting (industrial) policies for early deployment of low-carbon technologies directly support these goals, in part by addressing knowledge and cost-saving spillovers from creating markets for low-carbon alternatives (Chapter 4). They also help overcome missing markets for emissions prices and potential time-inconsistencies in implementation (Chapter 5). These polices are particularly important in countries that specialize in innovations and production of low-carbon alternatives, where they also foster new interests in sustaining energy reforms (Chapter 8). When well sequenced with industrial policies, cost-effective emissions pricing consistent with this quantitative limit on emissions brings forward in time its long-run shadow price and promotes relative price shifts economy-wide to reflect the constraint, especially in hard-to-transform sectors and those with market linkages to them (Chapter 5). Emissions pricing also creates a market-based exit from industrial policies and, if differentiated between easy- and hard-to decarbonize sectors, can help manage adverse distributional impacts, especially on house-holds. Adaptations of government-designed and government-regulated energy markets and infrastructures are necessary to advance alternative technologies and choices, especially in electric power, which is the expected backbone of low-carbon systems (Chapter 6). Energy- and material-efficiency measures as warranted by market imperfections also create economic benefits that help offset the costs of advancing low-carbon technologies (Chapter 7).

Such a heterodox energy-reform strategy is no more expansive than warranted by careful consideration of domestic energy-reform contexts. Moreover, its design manages distributional impacts, especially on house-holds, and brings to the fore self-reinforcing mechanisms that sustain and deepen energy-system transformations. They include effectively sequenced policies and substantial scale economies and network effects from increasing deployment of low-carbon alternatives. Such mechanisms, together with adapting market-supporting institutions to low-carbon alternatives and fos-tering arrangements that promote policy consistency, build the credibility of energy-reform strategies and the climate goals they serve to implement. Over time, they lock-in low-carbon alternatives to incumbent technologies and traditional use of fossil fuels.

The bases for international coordination of the energy reforms to achieve net zero emissions from all energy systems also require careful consideration if the Paris Agreement goals are to be achieved. Achieving net zero emissions from energy systems requires almost all countries to transform their systems to low-carbon alternatives. Moreover, country specializations in innovations and manufacturing low-carbon technologies mean that acting alone in reforming energy systems is costlier than coordinated actions. Some self-reinforcing mechanisms of domestic energy reforms also facilitate international policy coordination and competition in manufacturing tradable low-carbon technologies, such as scale economies and network effects. These mechanisms operated in now easy-to-transform energy sectors in those countries that initiated the technological disruptions. Moreover, the sectors in which these technologies deploy produce largely non-traded energy (electric power) and energy services (surface transport and activities in buildings), so early actions in initiating countries give rise to few concerns about competitiveness losses. The mechanisms necessary to advance energy reforms in hard-to-decarbonize sectors in contrast are likely to require additional instruments, such as adequate emissions pricing and carbon border adjustments. These sectors produce tradable low-carbon fuels, materials and goods (heavy industries) and energy services (commercial aviation and shipping).

The constituent elements of comprehensive, coherent and credible domestic reform strategies that can be effectively coordinated across countries are thus emerging from experiences in transforming energy systems. Looking forward, it is important too to consider bilateral and multilateral technical cooperation on capacity-building and development financing for developing countries, enabling them to adapt and adopt low-carbon alternatives to facilitate their widespread diffusion. This is especially important in the electric power sector as low-carbon alternatives become more cost-competitive with incumbent ones. These elements are considered in turn.

COMPREHENSIVE AND COHERENT ENERGY-REFORM STRATEGIES

While there is consensus on many constituent policies of comprehensive energy reforms, primarily on economic-efficiency grounds, some policy elements, effective policy sequencings and complementary combinations remain subject to debate. Government support for R&D investments to account for knowledge spillovers and emissions pricing to internalize the environmental externalities are widely accepted in principle and increasingly applied in practice (Chapters 4 and 5). However, they are not yet implemented with the intensity necessary to achieve government climate goals. For example, emissions pricing in most countries remains well below the level and scope deemed adequate (Chapter 5). Policies to promote energy efficiencies, especially in buildings, appliances and automobiles, to overcome market imperfections and behavioural anomalies are also widely accepted and applied in practice

(Chapter 7). The economic case for adapting government-designed energy markets and network infrastructures to low-carbon technologies is widely accepted, but implementation remains subject to policy experimentation, learning and improvement (Chapter 6). Given complex trade-offs involved in their design, as around capacity remuneration mechanisms, there is no single reform model for electric power markets, their government design and supporting institutions and regulations. Addressing the difficult investment-coordination challenges in adapting natural gas infrastructures and buildings to low-carbon alternatives mostly remains on policy to-do lists, as do material-efficiency measures (Chapter 6 and 7).

The more pressing policy challenge, though, relates to the role of market-creating and industry-supporting (industrial) policies in early deployment of low-carbon technologies. They are, together with R&D activities, the foundational investments in transforming energy systems. The role of industrial policies in early deployment of low-carbon technologies is an issue primarily for countries that specialize—or seek to specialize—in innovations and manufacturing low-carbon technologies (Chapter 8). But importantly, countries demonstrating leadership in tackling climate change also contribute to advancing low-carbon technologies by participating in their market creation. Most such industrial policies in advanced industrialized countries and China create markets for low-carbon technologies by subsidizing investments in them by early-adopting firms and households. This policy approach uses investment and/or operating subsidies to direct the upfront investments in innovation and market creation by firms towards low-carbon energy alternatives. This approach can be designed to promote competition in producing low-carbon technologies as a further spur to firms' investments in innovations and early-stage deployments. Their industry-supporting effects to the extent that they arise do so indirectly through home-market effects of market-creating policies.

Long-standing economic critiques of such industrial policies are that they are unnecessary, ineffective and vulnerable to capture by interest groups. That they are unnecessary in transforming energy systems assumes that markets adequately remunerate investments in market creation through temporary market power from product-differentiation and cost advantages. Markets that are closer to this Schumpeterian ideal of 'creative-destruction' to create and manage such market rents are those for automobiles. Those farther away are government-designed and government-regulated electricity markets. This critique also typically glosses over other material market imperfections, distributional impacts, missing markets for future emissions pricing, and potential time-inconsistencies in its implementation.

The argument that industrial policies are ineffective and vulnerable to capture assumes 'losing' technologies select governments rather than governments select 'winning' technologies. But evidence on industrial policies in general points to their potential effectiveness in fostering enduring commercial successes even after the market-creating and industry-supporting props are removed (Chapter 4). Moreover, the bulk of real resources directed by government towards innovations and early deployment of low-carbon alternatives appears to have supported ones that will endure the market-selection test with only modest or no emissions pricing in the long run. The argument that industrial policies turned these otherwise-losing technologies into market winners would need to provide plausible counterfactuals that better technologies were overlooked by government policies in timeframes relevant to achieving climate goals.

That said, industrial policies for low-carbon technologies could have been better designed—more cost-effective and fairer in their distributional impacts. For example, inefficiencies include relatively slow transitions in Europe and elsewhere to auctioning of government-supported contracts with feed-in tariffs for investments in renewables generation and untargeted purchase subsidies of alternative-drivetrain automobiles. Supporting early deployment of renewables generation through tax credits in the United States had fairer distributional impacts than regressive feed-in tariff obligations in Europe funded through electricity bill surcharges. Also, it is important to emphasize that not all industrial policies have been necessarily effective in promoting industry development. For example, the jury remains out on cost reductions of new large-scale nuclear reactors in advanced industrialized countries, although levelized costs of new nuclear generation in China appear more competitive. Nevertheless, many governments have learned from experience and improved the efficiency and effectiveness of these energy policies over time. For example, the transition to auctioning of contracts with feed-in tariffs for renewables generation across advanced industrialized and developing countries was rapid and extensive. This shift occurred as supply chains developed, the number of competing investment projects increased and technology costs declined, enabling more competitive allocation of these long-term investment contracts for investments in renewable generation.

Associated with the role of industrial policies in transforming energy systems is a debate around policy sequencing, in particular between market-creating policies and emissions pricing. The orthodox policy prescription is to use emissions pricing to direct innovations towards low-carbon technologies and government supports for investments in R&D to internalize the knowledge spillovers. The heterodox policy approach uses market-creating and industry-supporting policies to direct innovations towards low-carbon technologies combined with R&D supports. Emissions pricing in this policy sequence supports widespread deployment and diffusion of low-carbon technologies as necessary and consistent with long-run emissions constraints but does not play an earlier role in directing innovations and fostering early-adopter demands. There are several potential benefits from the heterodox approach. It addresses missing markets for future emissions prices and reduces investment risks in advancing low-carbon alternatives, many of which are highly capital-intensive, and compensates for positive cost spillovers from early adopters. This policy sequencing also fosters interests in low-carbon technologies and creates opportunities for making alternative choices before turning to more adequate and effective emissions pricing.

Potential inefficiencies from this sequencing are the extent to which a market-creating policy is second best to emissions pricing at early stages of deployment, government selection of low-carbon technologies for support is misguided and its management of rent-seeking is ineffective. An efficiency benefit of a heterodox approach is that it avoids over-reliance on emissions pricing early in system transformations (Acemoglu et al., 2012 and 2016; Stiglitz, 2019; Stern and Stiglitz, 2021). An emissions-pricing-led approach also risks becoming stuck at a low level of policy ambition for want of self-reinforcing mechanisms. That said, well-sequenced cost-effective emissions pricing based on a time-bound net zero emission constraint is an important complement to industrial policies, providing a market-based exit to them. Such emissions pricing brings forward in time the shadow price of the binding emissions constraint, promoting economy-wide shifts in relative prices in anticipation of the binding constraint, especially in hard-to-abate sectors and those with market linkages to them (Chapter 5). Moreover, some sector differentiation in cost-effective emissions pricing between easy- and hard-to-decarbonize sectors can help manage adverse distributional impacts of emissions pricing, especially on households.

In practice, the heterodox policy combination and sequencing has been widely applied in transforming the electric power and automobiles sectors (Meckling et al., 2015 and 2017). A majority of countries that have taken policy actions in these sectors led with market-creating and industry-supporting policies. A minority led with emissions pricing. Most countries that led in advancing low-carbon technologies—China, Germany, Japan, South Korea and the United States—had at most modest emissions-pricing schemes in the 2000s and 2010s (Chapters 5 and 8). Moreover, those that achieved high emissions-pricing levels and wide sector coverage have tended to benefit from significant renewable resources (hydro) and low-carbon technologies (nuclear) legacies and had high transport fuel taxes motivated by energy security concerns (Chapter 5). Among these countries, only Denmark invested heavily in developing and deploying a variable renewable technology—wind turbines.

The countries and regions that have made the greatest progress in transforming their electric power and automobile sectors led primarily with government industrial policies, which have two self-reinforcing mechanisms that help sustain system transformations (Meckling et al., 2017). One is the positive feedback from scale economies—both internal and external to firms—as demand for low-carbon technologies grow and they gain increasing shares of energy systems. Examples include declining low-carbon technologies costs from mass manufacturing, such as with solar photovoltaic units and lithium-ion battery packs, and network effects from shared infrastructure such as electric vehicle charging facilities. Low-carbon technologies also benefit from new cost complementarities among them as they integrate into energy systems, as with variable renewables and energy storage in electric power systems. A second is that institutional and infrastructural changes to support adoption of low-carbon technologies tend to lock them in over time. Adapting electricity markets, building thermal efficiencies and energy infrastructures to low-carbon technologies are examples. In contrast, leading with emissions prices appears to run the risk of getting stuck at low levels of policy ambition for want of sufficiently timely self-reinforcing mechanisms—except where countries already benefited from favourable renewable resource endowments and low-carbon-technology legacies.

It is important to emphasize, moreover, that the observed policy sequencing benefited from initial spurs to energy innovations unrelated to climate change—a sequence of energy prices shocks and energy security issues arising from nuclear accidents (Chapter 4). As a result, alternative technologies were already in development in innovation systems in the United States, Europe and Japan as governments increasingly prioritized climate actions—by happenstance more than energy reform strategy. Looking ahead, time-bound commitments to net zero emissions to the extent credible could impose a similar 'shadow emissions price shock' in directing private investments towards innovations and market creation for low-carbon alternatives (Chapter 5). But credible energy reform strategies are necessary to strengthen such price signals and market-based expectations.

In addition to effective policy sequencing, their combinations are also important to manage policy costs because of cost complementarities and spillovers to the extent that they are not adequately internalized by markets and market-based relationships among firms. For example, the early focus on energy efficiency was important in its own right because for some technologies such as buildings, appliances and automobiles this technical performance characteristic is at least partly neglected by customers (Chapter 7). Cost-effective energy efficiency policies also helped counter overall increases in energy bills of firms and households from higher electricity policy costs (Committee on Climate Change, 2017, pp. 33–49). Material efficiency policies hold similar potential to complement emissions pricing emissions in heavy industries, or comparable policy measures. In general, cost-effective efficiency measures counter impacts of other energy reforms that increase unit energy and material costs to keep down overall energy and material service costs to households and firms.

An example of potential technology spillovers across sectors arises from advancing low-carbon-technology alternatives in road transport, for which the unit energy cost of fossil fuels is relatively high. The real resource costs of early deployment of low-carbon alternatives such as lithium-ion battery or hydrogen-fuel-cell electric vehicles are lowered by relatively high transport fuel costs, helping to manage their related policy costs. For such general-purpose technologies, cost declines from increasing scale and experience-based learning in transport facilitates their deployment in stationary applications, such as battery storage for system-balancing and ancillary services in electric power. In stationary applications, alternative low-carbon technologies displace thermal fossil fuels such as coal and natural gas, which have relatively low unit energy costs.

CREDIBILITY OF REFORM STRATEGIES

While comprehensive and coherent policy strategies are necessary to address material market imperfections and advance energy-system transformations, even well-crafted strategies remain subject to substantial uncertainties in their development and implementation. This lack of policy predictability and credibility increases uncertainties that firms and households face in their long-run investment decisions, including those for developing and creating markets for low-carbon alternatives. Policy risks shorten investment horizons and increase the costs of transforming energy systems (Bosetti and Victor, 2011; Fæhn and Isaksen, 2016). Policy frameworks that promote greater policy predictability reduce this source of uncertainty for investments and thus advance energy transformations at lower cost.

There are several potential sources of such policy unpredictability (Brunner et al., 2011). One is that governments and firms interact strategically. For example, private firms can make investments in low-carbon technologies in anticipation of future government policies, but which are not fully forthcoming once investments are made (Kydland and Prescott, 1977). This is a particular concern among energy utilities with large sunk and immovable investments (Blackmon and Zeckhauser, 1992). Another concern is maintenance of emission-intensive activities by firms and households when governments retain discretion over future policies, potentially raising future switching costs unless countered by declining low-carbon-technology costs. A second source is the high uncertainties around estimates of the social benefits and costs of climate actions (Chapter 5). Advances in climate science and alternative technologies improve over time, shifting projections of the social benefits and costs of climate actions and in turn potentially policies. A third is shifting societal values and political volatility, such as growing antipathy towards science and rising economic nationalism, which can stand in the way of domestic and internationally coordinated policy actions to limit climate change (Chapter 8).

Drawing on approaches to promote credibility in other policy areas, several framework elements are important: (1) clear goals, reform strategies and implementation; (2) transparency and reputation; (3) mitigation of distributional impacts; and (4) policy robustness (Nemet et al., 2017). Clear climate goals, reform strategies and implementation help firms and households to form expectations of future policy developments. For example, the UK government has a well-developed institutional framework for goal-driven climate-policy development and experience with policy-implementation rules. Transparency and reputation involve clear and systematic reporting on progress towards achieving climate goals along with policy development and implementation. Independent climate change committees or councils in several countries, for example, perform this role. Providing transparency can promote policy self-discipline and reputations for policy predictability. Addressing distributional impacts of energy reforms can also enhance reform credibility. In addition to galvanizing economic interests in energy reforms, low-carbon technologies and renewable resources, compensating those who face costs from structural changes can avoid their efforts to weaken or block policies. For example, German government support for coal-mine workers and communities facilitated plans to phase-out coal-fired electricity generation. Lastly, policy robustness—resiliency to reversals—can be strengthened in several ways. Ensuring that all relevant market imperfections and institutional reforms are adequately addressed is essential. Within countries, recognizing the diversity of interests in energy system transformations and tapping those at subnational levels, such as in cities, can help advance reforms in their early stages. In addition, international coordination of reforms is necessary for their overall effectiveness and resilience. Each of these aspects to promoting reform credibility are considered in turn.

Clear Climate Goals, Reform Strategies and Implementation

The Paris Agreement committed 185 countries to achieving net zero emissions of greenhouse gas emissions in the second half of this century and by the end of 2020 six countries had adopted legally binding targets to achieve net zero emissions by 2050 or sooner. These countries—Denmark, France, Hungary, New Zealand, Sweden and the United Kingdom—expressed their climate-policy goals in terms of a time-bound commitment to stop net emissions of greenhouse gases. In addition, the European Union (EU) and five countries (Canada, Chile, Fiji, South Korea and Spain) had proposed legislation to achieve net zero emissions by mid-century, while the governments of Japan and China announced such goals by 2050 and 2060, respectively. A number of other countries include net zero emission goals in their NDCs as part of the Paris Agreement pledge-and-review process, and more are anticipated by the 2021 United Nations Framework Convention on Climate Change review point. These commitments and goals are expressed in terms of the variable that countries must control and cap—their cumulative emissions of carbon dioxide and other greenhouse gases—to limit their contributions to climate change (Chapter 3).

The UK Climate Change Act of 2008 and its application over time is a well-developed policy framework for goal-driven, economy-wide emissions reductions. Under this Act, the independent Committee on Climate Change advises on and the government adopts in law five-yearly carbon budgets on a rolling basis at least 12 years in advance (for example, Committee on Climate Change, 2015; the Carbon Budget Order 2016). Accompanying these carbon budgets are government policy strategies that set out how it intends to meet the carbon budget through its policies and programmes (HM Government, 2017). The carbon budgets framework, including independent advice of the Committee on Climate Change, helps shape the design of domestic reform strategies aimed at delivering substantial emissions reductions (Averchenkova et al., 2018). In 2019–20, however, the government's strategy was only partly on track to deliver the reductions required to meet the carbon budgets in the late 2020s and early 2030s (Committee on Climate Change, 2019, pp. 55–78). In some sectors there are substantial policy-driven emissions cuts, such as reductions from variable renewables generation and coal to natural gas switching in power generation as well as renewable heating in buildings. There are also structural changes in heavy industry that contribute to emissions reductions that appear unrelated to climate-policy actions. But in other sectors, government policies fall short of those required to achieve the interim budgets and long-run emissions-reduction goals.

While the policy-driven reductions reflect UK domestic policy implementation, two were implemented in the context of the 2009 EU Renewables Directive: renewables generation of electric power and renewable heating. This Directive, agreed by each EU member country, including then the United Kingdom, obligated it to increase the share of renewables in total energy consumption to a target level by 2020, a legal obligation subject to EU enforcement mechanisms. As a commitment device, albeit imperfect, this directive is associated with a substantial overall increase in the share of renewables in EU electric power generation and was a significant consideration in UK policy development. But about one-third of EU member states are falling short of their obligations, France and the Netherlands in particular. However, they are making significant if belated progress towards them, while benefiting from upfront investments in cost reductions by others.[2]

The other source of significant UK emissions reduction reflects partial operation of a domestic policy rule. The UK government introduced a carbon-price
floor for the electric power sector in 2013, following the collapse of the EU
Emissions Trading System allowance price in the aftermath of the 2008–09
global recession and subsequent European sovereign debt stresses. This
carbon-price floor added a price-support payment to HM Treasury on top
of the EU allowance price to raise the actual and expected future emissions
price in the sector. The target for the effective carbon price for electricity was
£30/tCO_2 in 2020 and £70/tCO_2 in 2030 (HM Treasury, 2011). These target
emissions prices reflected the projected marginal cost of cutting emissions in
the UK economy in line with both interim carbon budgets and the then 2050
goal of reducing emissions by 80 per cent from its 1990 baseline set under the
2008 Climate Change Act. HM Treasury escalated the carbon-price support
in line with the announced policy for three years, and then paused further
increases with growing resistance from electricity-intensive industries over
competitiveness losses and concerns about adverse distributional impacts on
household budgets (Hirst, 2018). While falling short of the original policy
goal of reflecting the cost-effective emissions price implied by UK carbon
budgets, the level of sector-based emissions pricing achieved was nevertheless
sufficient to encourage power generators to switch to natural gas from coal as
their thermal fuel.

While the carbon-price support incentivized significant switching of thermal
generation of electricity to natural gas from coal, it failed to inspire the confidence of investors in renewable generation projects in its rising long-run
trajectory to 2030 (Grubb and Newbery, 2018). Rather, these investments
benefited from contracts-for-difference feed-in tariffs introduced by a complementary Electricity Market Reform in 2013. These contracts for investments in
new low-carbon generation ensured a fixed wholesale electricity price over the
contract length for low-carbon electricity generated. They mitigated electricity-
and carbon-price risks and reduced financing costs for these capital-intensive
projects. In this case, legally enforceable long-term investment contracts were
the government commitment device that mobilized substantial investments
in low-carbon generation rather than the promised rise in the emissions
price, over which the government retained and exercised discretion. The
carbon-price-floor policy rule, shaped primarily by economy-wide efficiency
considerations, overlooked its distributional impacts on firms' competitiveness
and household budgets.

The UK's experience with economy-wide climate goals, reform strategies
and implementation rules illustrates well their strengths and weakness. They
help focus policy debates on long-run goals and promote reform strategies
in line with them. But the approach tends to rely on institutional commitment devices more than fostering domestic interests in change. Their formal

enforcement mechanism—domestic judicial reviews of government policies—has yet to be invoked even though policies are not yet in line with the carbon budgets. Other commitment devices for UK renewables supports, the prospect of EU infringement proceedings in developing and implementing renewables policies and legal enforcement of low-carbon investment contracts in policy implementation played more significant roles.

Attaining long-run climate goals, however, ultimately rests on broader domestic interests in and acceptance of low-carbon alternatives to sustain institutional frameworks, laws and policy strategies over time. For example, a key UK policy strategy decision was to direct its support for variable renewables towards offshore wind generation, a renewable resource with which it is abundantly endowed. This low-carbon technology also taps UK specializations in innovation, complex manufactured goods and offshore infrastructure. With investment supports for early-stage deployment of offshore wind, this strategic decision proved successful in substantially cutting technology costs and expanding the country's accessible renewable resource base. The policy continues, extending well beyond the EU Renewables Directive period (HM Government, 2019, pp. 4–8).

Transparency and Reputation

Transparency and reputation for consistent policy implementation over time are further means to impart policy credibility. For example, many central banks follow this approach (Barro and Gordon, 1983; Geraats, 2014). Similarly, independent monitoring agencies for government-designed and government-regulated energy markets provide transparency on their functioning with the aim of promoting more efficient market arrangements and detecting potential abuses of market power (Léautier, 2019, pp. 101–2). In the context of climate goals and energy reforms, transparency means governments providing the public with comprehensive information about their strategies, assessments and policy decisions as well as their implementation in an open, clear and timely manner. Transparency helps firms and households understand policy goals and implementation, which in turn can make them more credible and effective. A strong commitment to transparency can also impose self-discipline on policymakers and promote policy consistency over time. However, the effectiveness of transparency depends on the quality of information being disseminated, its inherent uncertainties and broad political support for the institutional arrangements.

Several countries with long-run climate goals or committed targets also have related advisory councils to provide independent expert advice on policy and its implementation that is open and transparent. Such arrangements exist in Canada, Denmark, Finland, France, Germany (environment), Ireland and the United Kingdom (Weaver et al., 2019). Yet other countries have within their governments advisory councils on climate change and related policies. The independent expert councils typically have a legal duty to advise the government on short- and long-run reform strategies and implementation plans. All of these councils have transparent working practices, consulting with the general public and policy makers, producing public reports and engaging with the media on their assessments. Their expert and transparent assessments help provide a basis for political consensus and wider societal support for policies to implement climate goals (Averchenkova et al., 2018). However, such independent bodies have limitations. They can help forge consensus where political divides are not wide. But where they have been, as in the United States and Australia, independent and expert councils were created and abolished amidst shifts in political power and wide swings in approach.

In addition, public expert assessments are not a substitute for firms and households making their own assessments of government policies, technology developments and behavioural changes that affect their investment decisions. This is especially so when public information has substantial inherent uncertainties, and there is value in multiple assessments to improve the overall precision of available information (Morris and Shin, 2002). Rigorous assessments of energy transformations—both public and private—help advance them cost-effectively. Financial investors too must make their own assessments in the context of their long-run financing decisions, supported by climate-related public information disclosures by firms (Task Force on Climate Related Financial Disclosures, 2017).

By late 2020, more than 1500 major businesses had set net zero emission goals, accounting for more than \$11 trillion in revenues and 19 million employees (Data-Driven EnviroLab and NewClimate Institute, 2020). Of these businesses, more than 1000 announced these ambitions in the past year. They take several forms, including specific targets and pledges to monitor and report on individual business performance. Others are more general, such as pledges to develop further targets and action plans. Sectors with the greatest participation as measured by the revenue of firms making these commitments are in the discretionary consumer goods, financial services, and information and computing technology sectors—relatively electricity-intensive sectors that are now 'easy to decarbonize'. The sector with the least participation is real estate and buildings. Many businesses setting net zero emission goals prioritize sustainability in the firms' identity and differentiate their products and services in terms of this characteristic, seeking to attract customers that value it. This product-market differentiation provides some economic underpinning to these firm-level ambitions. However, they also benefit substantially from now relatively low-cost abatement opportunities through renewable electric power enabled by past policy actions, as well as some low-cost bioresources and nature-based carbon dioxide removals from the atmosphere.

Fair and Just Transformations

As with any deep structural change of economies, the distributional impacts of transforming energy systems are significant, but manageable in principle. Consider, for example, potential impacts of an economy-wide emissions tax on household income groups, a policy approach amenable to extensive research but not implemented in practice. Such an emissions tax would affect the prices of goods and services purchased by households and firms as well as sources of household income: wages, capital income and any associated changes to government taxes and transfers. Assessments of the distributional impacts of such structural measures must take account of not only short-run impacts but also long-run economic responses to emissions pricing and changing technology costs. They include sectoral changes that could have significant and enduring costs that put adjustments out of reach for some households and communities.

The impacts of economy-wide emissions pricing on the prices of the goods and services that households consume would clearly be regressive. For example, an economy-wide emissions tax would significantly raise the price that consumers would face for transport fuels, electricity, thermal fuels for space heating (fuel oil and natural gas), travel services and to a lesser extent food. The household expenditure shares of relatively emissions-intensive goods and services are higher for low-income than high-income households. These potentially regressive distributional impacts are found in analyses of household expenditures in France, the United Kingdom and the United States (Rausch et al. 2011; Berry, 2019; Goulder et al., 2019; Burke et al., 2020). Similar regressive impacts are found for static policy costs of some regulations, including renewable portfolio standards for the electricity sector and energy-efficiency standards for appliances. These standards increase prices of electricity and electrical appliances, which weigh more heavily on low-income than high-income households (Rausch and Mowers, 2014; Allcott and Greenstone, 2017; McCoy and Kotsch, 2020). However, better policy design and dynamic efficiency gains could ease them.

One way to manage such regressive distributional impacts is through well-designed and sequenced industrial policies to advance low-carbon alternatives and emissions pricing consistent with a time-bound net zero emissions constraint to support their deployment at scale. The anchor for such emissions pricing is the expected long-run marginal cost of abatement consistent with the climate goal—either economy-wide or sector-specific. Discounting this long-run marginal cost to its present value using either a private or social discount rate would be one way to bring forward in time the shadow price of the emissions constraint. There is scope also to differentiate such emissions pricing between easy- and hard-to-decarbonize sectors to manage their distribution impacts in cost-effective emissions pricing. Given the significance of energy services demands for thermal comfort in dwellings, personal transport and other activities in household budgets, differentiated emissions pricing for these easy-to-transform sectors would ease some adverse distributional impacts of emissions pricing with few efficiency losses.

An alternative approach to managing distribution impacts of emissions pricing is recycling its revenues to households. In this approach, modelled equity and efficiency impacts appear best managed by reducing inefficiencies in existing tax systems (Klenert et al., 2018a and 2018b). Models of government taxation that capture its impacts on both economic efficiency and income distribution show that if the initial tax system is not economically optimal (none are), using proceeds from an emissions tax to reduce inefficiencies in existing taxes takes precedence. For example, a cut in labour taxes (payroll and income) can enhance both equity and efficiency. But for other types of changes, there are trade-offs between the two. Cuts in corporate taxation in modelled tax systems tend to reduce economic inefficiencies but increase inequality. Conversely, targeted transfers to households are the most equitable but least efficient. Uniform lump-sum recycling outperforms labour tax cuts only when the initial tax system is close to optimal (Klenert et al., 2018b).

The distributional impacts of economy-wide emissions pricing on industries and firms would broadly mirror those on households. Sectors exposed to higher costs, less demand and fewer profits as the economy adjusts to emissions pricing are those that are relatively emission-intensive in their production and product end-use (Goulder and Hafstead, 2019, pp. 182–200). They include thermal generation of electricity using fossil fuels; production, refining and distribution of fossil fuels; heavy industrial production—steel, cement and chemicals; and transport services. A key distributional issue for domestic producers would be with emission-intensive industries that produce internationally traded goods—primarily transport fuels, metals, chemicals, plastics and to a lesser extent cement. International aviation and shipping services are also affected by these policy coordination issues. A lack of international policy coordination in these sectors can create differences in emissions costs across countries and unlevel playing fields for internationally traded, emission-intensive goods and services. Carbon border adjustments for emission-intensive imports that do not face significant domestic emission prices are one way to address this issue (Morris, 2018; Evans et al., 2021). While difficult to design and implement, such adjustments could encourage wider adoption of emissions pricing across countries while levelling the international competitive playing field for tradable industrial goods and services (Helm et al., 2012; Nordhaus, 2015). Other energy-intensive sectors provide largely non-traded energy and energy services—electric power, domestic transport services and activities in buildings.

Looking ahead, there are two distributional impacts that would require particular focus beyond better design of existing policies to help ensure fair and just energy-system transformations (Chapter 8). This attention is necessary because of limited capacities of some households, firms and communities to adapt to these structural changes. One is that many coal-mining communities would be highly exposed to energy transformations with significant potential for 'stranded' employees and regions. Past examples of coal industry declines show that they can be associated with large and persistent economic and social dislocations. A second is the risk that poorer households in poor-quality housing become locked into fossil-fuel heating with diminishing capacity to adapt. There is a risk that low-income households in poor-quality private dwellings become locked into fossil-fuel heating as it becomes increasingly costly and these costs become capitalized in property values. Their capacity to adapt could thus become increasingly constrained.

Robustness

An important source of domestic policy robustness is ensuring that reform strategies and policies are sufficiently comprehensive and coherent to adequately address important market failures that could impede realization of climate goals. There are of course significant uncertainties around reform contexts and strategies, so reforms inevitably involve managing risks and trade-offs. For example, the orthodox reform strategy avoids the risks of potential market inefficiencies and government policy failures associated with heterodox industrial policies. But this narrow approach potentially glosses over key market imperfections and potential time-inconsistencies in government policies that could hold back low-carbon-technology advances. This approach to conserving the role of markets and individual choice in the short run thus holds longer-run risks of system transformations becoming stuck. A heterodox strategy that sequences industrial policies and emissions pricing perhaps introduces some short-run market inefficiencies and risks of government policy failures. But it holds the potential benefit of added robustness to domestic energy-reform strategies, including by ensuring that material market imperfections that could hold back advances of low-carbon alternatives are addressed.

In addition, subnational initiatives are important complements to national reform strategies. For example, by late 2020, more than 800 cities and 100 regions set net zero emission goals—many of them over the preceding year (Data-Driven EnviroLab and NewClimate Institute, 2020). These subnational ambitions benefit from within-country variation in political sentiments towards climate action as well as co-benefits from taking such actions, including improved urban air quality. Some low-carbon-technology alternatives, such as battery electric vehicles, also have performance characteristics more suited to urban than rural road travel, at least at their current stage of development. Moreover, some energy-related reforms are devolved to local and regional governments. They include adaptations to transport infrastructure, such as public transport networks, cycle lanes and pedestrian pathways, as well as for buildings' energy-efficiency measures and adaptations of heating systems and energy infrastructure to low-carbon alternatives.

Another important source of national energy-reform robustness is policy coordination across countries. Government market-creating and industry-supporting policies in the automobile sector provide one example. Ten countries, many with significant domestic markets and manufacturing industries, announced in quick succession future internal combustion engine bans in 2016–17 (Chapter 8). These bans were announced as cost uncertainties around alternative drivetrain technologies began to resolve. Another, earlier example of such tacit coordination was the introduction of market-creating policies for renewable generation technologies in the economic recovery from the 2008–09 financial crisis and global recession (Chapter 4). In this case, the EU (2009 Renewables Directive), the United States (American Recovery and Reinvestment Act of 2009) and China's Twelfth Five-Year Plan (2011–15) directed significant domestic resources to support the early deployment of variable renewables. In both cases of tacitly coordinated industrial policies, the increasingly evident cost reductions and scale economies of these low-carbon technologies helped create not only self-reinforcing mechanisms for domestic energy-reform strategies, but also a basis for their coordination across countries. Much more extensive coordination and cooperation among countries, though, would be necessary to achieve the Paris Agreement goals.

COORDINATING ENERGY REFORMS ACROSS COUNTRIES

The Paris Agreement commits its signatory governments to achieving net zero emissions of greenhouse gases in the second half of this century but leaves open precisely how and by when this is achieved.[3] Country reporting of NDCs and their monitoring under the Paris Agreement rule book provides transparency on country goals, reform strategies and policy implementation, with peer reviews of country climate ambitions and actions. The robustness of the Paris Agreement relies on both domestic reform strategies and implementation as well as their international coordination—a two-level game between domestic interests and international actions (Putnam, 1988; Barrett, 2003, pp. 218–9). International actions are necessary because almost all countries would need to transform their energy systems to achieve net zero emissions from energy systems. Also, countries have comparative advantages and specializations in innovations and manufacturing low-carbon technologies, so acting alone on climate change would be costlier than coordinated actions, including through markets.

International climate action is typically framed as providing a safe climate—a global public good of benefit to all, albeit to differing degrees because climate change impacts vary across countries. As with all public goods, any one person or country benefiting from it does not exclude or prevent others from doing so. But because actions to limit climate change are costly, a safe climate suffers from global under-provision of climate actions—a tragedy of the commons. Governments, firms and households have an incentive to free-ride on the actions of others (Ostrom, 1990, 30–33). Moreover, even if a country has an incentive to act unilaterally—for example, from strong domestic support for climate leadership—few countries could achieve net zero emissions from energy through domestic preferences, capabilities and resources alone. Because of such differences in societal values as well as economic costs and benefits of climate actions, many such actions are not set simply in the context of global collective actions and potential free-riding problems (Ostrom, 2009).

There are in particular three aspects to climate actions that depart from the classic collective-action problem (T. Hale, 2018). One is joint goods to the extent that climate actions towards the global public good also yield private benefits to those who act. For example, many climate actions give rise to co-benefits such as improving local air quality, strengthening domestic energy security and developing new sources of economic growth and comparative advantages (Chapter 8).

A second is heterogenous preferences to the extent that the costs and benefits of actions towards the common good vary across governments, firms and households. Some customers, for example, may value a premium battery electric vehicle or electricity from renewables more than others. Popular political sentiments in some countries, and at certain times, may tilt towards global environmental leadership and in other contexts towards the 'national' interest and antipathy towards global cooperation.

A third is increasing returns whereby past climate actions reduce the costs and increase the benefits of future actions. These increasing returns arise from experience-based learning from the development and early-stage deployment of low-carbon technologies, scale economies, and integration of complementary low-carbon technologies in energy systems (Chapters 4, 6 and 7). These aspects to climate actions increase their benefits and reduce their costs, with upfront investments in innovations and market creation for low-carbon technologies providing quasi-public goods.

These three aspects of climate actions can make their coordination or more formal government cooperation on them more likely (Keohane and Victor, 2016). Coordination of climate actions arises from a shared understanding of the benefits and costs of actions which become self-reinforcing. Such shared understandings—or focal points—in transforming energy systems are the long-run role of renewable resources in displacing fossil primary-energy resources and the growing role of electric power, including through electrification of surface transport and most activities in buildings and manufacturing. Specific examples of tacit government coordination of climate actions in these sectors by China, EU member states and the United Kingdom, Japan, South Korea and the United States are R&D supports for innovations and early-stage deployment supports for variable renewables and electric vehicles. These countries have specializations in innovations and manufacturing, and the costs of these climate actions were lower for them given these capabilities than for other countries. In addition, they could reasonably have expected to gain co-benefits from new sources of growth and comparative advantages in manufacturing tradable low-carbon technologies that require at most moderate emissions pricing to sustain them in the long run. In these sectors, tacit coordination of market-creating policies among these countries can become increasingly self-reinforcing over time (Meckling et al., 2017; T. Hale, 2018). Increasingly, international competition in producing tradable low-carbon technologies that are competitive with incumbent ones is widening diffusion of these new technologies.

A challenge ahead is to extend coordinated climate actions to advance more low-carbon technologies across more sectors. Low-carbon technologies with product characteristics that lend themselves to coordinated actions are those with wide potential applications across sectors and countries, and amenable to mass manufacturing. These product characteristics allow varied interests to participate in coordinated actions and become self-reinforcing through scale economies and cost complementarities in system integration. Other technologies with similar potentials for mass manufacturing and widespread deployments include hydrogen fuel cells and electrolysers for low-carbon hydrogen production and use. They could deploy initially in road freight transport and rail as alternatives to diesel engines where unit energy costs of transport fuels are relatively high. They could also play a role in decarbonizing production processes in heavy industries; for example, through end-products that use the low-carbon materials and are sufficiently differentiated in product markets, such as high-quality, specialist steel. Such initiatives could centre initially on countries that specialize in innovations related to fuel cells and electrolysers, truck manufacturing and some value chains that include heavy industries.

Not all low-carbon technologies, however, appear amenable to such approaches. For example, low-carbon technologies for steel, cement, chemicals and plastics are largely sector-specific and implemented through large-scale investment projects rather than mass manufacturing. There are fewer shared bases on which to coordinate actions and less scope for them to become self-reinforcing because of their expected relatively high costs compared with market incumbents. Another approach is likely required in heavy industry sectors and their internationally traded goods. Emissions pricing with carbon border adjustments are feasible and efficient policies in countries initiating technological disruptions, and they impart incentives for following countries to price industrial emissions as well. However, such border adjustments are difficult to design and implement in practice.

In yet other sectors, transformation of energy systems would likely require formal international cooperation, such as international aviation and shipping. The elimination of emissions from these cross-border activities through relatively expensive low-carbon fuels, such as sustainable aviation fuel, hydrogen and ammonia, would require international agreement on emissions pricing or standards and their enforcement. Countries and firms taking a lead in demonstrating low-carbon alternatives could help reduce uncertainties around costs and benefits of low-carbon fuels and alternative prime movers, laying foundations for broader agreement. But formal international cooperation on eliminating net emissions from these cross-border sectors would likely require a few major countries providing leadership on the issue, potentially those with aviation manufacturing and transport service industries.

In contrast to tacit coordination, formal cooperation among governments aims to promote mutual policy adjustments by conditioning the actions of any one government on those of others through negotiation (Keohane, 1984, pp. 51–5). Formal cooperation to provide a global public good is more likely when a dominant country or small group of countries takes the lead, not only incurring a disproportionate share of the costs but also realizing much of the benefits (Olson, 1971, pp. 33–6; Keohane, 1984, pp. 177–81). The 1997 Kyoto Protocol is an example of such a formal agreement on climate actions. The protocol was calibrated to encourage low-cost emissions reductions such as those from energy-efficiency gains, facilitating cooperation on a goal but not promoting deep emissions reductions. It also offered developing countries a low-carbon development mechanism to help fund more sustainable development, which was to draw resources primarily from emissions-allowance trading systems. However, it suffered from higher-than-expected compliance costs, adverse domestic distributional impacts, contentious international transfers and little impact on the global public good because of rapid growth in developing countries' emissions (Victor, 2001; Schelling, 2002). The partial unravelling of the Kyoto Protocol and failure to reach agreement on its successor at the 2009 Copenhagen Conference of UN Framework Convention on Climate Change reflected both by then wide dispersion of greenhouse gas emissions across countries and fragmentation of power in international affairs away from the United States (Victor, 2011, pp. 204–8).

The 2015 Paris Agreement pivoted away from top-down targets and timetables to NDCs in which governments pledge what national climate actions they intend, including those that can be reasonably expected of their firms and households. The strengths of this approach are that it reflects the diverse capabilities and interests across countries, includes not only advanced industrialized countries but also developing countries, and does not necessarily require a dominant geopolitical actor to forge agreements. But how these NDCs are assessed remains remarkably similar to the targets and timetables of the Kyoto Protocol. They are scored primarily in terms of their emissions-reduction pathways over the next decade (Victor et al., 2017). There is a logic to this approach, since limiting cumulative emissions of carbon dioxide is necessary to limit climate change, but it underplays those climate actions that reduce costs and increase benefits of future actions.

The tacit coordination of country actions in advancing low-carbon technologies where possible, however, is not solely dependent on the formal NDC review mechanism. Coordinated actions arise from shared understandings or focal points in transforming energy systems that become self-reinforcing. There is potential for two such self-reinforcing mechanisms—scale economies in easy-to-decarbonize sectors and emissions pricing with carbon border adjustments in hard-to-transform heavy industries. Easy-to-transform sectors are electricity, surface transport and activities in buildings—largely non-traded energy and energy services. Their reform requires little international coordination of sector policies apart from innovation and industrial policies to advance low-carbon alternatives for them. Important complements to these efforts are bilateral and multilateral technical cooperation and financing among countries to facilitate widespread diffusion of low-carbon technologies in developing countries as they become increasingly commercially viable (Chapter 8). Hard-to-decarbonize sectors include heavy industries that produce tradable materials and goods. These sectors would likely require emissions pricing to support their reform. This policy could either be implemented through formal cooperation on emissions pricing across countries, or countries that are initiating change complementing such pricing with carbon border adjustments to ensure a level domestic playing field. Such adjustments in initiating countries would provide following countries with an incentive to adopt emissions pricing as well.

The sectors that would likely require formal international cooperation to reform are international aviation and shipping, given the cross-border nature of the activities. International cooperation would also be necessary to agree rules for international cooperation mechanisms as envisaged in Article 6 of the Paris Agreement. Such rules would make it possible for emissions-reduction measures to be implemented in one country and the resulting emission reductions to be transferred to another and counted towards its NDC. This would require a transparent process and accurate accounting of the emissions reductions achieved to avoid such reductions being counted more than once. Such mechanisms could enable, for example, use of nature-based management of emissions through sustainable afforestation and reforestation measures and exchange of such emissions reductions across countries. They would help manage upward pressure on emissions-abatement costs, especially in hard-to-decarbonize sectors, and support the attainment of net zero emissions commitments and goals (Chapter 5).

ACCELERATING CHANGE

This book draws lessons from experiences in transforming energy systems over recent decades and uses them to explore how energy transformations could be accelerated and expanded in pursuit of the Paris Agreement goals. Time-bound commitments to achieve net zero emissions from energy are a potentially powerful spur to accelerating these structural changes. But this galvanizing effect requires comprehensive, coherent and credible domestic energy-reform strategies consistent with these climate goals. Comprehensive reforms are more expansive than the orthodox economic prescription of government R&D supports and emissions pricing based on the SC CO_2—but no more so than warranted by market imperfections and institutional requirements adequately assessed.

The heterodox energy-reform strategy advanced in this book rests on time-bound goals and commitments to achieve net zero emissions and include market-creating and industry-supporting policies for early deployment of low-carbon alternatives. Such goals and commitments create relatively stable and informative focal points for coordinated action by governments, firms and households to eliminate net carbon dioxide and other greenhouse gas emissions within and across countries. Market-creating policies for low-carbon alternatives, especially in countries with existing or aspiring specializations in innovation and manufacturing, help counter investment disincentives from missing markets for future emissions prices and potential arbitrary shifts and time-inconsistencies in implementation. They also target external scale economies and network effects in transforming energy systems. These policies thus address directly and decisively the goals of climate action, although their design needs to aim at minimizing policy costs and allocating them fairly.

Cost-effective emissions pricing consistent with climate goals brings forward in time the shadow costs of the binding emissions constraint and promotes economy-wide relative price shifts consistent with the constraint. This pricing effect is especially important in hard-to-decarbonize sectors and those with market linkages to them. Effective sequencing of market-creating policies and emissions pricing can enhance the coherence and credibility of energy reforms by fostering interests in low-carbon alternatives and creating tangible opportunities for firms and households to embrace them. Substantial differences in the shadow costs of the binding emissions constraint between easy- and hard-to-decarbonize sectors also create scope for some differentiation in emissions pricing to help manage its distributional impacts, especially on households, without significantly sacrificing efficiency of emissions abatement.

Adapting government-designed wholesale electricity markets and network regulations in liberalized systems to variable renewables and complementary flexibility technologies is also necessary. Reforms are necessary too in adapting planning-led approaches to electricity system development in traditionally organized systems especially in developing countries. Bilateral and multilateral technical cooperation and financing to these countries should prioritize electricity sector reforms and capacity building to foster the widespread deployment of increasingly cost-effective renewable generation technologies. Many developing countries are advantaged by these technological advances because of their abundant renewable resource endowments, and together with international partners they should act decisively to enable widespread deployment of renewable technologies. In countries with cooler climates and extensive natural gas networks, early prioritization of buildings' thermal-efficiency measures can help ease the lock-in of fossil-fuel boilers for heating in buildings and facilitate the coordination of investments necessary for their transformation to low-carbon alternatives.

In addition, energy- and material-efficiency measures are important in directing a focus on technology and product characteristics that customers have at least partly neglected and fostering investments in innovations to improve efficiencies. These measures lower energy and material service costs and form an economic response to higher expected costs of low-carbon materials. An early focus on cost-effective measures helps ease the policy-cost impacts of other energy reforms on households and firms. These measures also reduce investments in low-carbon electric power, fuels and industrial processes necessary for those energy and material services demanded by societies.

Comprehensive, coherent and credible reform strategies must be attuned to the political-economy aspects of reform. Countries that specialize in innovation and manufacturing low-carbon technologies are more likely than others to incur the costs of disruptive technological changes, such as investments in innovation and market creation, and capture some economic benefits of change. A fair and just energy-system transformation also requires supporting households and communities with limited capacities to adapt to the deep structural changes of energy systems. They include, for example, coal-mining communities and low-income households living in thermally inefficient dwellings.

But the success of comprehensive, credible and coherent domestic energy reforms ultimately depends on international coordination of climate actions. The benefits of a safer climate by transforming energy systems can only be realized if almost all firms and households in advanced industrialized countries and many in developing countries invest in change. Given country specializations in innovations and manufacturing low-carbon technologies, the feasibility and cost-effectiveness of any one country's efforts to achieve net zero emissions would depend on actions of other countries. The range of interests in low-carbon technologies, renewable resource endowments and their environmental benefits create much scope for coordination of country actions. For example, the characteristics of some such low-carbon technologies help them to become self-reinforcing. Widely deployable technologies in low-carbon energy systems that can be mass-manufactured hold this potential, as is increasingly seen in easy-to-decarbonize sectors. The technologies that unlock substantial economic value from renewable resource endowments, especially in developing countries, can add further impetus to energy reforms. Among countries, official bilateral and multilateral technical cooperation and financing can play a key role in facilitating widespread diffusion to developing countries of low-carbon technologies that are increasingly competitive with incumbent ones.

But other low-carbon technologies appear less amenable to such approaches, such as those for heavy industries and commercial transport. Other routes to international coordination would likely be required for these sectors, such as emissions pricing and carbon border adjustments in countries initiating technological disruptions in heavy industry. Still others—such as international aviation and shipping—would require formal cooperation among countries because of the cross-border nature of their activities. However, leading countries and firms can initiate advances in such low-carbon alternatives for these sectors, de-risking technologies and lowering costs of potential formal cooperation in these sectors. International cooperation is also necessary to agree rules on accounting for emissions reductions and transferring them across countries—implementing rules for Article 6 of the Paris Agreement— that could support emissions management through nature-based solutions. Agreement on such rules would help manage long-run marginal abatement costs in hard-to-decarbonize sectors and strengthen the credibility of stretching goals to achieve net zero emissions, including through international cooperation.

This book provides evidence and examples of actions to decisively accelerate change to low-carbon alternatives, reflecting lessons from experiences of transforming energy systems. But time is short in the timescales of systemic transformations and market-based change (Chapter 1). Now is the time for these decisive coordinated actions of governments, businesses and households to reach the climate goals that governments adopt for their societies.

NOTES

1. On the role of focal points in coordinating behaviours and choices, see Schelling (1960, pp. 54–8) and Mehta et al. (1984a, 1984b).
2. Eurostat (2020), Energy Data SHARES (Renewables). Retrieved from https://ec .europa.eu/eurostat/web/energy/data/shares.
3. A goal of the Paris Agreement is "a balance between anthropogenic emissions by sources and removals by sinks of greenhouse gases in the second half of this century, on the basis of equity, and in the context of sustainable development and efforts to eradicate poverty". The balancing of emissions and removals is akin to net zero emissions. Paris Agreement, Article 4, paragraph 1. Retrieved from https://unfccc.int/process-and-meetings/the-paris-agreement/the-paris-agreement.

References

Abaluck, J., and Gruber, J. (2011). Heterogeneity in choice inconsistencies among the elderly: Evidence from prescription drug plan choice. *American Economic Review*, 101(3), 377–81. https://doi.org/10.1257/aer.101.3.377.

Abel, G.J., et al. (2019). Climate, conflict and forced migration. *Global Environmental Change*, 54, 239–49. https://doi.org/10.1016/j.gloenvcha.2018.12.003.

Acemoglu, D. (2002). Directed technical change. *Review of Economic Studies*, 69(4), 781–809. https://doi.org/10.1111/1467-937X.00226.

Acemoglu, D., and Robinson, J.A. (2012). *Why Nations Fail: The Origins of Power, Prosperity and Poverty*. New York, NY: Crown Publishers.

Acemoglu, D., et al. (2012). The environment and directed technical change. *American Economic Review*, 102(1), 131–66. https://doi.org/10.1257/aer.102.1.131.

Acemoglu, D., et al. (2016). Transition to clean technologies. *Journal of Political Economy*, 124(1), 52–104. https://doi.org/10.1086/684511.

ACER (2019a). *Annual Report on the Results of Monitoring the Internal Electricity and Natural Gas Markets in 2018: Electricity and Gas Retail Markets Volume*. Ljubljana: Agency for Cooperation of Energy Regulators. https://documents.acer.europa.eu/Official_documents/Publications.

ACER (2019b). *ACER Market Monitoring Report 2018: Gas Wholesale Market Volume*. Ljubljana: Agency for Cooperation of Energy Regulators. https://documents.acer.europa.eu/Official_documents/Publications.

Aghion, P., and Howitt, P. (1998). *Endogenous Growth Theory*. Cambridge, MA: MIT Press.

Aghion, P., et al. (2015). Industrial policy and competition. *American Economic Journal: Macroeconomics*, 7(4), 1–32. https://doi.org/10.1257/mac.20120103.

Aghion, P., et al. (2016). Carbon taxes, path dependency and directed technological change: Evidence from the automobile industry. *Journal of Political Economy*, 124(1), 1–51. https://doi.org/10.1086/684581.

Akerlof, G.A. (2020). Sins of omission and the practice of economics. *Journal of Economic Literature*, 8(2), 404–18. https://doi.org/10.1257/jel.20191573.

Aklin, M., and Urpelainen, J. (2018). *Renewables: The Politics of a Global Energy Transition*. Cambridge, MA: MIT Press.

Alberini, A., Gans, W., and Towe, C. (2016). Free riding, upsizing, and energy efficiency incentives in Maryland homes. *Energy Journal*, 37(1), 259–90. https://doi.org/10.5547/01956574.37.1.aalb.

Alberini, A., and Towe, C. (2015). Information v. energy efficiency incentives: Evidence from residential electricity consumption in Maryland. *Energy Economics*, 52(S1), S30–40. https://doi.org/10.1016/j.eneco.2015.08.013.

Allcott, H. (2011). Consumers' perceptions and misperceptions of energy costs. *American Economic Review*, 101(3), 98–104. https://doi.org/10.1257/aer.101.3.98.

Allcott, H. (2013). The welfare costs of misperceived product costs: Data and calibrations from the automobile market. *American Economic Journal: Economic Policy*, 5(3), 30–66. https://doi.org/10.1257/pol.5.3.30.

Allcott, H., and Greenstone, M. (2012). Is there an energy efficiency gap? *Journal of Economic Perspectives*, 26(1), 3–28. https://doi.org/10.1257/jep.26.1.3.

Allcott, H., and Greenstone, M. (2017). Measuring the welfare effects of residential energy efficiency programs (NBER Working Paper No 23386). Cambridge, MA: National Bureau of Economic Research. www.nber.org/papers/w23386.

Allcott, H., and Knittel, C. (2019). Are consumers poorly informed about fuel economy? Evidence from two experiments. *American Economic Journal: Economic Policy*, 11(1), 1–37. https://doi.org/10.1257/pol.20170019.

Allcott, H., and Rogers, T. (2014). The short-run and long-run effects of behavioral interventions: Experimental evidence from energy conservation. *American Economic Review*, 104(10), 3003–37. https://doi.org/10.1257/aer.104.10.3003.

Allcott, H., and Sweeney, R.L. (2017). The role of sales agents in information disclosure: Evidence from a field experiment. *Management Science*, 63(1), 21–39. https://doi.org/10.1287/mnsc.2015.2327.

Allcott, H., and Taubinsky, D. (2015). Evaluating behaviorally motivated policy: Experimental evidence from the lightbulb market. *American Economic Review*, 105(8), 2501–38. https://doi.org/10.1257/aer.20131564.

Allcott, H., and Wozny, N. (2014). Gasoline prices, fuel economy and the energy paradox. *Review of Economic and Statistics*, 96(5), 779–95. https://doi.org/10.1162/REST_a_00419.

Allen, M.R., et al. (2009). Warming caused by cumulative carbon emissions towards the trillionth tonne. *Nature*, 458, 1163–6. https://doi.org/10.1038/nature08019.

Allen, M.R., et al. (2018). Framing and context. In V. Masson-Delmotte et al. (Eds). *Global Warming of 1.5°C: An IPCC Special Report on the Impacts of Global Warming of 1.5°C above Pre-Industrial Levels and Related Global Greenhouse Gas Emission Pathways, in the Context of Strengthening the Global Response to the Threat of Climate Change, Sustainable Development, and Efforts to Eradicate Poverty* (pp. 49–91). www.ipcc.ch/site/assets/uploads/sites/2/2019/06/SR15_Full_Report_Low_Res.pdf.

Allen, R.C. (2009). *The British Industrial Revolution in Global Perspective*. Cambridge: Cambridge University Press.

Allwood, J.M., and Cullen, J.M. (2015). *Sustainable Materials Without the Hot Air*. Cambridge: UIT.

Allwood, J.M., et al. (2011). Material efficiency: A white paper. *Resources, Conservation and Recycling*, 55(3), 362–81. https://doi.org/10.1016/j.resconrec.2010.11.002.

Alvaredo, F., et al. (2018). *World Inequality Report 2018*. Paris: World Inequality Lab, Paris School of Economics. https://wir2018.wid.world.

Arapostathis, S., et al. (2013). Governing transitions: Cases and insights from two periods in the history of the UK gas industry. *Energy Policy*, 52, 25–44. https://doi.org/https://doi.org/10.1016/j.enpol.2012.08.016.

Archer, D. (2010). *The Global Carbon Cycle*. Princeton, NJ: Princeton University Press.

Aroonruengsawat, A., Auffhammer, M., and Sanstad, A.H. (2012). The impact of state level building codes on residential electricity consumption. *Energy Journal*, 33(1), 31–52. https://doi.org/10.5547/ISSN0195-6574-EJ-Vol33-No1-2.

Arrow, K.J. (1962a). The economic implications of learning by doing. *Review of Economic Studies*, 29(3), 155–73. https://doi.org/10.2307/2295952.

Arrow, K.J. (1962b). Economic welfare and the allocation of resources for invention. In R. Nelson (Ed.). *The Rate and Direction of Inventive Activity* (pp. 609–26). Princeton, NJ: Princeton University Press.

Arthur, W.B. (1989). Competing technologies, increasing returns, and lock-in by historical events. *Economic Journal*, 99(394), 116–31. https://doi.org/10.2307/2234208.

Arvizu, D., et al. (2011). Direct solar energy. In O. Edenhofer et al. (Eds). *IPCC Special Report: Renewable Energy Sources and Climate Change Mitigation* (pp. 333–400). Cambridge: Cambridge University Press.

Atalla, T., Gualdi, S., and Lanza, A. (2018). A global degree days database for energy-related applications. *Energy*, 143, 1048–55. https://doi.org/10.1016/j.energy.2017.10.134.

Audretsch, D.B., and Feldman, M.P. (2004). Knowledge spillovers and the geography of innovation. In J.V. Henderson and J.-F. Thisse (Eds). *Handbook of Regional and Urban Economics, Volume 4* (pp. 2713–39). Amsterdam: North-Holland. https://doi.org/10.1016/S1574-0080(04)80018-X.

Auffhammer, M. (2018). Quantifying economic damages from climate change. *Journal of Economic Perspectives*, 32(4). 33–52. https://doi.org/10.1257/jep.32.4.33.

Averchenkova, A., Fankhauser, S., and Finnegan, J. (2018). *The Role of Independent Bodies in Climate Governance: The UK's Committee on Climate Change*. London: Grantham Research Institute on Climate Change, London School of Economics. www.lse.ac.uk/granthaminstitute/publication/role-independent-bodies-in-climate-governance-uk-committee-on-climate-change.

Bailer, S. (2012). Strategies in climate negotiations: Do democracies negotiate differently? *Climate Policy*, 12(5), 543–51. https://doi.org/10.1080/14693062.2012.691224.

Ballasa, B. (1965). Trade liberalisation and 'revealed' comparative advantage. *Manchester School*, 33(2), 99–123. https://doi.org/10.1111/j.1467-9957.1965.tb00050.x.

Baptista, R. (2000). Do innovations diffuse faster within geographical clusters? *International Journal of Industrial Organization*, 18(3), 515–35. https://doi.org/10.1016/S0167-7187(99)00045-4.

Barber, J. (2009). Photosynthetic energy conversion: Natural and artificial. *Chemical Society Reviews*, 38(1), 185–96. https://doi.org/10.1039/B802262N.

Barrett, S. (2003). *Environment and Statecraft: The Strategy of Environmental Treaty-Making*. Oxford: Oxford University Press.

Barro, R.J., and Gordon, D.B. (1983). Rules, discretion and reputation in a model of monetary policy. *Journal of Monetary Economics*, 12(1), 102–21. https://doi.org/10.1016/0304-3932(83)90051-X.

Bartlett, J. (2019). Reducing risk in merchant wind and solar projects through financial hedges (RFF Working Paper 19-06). Washington, DC: Resources for the Future. https://media.rff.org/documents/WP_19-06_Bartlett.pdf.

Bättig, M.B., and Bernauer, T. (2009). National institutions and global public goods: Are democracies more cooperative in climate change policy? *International Organization*, 63(2), 281–308. https://doi.org/10.1017/S0020818309090092.

Bayer, P., and Aklin, M. (2020). The European Union Emissions Trading System reduced CO_2 emissions despite low prices. *Proceedings of the National Academy of Sciences*, 117(16), 8804–81. https://doi.org/10.1073/pnas.1918128117.

Bell, M., and Pavitt, K. (1995). The development of technological capabilities. In I. ul Haque et al. (Eds). *Trade, Technology and International Competitiveness*

(pp. 69–101). Washington, DC: World Bank. https://documents.worldbank
.org/curated/en/265331468765926233/Trade-technology-and-international
-competitiveness.

Benson, C.L., and Magee, C.L. (2015). Quantitative determination of technological
improvement from patent data. *PLoS One*, 10(4), e0121635. https://doi.org/10.1371/
journal.pone.0121635.

Bento, A.M., Li, S., and Roth, K. (2012). Is there an energy efficiency paradox in fuel
economy? A note on the role of consumer heterogeneity and sorting bias. *Economic
Letters*, 115(1), 44–8. https://doi.org/10.1016/j.econlet.2011.09.034.

Bento, N., and Wilson, C. (2016). Measuring the duration of formative phases
for energy technologies. *Environmental Innovations and Social Transitions*, 21,
95–112. https://doi.org/10.1016/j.eist.2016.04.004.

Bento, N., Wilson, C., and Anadon, L.D. (2018). Time to get ready: Conceptualizing
the temporal and spatial dynamics of formative phases for energy technologies.
Energy Policy, 119, 282–93. https://doi.org/10.1016/j.enpol.2018.04.015.

Bernini, C., and Pellegrini, G. (2011). How are growth and productivity in private firms
affected by public subsidy? Evidence from a regional policy. *Regional Science and
Urban Economics* 41, 253–65. https://doi.org/10.1016/j.regsciurbeco.2011.01.005.

Berry, A. (2019). The distributional effects of a carbon tax and its impact on fuel
poverty: A microsimulation study in the French context. *Energy Policy*, 124, 81–94.
https://doi.org/10.1016/j.enpol.2018.09.02.

Bettencourt, L.M.A., Trancik, J.E., and Kaur, J. (2013). Determinants of the pace of
global innovation in energy technologies. *PLoS ONE*, 8(10), e67864. https://doi.org/
10.1371/journal.pone.0067864.

Blackmon, G., and Zeckhauser, R. (1992). Fragile commitments and the regulatory
process. *Yale Journal of Regulation*, 9(1), 73–105. https://digitalcommons.law.yale
.edu/yjreg/vol9/iss1/3.

Blankenship, R.E., et al. (2011). Comparing the photosynthetic and photovoltaic efficiencies and recognising the potential for improvement. *Science*, 332(805), 805–9.
https://doi.org/10.1126/science.1200165.

Bloom, N., Griffith, R., and Van Reenen, J. (2002). Do R&D tax credits work?
Evidence from a panel of countries 1979–1997. *Journal of Public Economics*, 85(1),
1–31. https://doi.org/10.1016/S0047-2727(01)00086-X.

Bloom, N., et al. (2019). Are ideas getting harder to find? *American Economic Review*,
110(4), 1104–44. https://doi.org/10.1257/aer.20180338.

BNEF (2019). *2019 Lithium-ion Battery Price Survey*, 3 December 2019. https://bnef
.com.

Bogdanov, D., et al. (2019). Radical transformation pathway towards sustainable electricity via evolutionary steps. *Nature Communication*, 10, 1077. https://doi.org/10
.1038/s41467-019-08855-1.

Boiteux, M. (1960). Peak load pricing. *Journal of Business*, 33(2), 157–79. www.jstor
.org/stable/2351015.

Bolt, J., et al. (2018). Rebasing Maddison: New income comparisons and the shape of
long-run economic development (GGDC Research Memorandum 174). Groningen:
University of Groningen. www.rug.nl/ggdc/html_publications/memorandum/gd174
.pdf.

Borenstein, S. (2002). The trouble with electricity markets: Understanding California's
restructuring disaster. *Journal of Economic Perspectives*, 16(1), 191–211. https://doi
.org/10.1257/0895330027175.

Borenstein, S. (2012). The private and public economics of renewable electricity generation. *Journal of Economic Perspectives*, 26(1), 67–92. https://doi.org/10.1257/jep.26.1.67.

Bosetti, V., and Victor, D.G. (2011). Politics and economics of second-best regulation of greenhouse gases: The importance of regulatory credibility. *Energy Journal*, 32(1), 1–24. https://doi.org/10.5547/ISSN0195-6574-EJ-Vol32-No1-1.

Bradford, D., et al. (2017). Time preferences and consumer behaviour. *Journal of Risk and Uncertainty*, 55(2–3), 119–45. https://doi.org/10.1007/s11166-018-9272-8.

Bramoullèa, Y., and Olson, L.J. (2005). Allocation of pollution abatement under learning by doing. *Journal of Public Economics*, 89(9–10), 1935–60. https://doi.org/10.1016/j.jpubeco.2004.06.007.

Briceño-Garmendia, C., and Shkaratan, M. (2011). Power tariffs: Caught between cost recovery and affordability (World Bank Policy Research Working Paper 5904). Washington, DC: World Bank. https://documents1.worldbank.org/curated/en/234441468161963356/pdf/WPS5904.pdf.

Bronski, P., et al. (2015). *The Economics of Demand Flexibility: How 'Flexiwatts' Create Certifiable Value for Customers and the Grid*. Boulder, CO: Rocky Mountain Institute. https://rmi.org/insight/the-economics-of-demand-flexibility.

Broomhower, J., and Davis, L.W. (2014). A credible approach for measuring inframarginal participation in energy efficiency programs. *Journal of Public Economics*, 113, 67–79. https://doi.org/10.1016/j.jpubeco.2014.03.009.

Brucal, A., and Roberts, M.J. (2019). Do energy efficiency standards hurt consumers? Evidence from household appliance sales. *Journal of Environmental Economics and Management*, 96, 88–107. https://doi.org/10.1016/j.jeem.2019.04.005.

Bruckner, T., et al. (2014). Energy systems. In O. Edenhofer et al. (Eds). *Climate Change 2014: Mitigation of Climate Change—Contribution of Working Group III to the Fifth Assessment Report of the Intergovernmental Panel on Climate Change* (pp. 511–97). Cambridge: Cambridge University Press.

Brunner, S., Flachsland, C., and Marschinski, R. (2011). Credible commitment in carbon policy. *Climate Policy*, 12(2), 255–71. https://doi.org/10.1080/14693062.2011.582327.

Budinis, S., et al. (2018). An assessment of CCS costs, barriers and potential. *Energy Strategy Reviews*, 22, 61–81. https://doi.org/10.1016/j.esr.2018.08.003.

Burck, J., et al. (2019). *Climate Change Performance Index: Background and Methodology*. Bonn: Germanwatch. https://ccpi.org/methodology.

Bürger, V., et al. (2016). *Klimaneutraler Gebäudebestand 2050*, Dessau-Roßlau: Umweltbundesamt. www.umweltbundesamt.de/publikationen/klimaneutraler-gebaeudebestand-2050.

Burke, J., et al. (2020). *Distributional Impacts of a Carbon Tax in the UK: Report 2—Analysis by Income Decile*. London: Grantham Research Institute on Climate Change, London School of Economics, and Vivid Economics. www.lse.ac.uk/granthaminstitute/publication/distributional-impacts-of-a-carbon-tax-in-the-uk.

Busse, M.R., Knittel, C.R., and Zettelmeyer, F. (2013). Are consumers myopic? Evidence from new and used car purchases. *American Economic Review*, 103(1), 220–56. https://doi.org/10.1257/aer.103.1.220.

Cai, W., et al. (2015). Short-lived buildings in China: Impacts on water, energy, and carbon emissions. *Environmental Science & Technology*, 49(24), 13921–8. https://doi.org/10.1021/acs.est.5b02333.

Cain, L.P. (2006). Motor vehicle registrations, by vehicle type: 1900–1995. Series Df339–42. In S.B. Carter et al. (Eds). *Historical Statistics of the United States,*

Earliest Times to the Present: Millennial Edition. New York, NY: Cambridge University Press. https://doi.org/10.1017/ISBN-9780511132971.Df184-577.

Calcott, P., and Walls, M. (2000). Can downstream waste disposal policies encourage upstream 'design for environment'? *American Economic Review*, 90(2), 233–7. https://doi.org/10.1257/aer.90.2.233.

Calcott, P., and Walls, M. (2005). Waste, recycling, and 'design for environment': Roles for markets and policy instruments. *Resource and Energy Economics*, 27(4), 287–305. https://doi.org/10.1016/j.reseneeco.2005.02.001.

Calel, R. (2020). Adopt or innovate: Understanding technological responses to cap-and-trade. *American Economic Journal: Economic Policy*, 12(3), 170–201. https://doi.org/10.1257/pol.20180135.

Calel, R., and Dechezleprêtre, A. (2016). Environmental policy and directed technological change: Evidence from the European carbon market. *Review of Economics and Statistics*, 98(1), 173–91. https://doi.org/10.1162/REST_a_00470.

Cambridge Economic Research Associates (2018). *Study on the Estimation of the Value of Lost Load of Electricity Supply in Europe.* Ljubljana: Agency for the Cooperation of Energy Regulation.

Cao, Z., et al. (2017). Elaborating the history of our cementing societies: An in-use stock perspective. *Environmental Science and Technology*, 51(19), 11468–75. https://doi.org/10.1021/acs.est.7b03077.

Carbon Budget Order 2016, No. 785 (United Kingdom). www.legislation.gov.uk/uksi/2016/785/contents/made.

Carleton, T., et al. (2020). Valuing the global mortality consequences of climate change accounting for adaptation costs and benefits (Becker Friedman Institute for Economics, University of Chicago, Working Paper 2018–51, Version 19 June 2020). Chicago, IL: University of Chicago. https://doi.org/10.2139/ssrn.3224365.

Carneiro, R.L. (1970). A theory of the origin of the state. *Science*, 169(3967), 733–8. https://doi.org/10.1126/science.169.3947.733.

Carruth, M.A., Allwood, J.M., and Moynihan, M.C. (2011). The technical potential for reducing metal requirements through lightweight product design. *Resources, Conservation and Recycling*, 57, 48–60. https://doi.org/10.1016/j.resconrec.2011.09.018.

Carvalho, M., Dechezleprêtre, A., and Glachant, M. (2017). Understanding the dynamics of global value chains for solar photovoltaic technologies (WIPO Economic Research Working Paper No. 40). Geneva: World Intellectual Property Organization. www.wipo.int/edocs/pubdocs/en/wipo_pub_econstat_wp_40.pdf.

Casey, J., and Koleski, K. (2011). *Backgrounder: China's 12th Five-Year Plan.* Washington, DC: US–China Economic and Security Review Commission. www.uscc.gov/sites/default/files/Research/12th-FiveYearPlan_062811.pdf.

Castle, J., and Hendry, D. (2020, 4 June). Decarbonising the future UK economy. VoxEU. https://voxeu.org/article/decarbonising-future-uk-economy.

Cattaneo, C. (2019). Internal and external barriers to energy efficiency: Which role for policy interventions? *Energy Efficiency*, 12, 1293–311. https://doi.org/10.1007/s12053-019-09775-1.

Cerqua, A., and Pellegrini, G. (2014). Do subsidies to private capital boost growth? A multiple regression discontinuity design approach. *Journal of Public Economics*, 109, 114–26. https://doi.org/10.1016/j.jpubeco.2013.11.005.

Chastas, P., Theodosiou, T., and Bikas, D. (2016). Embodied energy in residential buildings—towards the nearly zero energy building: A literature review. *Building and Environment*, 105, 267–82. https://doi.org/10.1016/j.buildenv.2016.05.040.

Christiansen, P., Gillingham, K., and Nordhaus, W. (2018). Uncertainty in forecasts of long-run economic growth. *Proceedings of the National Academy of Sciences*, 115(21), 5409–14. https://doi.org/10.1073/pnas.1713628115.

Chum, H., et al. (2011). Bioenergy. In O. Edenhofer et al. (Eds). *IPCC Special Report: Renewable Energy Sources and Climate Change Mitigation* (pp. 209–331). Cambridge: Cambridge University Press.

Chyong, C.K., Pollitt, M., and Cruise, R. (2019). Can wholesale electricity prices support 'subsidy-free' generation investment in Europe (Cambridge Working Papers in Economics 1955). Cambridge: University of Cambridge. www.econ.cam.ac.uk/research-files/repec/cam/pdf/cwpe1955.pdf.

Ciacci, L., et al. (2016). Metal dissipation and inefficient recycling intensify climate forcing. *Environmental Science & Technology*, 50(20), 11394–402. https://doi.org/10.1021/acs.est.6b02714.

Ciais, P., et al. (2013). Carbon and other biogeochemical cycles. In T.F. Stocker et al. (Eds). *Climate Change 2013: The Physical Science Basis—Contribution of Working Group I to the Fifth Assessment Report of the Intergovernmental Panel on Climate Change* (pp. 465–570). Cambridge: Cambridge University Press.

ClimateWorks Australia and Vivid Economics (2019). *A Low-Carbon Industrial Strategy for Vietnam*. London: Vivid Economics. www.vivideconomics.com/wp-content/uploads/2019/08/discussion_paper_a_low-carbon_industrial_strategy_for_vietnam-.pdf.

Cline, W.R. (1992). *The Economics of Global Warming*. Washington, DC: Institute for International Economics.

Clò, S., Cataldi, A., and Zoppoli, P. (2015). The merit-order effect in the Italian power market: The impact of solar and wind generation on national wholesale electricity prices. *Energy Policy*, 77, 79–88. https://doi.org/10.1016/j.enpol.2014.11.038.

Cohen, F., Glachant, M., and Söderberg, M. (2017). Consumer myopia, imperfect competition and the energy efficiency gap: Evidence from the UK refrigerator market. *European Economic Review*, 93, 1–23. https://doi.org/10.1016/j.euroecorev.2017.01.004.

Collins, M., et al. (2013). Long-term climate change: projections, commitments and irreversibility. In T.F. Stocker et al. (Eds). *Climate Change 2013: The Physical Science Basis—Contribution of Working Group I to the Fifth Assessment Report of the Intergovernmental Panel on Climate Change* (pp. 1028–136). Cambridge: Cambridge University Press.

Combes, P., Duranton, G., and Gobillon, L. (2008). Spatial wage disparities: Sorting matters! *Journal of Urban Economics*, 63(2), 723–42. https://doi.org/10.1016/j.jue.2007.04.004.

Combes, P., et al. (2012). The productivity advantages of large cities: Distinguishing agglomeration from firm selection. *Econometrica*, 80(6), 2543–94. https://doi.org/10.3982/ECTA8442.

Comin, D.A., Dmitriev, M., and Rossi-Hansberg, E. (2012). The spatial diffusion of technology (NBER Working Paper No. 18534). Cambridge, MA: National Bureau of Economic Research. www.nber.org/papers/w18534.

Comin, D., and Hobijn, B. (2010). An exploration of technology diffusion. *American Economic Review*, 100(5), 2031–59. https://doi.org/10.1257/aer.100.5.2031.

Committee on Climate Change (2015). *The Fifth Carbon Budget: The Next Step Towards a Low-Carbon Economy*. London: Committee on Climate Change. www.theccc.org.uk/publication/the-fifth-carbon-budget-the-next-step-towards-a-low-carbon-economy.

Committee on Climate Change (2017). *Energy Prices and Bills: Impacts of Meeting Carbon Budgets*. London: Committee on Climate Change. www.theccc.org.uk/publication/energy-prices-and-bills-report-2017.

Committee on Climate Change (2019). *Reducing UK Emissions: 2019 Progress Report to Parliament*. London: Committee on Climate Change. www.theccc.org.uk/publication/reducing-uk-emissions-2019-progress-report-to-parliament.

Cook, E. (1971). The flow of energy in an industrial society. *Scientific American*, 225(3), 134–47. www.jstor.org/stable/24923122.

Cooper, D.R., and Allwood, J.M. (2012). Reusing steel and aluminum components at end of product life. *Environmental Science & Technology*, 46(18), 10334–40. https://doi.org/10.1021/es301093a.

Cordon, W.M. (1974). *Trade Policy and Economic Welfare*. Oxford: Oxford University Press.

Costinot, A., et al. (2019). The more we die, the more we sell? A simple test of the home-market effect. *Quarterly Journal of Economics*, 134(2), 843–94. https://doi.org/10.1093/qje/qjz003.

Council of Economic Advisors (2016). *A Retrospective Assessment of Clean Energy Investments in the Economic Recovery Act*. Washington, DC: Executive Office of the President of the United States. https://obamawhitehouse.archives.gov/sites/default/files/page/files/20160225_cea_final_clean_energy_report.pdf.

Crabb, J.M., and Johnson, D.K.N. (2010). Fueling innovation: The impact of oil prices and CAFE standards on energy-efficient automobile technology. *Energy Journal*, 31(1), 199–216. https://doi.org/10.5547/ISSN0195-6574-EJ-Vol31-No1-9.

Cramton, P., Ockenfels, A., and Stoft, S. (2013). Capacity market fundamentals. *Economics of Energy & Environmental Policy*, 2(2), 27–46. https://doi.org/10.5547/2160-5890.2.2.2.

Cramton, P., and Stoft, S. (2006). *The Convergence of Market Designs for Adequate Generation with Special Attention to the CAISIO's Resource Adequacy Problem* (Center for Energy and Environmental Policy Research Working Paper 2006-007). Cambridge, MA: Massachusetts Institute of Technology. https://dspace.mit.edu/handle/1721.1/45053.

Cramton, P., and Stoft, S. (2008). Forward reliability markets: Less risk, less market power, more efficiency. *Utility Policy*, 16(3), 194–201. https://doi.org/10.1016/j.jup.2008.01.007.

Cretì, A., and Fontini, F. (2019). *Economics of Electricity: Markets, Competition and Rules*. Cambridge: Cambridge University Press.

Criscuolo, C., et al. (2019). Some causal effects of an industrial policy. *American Economic Review*, 109(1), 48–85. https://doi.org/10.1257/aer.20160034.

Cullen, J.M., and Allwood, J.M. (2010a). The efficient use of energy: Tracing the global flow of energy from fuel to service. *Energy Policy*, 38, 75–81. https://doi.org/10.1016/j.enpol.2009.08.054.

Cullen, J.M., and Allwood, J.M. (2010b). Theoretical efficiency limits for energy conversion devices. *Energy*, 35(5), 2059–69. https://doi.org/10.1016/j.energy.2010.01.024.

Cullen, J.M., Allwood, J.M., and Borgstein, E.H. (2010). Reducing energy demand: What are the practical limits? *Environmental Science and Technology*, 45(4), 1711–8. https://doi.org/10.1021/es102641n.

Czarnitzki, D., and Hussinger, K. (2018). Input and output additionality of R&D subsidies. *Applied Economics*, 50(12), 1324–41. https://doi.org/10.1080/00036846.2017.1361010.

Daehn, K.E., Cabrera Serrenho, A., and Allwood, J.M. (2017). How will copper contamination constrain future global steel recycling? *Environmental Science & Technology*, 51(11), 6599–606. https://doi.org/10.1021/acs.est.7b00997.

Dargay, J., Gately, D., and Sommer, M. (2007). Vehicle ownership and income growth, worldwide: 1960–2030. *Energy Journal*, 28(4), 143–70. https://doi.org/10.5547/ISSN0195-6574-EJ-Vol28-No4-7.

Dasgupta, P. (2008). Discounting and climate change. *Journal of Risk and Uncertainty*, 37(2–3), 141–69. https://doi.org/10.1007/s11166-008-9049-6.

Data-Driven EnviroLab and NewClimate Institute (2020). *Accelerating Net Zero: Exploring Cities, Regions, and Companies' Pledges to Decarbonise*. Research report prepared by A. Hsu et al. https://datadrivenlab.org/publications.

Davies, J.H., and Davies, D.R. (2010). Earth's surface heat flux. *Solid Earth*, 1(1), 5–24. https://doi.org/10.5194/se-1-5-2010.

Davis, L.W. (2012). Evaluating the slow adoption of energy efficient investments: Are renters less likely to have energy efficient appliances? In D. Fullerton and C. Wolfram (Eds), *The Design and Implementation of U.S. Climate Policy* (pp. 301–16). Chicago, IL: Chicago University Press.

Davis, L.W., Martinez, S., and Taboada, B. (2020). How effective is energy-efficient housing? Evidence from a field trial in Mexico. *Journal of Development Economics*, 143, 102390. https://doi.org/10.1016/j.jdeveco.2019.102390.

Davis, L.W., and Metcalf, G.E. (2016). Does better information lead to better choices? Evidence from energy-efficiency labels. *Journal of the Association of Environmental and Resource Economists*, 3(3), 589–625. https://doi.org/10.1086/686252.

Davis, S.J., et al. (2018). Net zero emission energy systems. *Science*, 360(6396), eaas9793. https://doi.org/10.1126/science.aas9793.

De la Tour, A., Glachant, M., and Ménière, Y. (2011). Innovation and international technology transfer: The case of the Chinese photovoltaic industry. *Energy Policy*, 39(2), 761–70. https://doi.org/10.1016/j.enpol.2010.10.050.

DECC (2012). *The Future of Heating: A Strategic Framework for Low Carbon Heat in the UK*. London: Department of Energy and Climate Change. https://assets.publishing.service.gov.uk/government/uploads/system/uploads/attachment_data/file/48574/4805-future-heating-strategic-framework.pdf.

Dechezleprêtre, A., and Glachant, M. (2014). Does foreign environmental policy influence domestic innovation? Evidence from the wind industry. *Environmental and Resource Economics*, 58(3), 391–413. https://doi.org/10.1007/s10640-013-9705-4.

Dechezleprêtre, A., et al. (2019). Do tax incentives for research increase firm innovation? An RD design for R&D (Centre for Economic Performance Discussion Paper No. 1413). London: London School of Economics. https://cep.lse.ac.uk/pubs/download/dp1413.pdf.

Dechezleprêtre, A., Martin, R., and Mohnen, M. (2017). Knowledge spillovers from clean and dirty technologies (Grantham Research Institute on Climate Change and the Environment Working Paper No. 135). London: London School of Economics. www.lse.ac.uk/granthaminstitute/publication/knowledge-spillovers-from-clean-and-dirty-technologies-a-patent-citation-analysis-working-paper-135.

DeShazo, J.R., Sheldon, T.L., and Carson, R.T. (2017). Designing policy incentives for cleaner technologies: Lessons from California's plug-in electric vehicle rebate program. *Journal of Environmental Economics and Management*, 84, 18–43. https://doi.org/10.1016/j.jeem.2017.01.002.

Diamond, J. (1997). *Guns, Germs and Steel: The Fate of Human Societies*. New York, NY: Norton.

Dietz, S., and Venmans, F. (2019). Cumulative carbon emissions and economic policy: In search of general principles. *Journal of Environmental Economics and Management*, 96, 108–29. https://doi.org/10.1016/j.jeem.2019.04.003.

Dixit, A.K., and Stiglitz, J.E. (1977). Monopolistic competition and optimum product diversity. *American Economic Review*, 67(3), 297–308. www.jstor.org/stable/1831401.

Dobbs, R., et al. (2012). *Urban World: Cities and the Rise of the Consuming Class*. Seoul: McKinsey Global Institute. www.mckinsey.com/featured-insights/urbanization/urban-world-cities-and-the-rise-of-the-consuming-class.

Dutta, S., et al. (2019). The Global Innovation Index 2019. In S. Dutta, B. Lanvin and S. Wunsch-Vincent (Eds). *The Global Innovation Index 2019: Creating Healthy Lives—The Future of Medical Innovation* (pp. 1–40). Ithaca, NY, Fontainebleau and Geneva: Cornell University, Institut Européen d'Administration des Affaires and World Intellectual Property Organization.

EBRD (1999). *Transition Report 1999: Ten Years of Transition*. London: European Bank for Reconstruction and Development.

Einstein, A. (1905). Does the inertia of a body depend on its energy content? *Annalen der Physik*, 18(13), 639–41. In *The Collected Papers of Albert Einstein, Volume 2: The Swiss Years: Writings, 1900–1909* (English translation supplement, translated by A. Beck). Princeton, NJ: Princeton University Press. https://einsteinpapers.press.princeton.edu/vol2-trans/186.

Ellen MacArthur Foundation (2019). *Completing the Picture: How the Circular Economy Tackles Climate Change*. Cowes: Ellen MacArthur Foundation. www.ellenmacarthurfoundation.org/publications.

Energy Transitions Commission (2018). *Mission Possible: Reaching Net Zero Emissions from Hard-to-Abate Sectors by Mid-Century*. London: Energy Transitions Commission. www.energy-transitions.org/mission-possible.

Ericsson, K. (2017). *Biogenic Carbon Dioxide as Feedstock for Production of Chemicals and Fuels: A Techno-Economic Assessment with a European Perspective*. Environmental and Energy Systems Study (IMES/EESS Report No. 103). Lund: Lund University.

Escobar Rangel, L., and Lévêque, F. (2015). Revisiting the cost escalation curse of nuclear power: New lessons from the French experience. *Economics of Energy & Environmental Policy*, 4(2), 103–26. https://doi.org/10.5547/2160-5890.4.2.lran.

European Environment Agency (2020). *Construction and Demolition Waste: Challenges and Opportunities in a Circular Economy*. Copenhagen: European Environment Agency. www.eea.europa.eu/themes/waste/waste-management/construction-and-demolition-waste-challenges.

European Patent Office and UN Environment Program (2014). *Climate Change Mitigation Technologies in Europe: Evidence from Patent and Economic Data*. Munich: European Patent Office. https://documents.epo.org/projects/babylon/eponet.nsf/0/6A51029C350D3C8EC1257F110056B93F/$File/climate_change_mitigation_technologies_europe_en.pdf.

Evans, S., et al. (2021). Border carbon adjustments and industrial competitiveness in a European Green Deal. *Climate Policy*, 21(3), 307–17. https://doi.org/10.1080/14693062.2020.1856637.

Fæhn, T., and Isaksen, E.T. (2016). Diffusion of climate technologies in the presence of commitment problems. *Energy Journal*, 37(2), 155–80. https://doi.org/10.5547/01956574.37.2.tfae.

Fankhauser, S., et al. (2013). Who will win the green race? In search of environmental competitiveness and innovation. *Global Environmental Change*, 23(5), 902–13. https://doi.org/10.1016/j.gloenvcha.2013.05.007.

FAO (2004). *Human Energy Requirements: A Joint Expert Report of a Joint FAO/ WHO/UNU Expert Consultation*. Rome: Food and Agriculture Organization of the United Nations. www.fao.org/3/a-y5686e.pdf.0.

Farmer, J.D., and Lafond, F. (2016). How predictable is technological progress? *Research Policy*, 45(3), 647–65. https://doi.org/10.1016/j.respol.2015.11.001.

Farsi, M., Fetz, A., and Filippini, M. (2008). Economies of scale and scope in multi-utilities. *Energy Journal*, 29(4), 123–43. https://doi.org/10.5547/ISSN0195 -6574-EJ-Vol29-No4-6.

Fasel, J. (2012). Timeline of the power market. Wikimedia Commons. https://commons .wikimedia.org/wiki/File:Timeline_of_the_power_market.jpg.

Fasihi, M., and Breyer, C. (2020). Baseload electricity and hydrogen supply based on hybrid PV-wind power plants. *Journal of Cleaner Production*, 243, 118466. https:// doi.org/10.1016/j.jclepro.2019.118466.

FERC (2004). *Current State of Issues Concerning Underground Natural Gas Storage*. Washington, DC: Federal Energy Regulatory Commission. www.ferc.gov/sites/ default/files/2020-05/UndergroundNaturalGasStorageReport.pdf.

FERC (2020). *Energy Primer: A Handbook of Energy Market Basics*. Washington, DC: Federal Energy Regulatory Commission. www.ferc.gov/sites/default/files/2020-06/ energy-primer-2020.pdf.

Fernihough, A., and O'Rourke, K.H. (2014). Coal and the European industrial revolution. (NBER Working Paper No. 19802). Cambridge, MA: National Bureau for Economic Research. www.nber.org/papers/w19802.

Figueiredo, N.C., and da Silva, P.P. (2019). The 'merit-order effect' of wind and solar power: Volatility and determinants. *Renewable and Sustainable Energy Reviews*, 102, 54–62. https://doi.org/10.1016/j.rser.2018.11.042.

Finnegan, J.J. (2019). Institutions, climate change and the foundations of long-term policymaking (Grantham Research Institute on Climate Change and the Environment Working Paper No. 321). London: London School of Economics. www.lse.ac .uk/GranthamInstitute/wp-content/uploads/2019/04/working-paper-321-Finnegan-1 .pdf.

Flomenhoft, G. (2018). Historical and empirical basis for communal title in minerals at the national level: Does ownership matter for human development? *Sustainability*, 10, 1958. https://doi.org/10.3390/su10061958.

Forsberg, C.W., et al. (2017). Converting excess low-price electricity into high-temperature stored heat for industry and high-value electricity production. *Electricity Journal*, 30(6), 42–52. https://doi.org/10.1016/j.tej.2017.06.009.

Fouquet, R. (2014). Long-run demand for energy services: Income and price elasticities over two hundred years. *Review of Environmental Economics and Policy*, 8(2), 186–207. https://doi.org/10.1093/reep/reu002.

Fouquet, R., and Broadberry, S. (2015). Seven centuries of European economic growth and decline. *Journal of Economic Perspectives*, 29(4), 227–44. https://doi.org/10 .1257/jep.29.4.227.

Fowler, T., et al. (2015). Excess winter deaths in Europe: A multi-country descriptive analysis. *European Journal of Public Health*, 25(2), 339–45. https://doi.org/10 .1093/eurpub/cku073.

Fowlie, M., Greenstone, M., and Wolfram, C. (2018). Do energy efficiency investments deliver? Evidence from the Weatherization Assistance Program. *Quarterly Journal of Economics*, 133(3), 1597–644. https://doi.org/10.1093/qje/qjy005.

Fremdling, R., and Solar, P. (2010). Industry. In S. Broadberry and K.H. O'Rourke (Eds). *The Cambridge Economic History of Modern Europe Volume 1: 1700–1870* (pp. 164–86). Cambridge: Cambridge University Press.

Friedlingstein, P., et al. (2019). Global carbon budget 2019. *Earth System Science Data*, 11, 1783–838. https://doi.org/10.5194/essd-11-1783-2019.

Frondel, M., Gerster, A., and Vance, C. (2020). The power of mandatory quality disclosure: Evidence from the German housing market. *Journal of the Association of Environmental and Resource Economists*, 7(1), 181–208. https://doi.org/10.1086/705786.

Frondel, M., and Vance, C. (2013). Heterogeneity in the effect of home energy audits: Theory and evidence. *Environmental and Resource Economics*, 55(3), 407–18. https://doi.org/10.1007/s10640-013-9632-4.

Fu, R., Remo, T., and Margolis, R. (2018). *2018 U.S. Utility-Scale Photovoltaics-Plus-Energy Storage System Costs Benchmark* (NREL/TP-6A20-71714). Golden, CO: National Renewable Energy Laboratory. www.nrel.gov/docs/fy19osti/71714 .pdf.

Fu, W., et al. (2018). Technology spillover effects of state renewable energy policy: Evidence from patent counts (NBER Working Paper 25390). Cambridge, MA: National Bureau of Economic Research. www.nber.org/papers/w25390.

Fuels Europe (2020). Fuel price breakdown. Retrieved from https://www.fuelseurope .eu/knowledge/refining-in-europe/economics-of-refining/fuel-price-breakdown/.

Fukuyama, F. (2011). *The Origins of Political Order*. New York, NY: Farrar, Straus and Giroux.

Fukuyama, F. (2014). *Political Order and Political Decay*. New York, NY: Farrar, Straus and Giroux.

Fulton, L., Cazzola, P., and Cuenot, F. (2009). IEA mobility model (MoMo) and its use in the ETP 2008. *Energy Policy*, 37(10), 3758-68. https://doi.org/10.1016/j.enpol .2009.07.065.

Fuss, S., et al. (2018). Negative emissions: Part 2—Costs, potentials and side effects. *Environmental Research Letters*, 13(6), 063002. https://doi.org/10.1088/1748-9326/ aabf9f.

Gaines, L. (2018). Lithium-ion battery recycling processes: Research towards a sustainable course. *Sustainable Materials and Technologies*, 17, e00068. https://doi.org/10 .1016/j.susmat.2018.e00068.

Galor, O. (2005). From stagnation to growth: Unified growth theory. In P. Aghion and S.N. Durlauf (Eds). *Handbook of Economic Growth, Volume 1, Part A* (pp. 171–293). Amsterdam: North-Holland.

Geraats, P.M. (2014). Monetary policy transparency. In J. Forssbaeck and L. Oxelheim (Eds). *The Oxford Handbook of Economic and Institutional Transparency* (pp. 68–97). Oxford: Oxford University Press.

Gerarden, T., Newell, R.G., and Stavins, R.N. (2015). Deconstructing the energy-efficiency gap: Conceptual frameworks and evidence. *American Economic Review*, 105(5), 183–6. https://doi.org/10.1257/aer.p20151012.

Gerarden, T.D, Newell, R.G., and Stavins, R.N. (2017). Assessing the energy efficiency gap. *Journal of Economic Literature*, 55(4), 1486–525. https://doi.org/10.1257/jel .20161360.

Geroski, P.A. (2000). Models of technology diffusion. *Research Policy*, 29(4–5), 603–25. https://doi.org/10.1016/S0048-7333(99)00092-X.

Gilbert, R.J., and Newbery, D.M.G. (1982). Preemptive patenting and the persistence of monopoly. *American Economic Review*, 72 (3), 514–26. www.jstor.org/stable/1831552.

Gillingham, K., Harding, M., and Rapson, D. (2012). Split incentives in residential energy consumption. *Energy Journal*, 33(2), 37–62. https://doi.org/10.5547/01956574.33.2.3.

Gillingham, K., and Huang, P. (2019). Is abundant natural gas a bridge to a low-carbon future or a dead-end? *Energy Journal*, 40(2), 1–26. https://doi.org/10.5547/01956574.40.2.kgil.

Gillingham, K., Newell, R.G., and Palmer, K. (2009). Energy efficiency economics and policy. *Annual Review of Resource Economics*, 1, 597–620. https://doi.org/10.1146/annurev.resource.102308.124234.

Gillingham, K., and Stock, J.H. (2018). The cost of reducing greenhouse gas emissions. *Journal of Economic Perspectives*, 32(4), 53–72. https://doi.org/10.1257/jep.32.4.53.

Gils, H.C. (2014). Assessment of the theoretical demand response potential in Europe. *Energy*, 67, 1–18. https://doi.org/10.1016/j.energy.2014.02.019.

Glaeser, E.L., and Maré, D.C. (2001). Cities and skills. *Journal of Labor Economics*, 19(2), 316–42. https://doi.org/10.1086/319563.

Gollier, C. (2013). *Pricing the Planet's Future: The Economics of Discounting in an Uncertain World*. Princeton, NJ: Princeton University Press.

Gordon, R.J. (2016). *The Rise and Fall of American Growth*. Princeton, NJ: Princeton University Press.

Goulder, L., and Hafstead, M. (2019). *Confronting the Climate Challenge: US Policy Options*. New York, NY: Columbia University Press.

Goulder, L.H., and Mathai, K. (2000). Optimal CO_2 abatement in the presence of induced technological change. *Journal of Environmental Economics and Management*, 39(1), 1–38. https://doi.org/10.1006/jeem.1999.1089.

Goulder, L.H., and Schneider, S.H. (1999). Induced technological change and the attractiveness of CO_2 abatement policies. *Resource and Energy Economics*, 21(3–4), 211–53. https://doi.org/10.1016/S0928-7655(99)00004-4.

Goulder, L.H., et al. (2019). Impacts of a carbon tax across US household income groups: What are the equity-efficiency trade-offs? *Journal of Public Economics*, 175, 44–64. https://doi.org/10.1016/j.jpubeco.2019.04.002.

Grant, N., et al. (2020). The appropriate use of reference scenarios in mitigation analysis. *Nature Climate Change*, 10(7), 605–10. https://doi.org/10.1038/s41558-020-0826-9.

Graus, W., et al. (2009). Global technical potentials for energy efficiency improvement (IAEE European Conference Paper). Utrecht: Ecofys Netherlands BV.

Green, F. (2018). Transition policy for climate change mitigation: Who, what, why and how (Centre for Climate Economics and Policy Working Paper 1807). Canberra: Australian National University. https://ccep.crawford.anu.edu.au/publication/ccep-working-paper/12850/transition-policy-climate-change-mitigation-who-what-why-and.

Gross, R., et al. (2018). How long does innovation and commercialisation in the energy sector take? Historical case studies of the timescales from innovation to widespread commercialisation in energy supply and end use technology. *Energy Policy*, 123, 682–99. https://doi.org/10.1016/j.enpol.2018.08.061.

Grubb, M., and Newbery, D. (2018). UK electricity market reform and the energy transition: Emerging lessons (CEEPR Working Paper No. 2018-004). Cambridge, MA: Massachusetts Institute of Technology. https://ceepr.mit.edu/files/papers/2018 -004.pdf.

Grubert, E. (2020). Fossil electricity retirement deadlines for a just transition. *Science*, 370(6521), 1171–3. https://doi.org/10.1126/science.abe0375.

Grübler, A. (2010). The cost of the French nuclear scale-up: A case of negative learning by doing. *Energy Policy*, 38(9), 5174–88. https://doi.org/10.1016/j.enpol.2010.05 .003.

Grübler, A., Nakićenović, N., and Victor, D.G. (1999). Dynamics of energy technologies and global change. *Energy Policy*, 27(5), 247–80. https://doi.org/10.1016/ S0301-4215(98)00067-6.

Grübler, A., Wilson, C., and Nemet, G. (2016). Apples, oranges, and consistent comparisons of the temporal dynamics of energy transitions. *Energy Research & Social Science*, 22, 18–25. https://doi.org/10.1016/j.erss.2016.08.015.

Hahn, R., and Metcalfe, R. (2016). The impact of behavioral science experiments on energy policy. *Economics of Energy & Environmental Policy*, 5(2), 27–44. https:// doi.org/10.5547/2160-5890.5.2.rhah.

Hale, E., et al. (2018). *Potential Roles for Demand Response in High-Growth Electric Systems with Increasing Shares of Renewable Generation* (NREL/TP-6A20-70630). Golden, CO: National Renewable Energy Laboratory. www.nrel.gov/docs/fy19osti/ 70630.pdf.

Hale, T. (2018). Catalytic cooperation (Blavatnik School of Government Working Paper 2018/-26). Oxford: University of Oxford. www.bsg.ox.ac.uk/research/publications/ catalytic-cooperation.

Hall, B.H., and Harhoff, D. (2012). Recent research on the economic effects of patents. *Annual Review of Economics*, 4, 541–65. https://doi.org/10.1146/annurev -economics-080511-111008.

Hall, B.H., Jaffe, A., and Trajtenberg, M. (2005). Market value and patent citations. *RAND Journal of Economics*, 36(1), 16–38. www.jstor.org/stable/1593752.

Hall, B.H., and Lerner, J. (2010). The financing of R&D and innovation. In B.H. Hall and N. Rosenberg (Eds). *Handbook of the Economics of Technological Innovation, Volume 1* (pp. 609–39). Amsterdam: North-Holland. https://doi.org/10.1016/S0169 -7218(10)01014-2.

Hall, B.H., Mairesse, J., and Mohnen P. (2010). Measuring the returns to R&D. In B.H. Hall and N. Rosenberg (Eds). *Handbook of the Economics of Technological Innovation, Volume 2* (pp. 1034–76). Amsterdam: North-Holland. https://doi.org/10 .1016/S0169-7218(10)02008-3.

Hall, B.H., and Van Reenen, J. (2000). How effective are fiscal incentives for R&D? A review of the evidence. *Research Policy*, 29(4–5), 449–69. https://doi.org/10 .1016/S0048-7333(99)00085-2.

Hall, R.E. (1997). The inkjet aftermarket: An economic analysis (unpublished manuscript). Palo Alto, CA: Stanford University. https://web.stanford.edu/~rehall/Inkjet %20Aftermarket%201997.pdf.

Han, J., et al. (2015). A comparative assessment of resource efficiency in petroleum refining. *Fuel*, 157, 292–8. https://doi.org/10.1016/j.fuel.2015.03.038.

Hänsel, M.C. (2020). Climate economics support for the UN climate targets. *Nature Climate Change*, 10, 781–9. https://doi.org/10.1038/s41558-020-0833-x.

Harari, Y.N. (2015). *Sapiens: A Brief History of Humankind*. New York, NY: HarperCollins.

Harding, M., and Hsiaw, A. (2014). Goal setting and energy conservation. *Journal of Economic Behavior & Organization*, 107, 209–27. https://doi.org/10.1016/j.jebo.2014.04.012.

Hashmi, A.R., and Van Biesebroeck, J. (2016). The relationship between market structure and innovation in industry equilibrium: A case study of the global automobile industry. *Review of Economics and Statistics*, 98(1), 192–208. https://doi.org/10.1162/REST_a_00494.

Hausfather, Z., and Peters, G.P (2020). Emissions: The 'business as usual' story is misleading. *Nature*, 577, 618–20. https://doi.org/10.1038/d41586-020-00177-3.

Hausmann, R., et al. (2014). *The Atlas of Economic Complexity: Mapping Paths to Prosperity*. Cambridge: MA: MIT Press.

Hausmann, R., et al. (2019). Implied comparative advantage (Center for International Development, Faculty Working Paper No. 276). Cambridge, MA: Harvard University. https://growthlab.cid.harvard.edu/files/growthlab/files/2019-09-cid-wp-276-revision-implied-comparative-advantage.pdf.

Heeren, N., et al. (2015). Environmental impact of buildings: What matters? *Environmental Science & Technology*, 49, 9832–41. https://doi.org/10.1021/acs.est.5b01735.

Hellebrandt, T., and Mauro, P. (2016). *World on the Move: Consumption Patterns in a More Equal Global Economy*. Washington, DC: Peterson Institute for International Economics.

Helm, D., Hepburn, C., and Ruta, G. (2012). Trade, climate change and the political game theory of border carbon adjustments. *Oxford Review of Economic Policy*, 28(2), 368–94. https://doi.org/10.1093/oxrep/grs013.

Hepburn, C., et al. (2019). The technological and economic prospects for CO_2 utilization and removal. *Nature*, 575, 87–97. https://doi.org/10.1038/s41586-019-1681-6.

Hepburn, C., Pless, J., and Popp, D. (2017). Encouraging innovation that protects environmental systems: Five policy proposals. *Review of Environmental Economics and Policy*, 12(1), 154–69. https://doi.org/10.1093/reep/rex024.

Hertwich, E.G., et al. (2019). Material efficiency strategies to reducing greenhouse gas emissions associated with buildings, vehicles and electronics: A review. *Environmental Research Letters*, 14(4), 043004. https://doi.org/10.1088/1748-9326/ab0fe3.

Hicks, J.R. (1932). *The Theory of Wages*. London: Macmillan.

Hidalgo, C.A., and Hausmann, R. (2009). The building blocks of economic complexity. *Proceedings of the National Academy of Sciences*, 106(26), 10570–75. https://doi.org/10.1073/pnas.0900943106.

Hidalgo, C.A., et al. (2007). The product space conditions the development of nations. *Science*, 317(5837), 482–7. https://doi.org/10.1126/science.1144581.

High-Level Commission on Carbon Prices (2017). *Report of the High-Level Commission on Carbon Prices*. Washington, DC: World Bank. www.carbonpricingleadership.org/report-of-the-highlevel-commission-on-carbon-prices.

Hirst, D. (2018). *Carbon Price Floor and the Price Support Mechanism* (Briefing Paper 05927). London: House of Commons Library. https://commonslibrary.parliament.uk/research-briefings/sn05927.

Hirth, L., and Steckel, J.C. (2016). The role of capital costs in decarbonizing the electricity sector. *Environmental Research Letters*, 11, 114010. https://doi.org/10.1088/1748-9326/11/11/114010.

Hirth, L., Ueckerdt, F., and Edenhofer, O. (2015). Integration costs revisited: An economic framework for wind and solar variability. *Renewable Energy*, 74, 925–39. https://doi.org/10.1016/j.renene.2014.08.065.

HM Government (2017). *Clean Growth Strategy: Leading the Way to a Low Carbon Future*. London: HM Government. www.gov.uk/government/publications/clean -growth-strategy.

HM Government (2019). *Offshore Wind Sector Deal*. London: HM Government. www .gov.uk/government/publications/offshore-wind-sector-deal.

HM Treasury (2011). *Carbon Price Floor Consultation: The Government Response*. London: HM Treasury. https://assets.publishing.service.gov.uk/government/uploads/ system/uploads/attachment_data/file/190279/carbon_price_floor_consultation_govt _response.pdf.

Hoegh-Guldberg, O., et al. (2018). Impacts of 1.5°C of global warming on natural and human systems. In V. Masson-Delmotte et al. (Eds). *Global Warming of 1.5°C: An IPCC Special Report on the Impacts of Global Warming of 1.5°C above Pre-Industrial Levels and Related Global Greenhouse Gas Emission Pathways, in the Context of Strengthening the Global Response to the Threat of Climate Change, Sustainable Development, and Efforts to Eradicate Poverty* (pp. 175–311). www .ipcc.ch/site/assets/uploads/sites/2/2019/06/SR15_Full_Report_Low_Res.pdf.

Hoekstra, M., Puller, S.L., and West, J. (2017). Cash for Corollas: When stimulus reduces spending. *American Economic Journal: Applied Economics*, 9(3), 1–35. https://doi.org/10.1257/app.20150172.

Hoffert, M.I., et al. (1998). Energy implications of future stabilization of atmospheric CO_2 content. *Nature*, 365, 881–4. https://doi.org/10.1038/27638.

Hogan, W.W. (1992). Contract networks for electric power transmission. *Journal of Regulatory Economics*, 4(3), 211–42. https://doi.org/10.1007/BF00133621.

Hogan, W.W. (1998). Nodes and zones in electricity markets: Seeking simplified congestion pricing. In Chao, H.P., and Huntington, H.G. (Eds). *Designing Competitive Electricity Markets* (pp. 33–62). New York, NY: Springer.

Hogan, W.W. (2005). *On an 'Energy Only' Electricity Market Design for Resource Adequacy* (Centre for Business and Government, John F. Kennedy School of Government, mimeo). Cambridge, MA: Harvard University. https://scholar.harvard .edu/whogan/files/hogan_energy_only_092305.pdf.

Hogan, W.W., and Pope, S.L. (2019). *PJM Reserve Markets: Operating Reserve Demand Curve Enhancements* (Harvard University and FTI Consulting mimeo). Cambridge, MA: Harvard University. https://hepg.hks.harvard.edu/files/hepg/files/ hogan_pope_pjm_report_032119.pdf.

Holmes, T.J., Levine, D.K., and Schmitz, D.K. (2012). Monopoly and the incentive to innovate when adoption involves switchover disruptions. *American Economic Journal: Microeconomics*, 4(3), 1–33. https://doi.org/10.1257/mic.4.3.1.

Hoppmann, J. (2018). The role of interfirm knowledge spillovers for innovation in mass-produced environmental technologies: Evidence from the solar photovoltaic industry. *Organization & Environment*, 31(1), 2–24. https://doi.org/10.1177/ 1086026616680683.

Houde, S. (2018a). How consumers respond to product certification and the value of energy information. *RAND Journal of Economics*, 49(2), 453–77. https://doi.org/10 .1111/1756-2171.12231.

Houde, S. (2018b). Bunching with the stars: How firms respond to environmental certification (Energy Institute at Haas E2e Working Paper 037). Berkeley, CA: University of California, Berkeley. https://e2e.haas.berkeley.edu/pdf/workingpapers/WP037 .pdf.

Houde, S., and Aldy, J.E. (2017). Consumers' response to state energy efficient appliance rebate programs. *American Economic Journal: Economic Policy*, 9(4), 227–55. https://doi.org/10.1257/pol.20140383.

Houde, S., and Spurlock, C.A. (2016). Minimum energy efficiency standards and appliances: Old and new rationales. *Economics of Energy & Environmental Policy*, 5(2), 65–83. https://doi.org/10.5547/2160-5890.5.2.shou.

Howard, P.H., and Sterner, T. (2017). Few and not so far between: A meta-analysis of climate damage estimates. *Environmental and Resource Economics*, 68, 197–225. https://doi.org/10.1007/s10640-017-0166-z.

Howell, S.T. (2017). Financing innovation: Evidence from R&D grants. *American Economic Review*, 107(4), 1136–64. https://doi.org/10.1257/aer.20150808.

Hsiang, S., et al. (2017). Estimating economic damages from climate change in the United States. *Science*, 356, 1362–69. https://doi.org/10.1126/science.aal4369.

Huber, R.A. (2020). The role of populist attitudes in explaining climate change skepticism and support for environmental protection. *Environmental Politics*, 29(6), 959–82. https://doi.org/10.1080/09644016.2019.1708186.

Huenteler, J., et al. (2016). Technology life-cycles in the energy sector: Technological characteristics and the role of deployment for innovation. *Technological Forecasting and Social Change*, 104, 102–21. https://doi.org/10.1016/j.techfore.2015.09.022.

Hughes, L., and Urpelainen, J. (2015). Interests, institutions, and climate policy: Explaining the choice of policy instruments for the energy sector. *Environmental Science & Policy*, 54, 52–63. https://doi.org/10.1016/j.envsci.2015.06.014.

Hughes, T.P. (1983). *Networks of Power: Electrification in Western Society, 1880–1930*. Baltimore, MD: Johns Hopkins University Press.

Hughes, T.P. (1989). The evolution of large technical systems. In W.E. Bijker, T.P. Hughes, and T. Pinch (Eds). *The Social Construction of Technological Systems: New Directions in the Sociology and History of Technology* (pp. 51–82). Cambridge, MA: MIT Press.

Hummels, D. (2007). Transportation costs and international trade in the second era of globalization. *Journal of Economic Perspectives*, 21(3), 131–54. https://doi.org/10 .1257/jep.21.3.131.

Hunt, S. (2002). *Making Competition Work in Electricity*. New York, NY: John Wiley and Sons.

Hunter, C., and Penev, M. (2019). Market segmentation analysis of medium- and heavy-duty trucks with a fuel cell emphasis (NREL/PR-5400-73491). Golden, CO: National Renewable Energy Laboratory. www.osti.gov/biblio/1511815.

ICAO (2017). *Annual Report of the Council*. Montreal: International Civil Aviation Organization. www.icao.int/annual-report-2017/Pages/default.aspx.

IEA (2013). *Transition to Sustainable Buildings*. Paris: International Energy Agency. www.iea.org/reports/transition-to-sustainable-buildings.

IEA (2016). *World Energy Outlook Special Report, Energy and Air Pollution*. Paris: International Energy Agency. www.iea.org/reports/energy-and-air-pollution.

IEA (2017a). Online Annex: Building Sector Model. In IEA, *Energy Technologies Perspectives 2017*. Paris: OECD/International Energy Agency.

IEA (2017b). *World Energy Outlook Special Report: Energy Access Outlook 2017— From Poverty to Prosperity*. Paris: International Energy Agency. www.iea.org/ reports/energy-access-outlook-2017.

IEA (2017c). *Technology Roadmap: Delivering Sustainable Bioenergy.* Paris: International Energy Agency.

IEA (2019a). *World Energy Outlook 2019.* Paris: International Energy Agency. https://iea.org/weo.

IEA (2019b). *The Future of Hydrogen: A Report Prepared for the G-20, Japan.* Paris: International Energy Agency. www.iea.org/reports/the-future-of-hydrogen.

IEA (2020a). *Tracking Clean Energy Progress: Cement.* www.iea.org/topics/tracking-clean-energy-progress.

IEA (2020b). *Tracking Clean Energy Progress: Iron and Steel.* www.iea.org/topics/tracking-clean-energy-progress.

IEA (2020c). *Tracking Clean Energy Progress: Chemicals.* www.iea.org/topics/tracking-clean-energy-progress.

IEA (2020d). *World Energy Outlook Special Report: Outlook for Biogas and Biomethane.* Paris: International Energy Agency. www.iea.org/reports/outlook-for-biogas-and-biomethane-prospects-for-organic-growth.

IEA (2020e). *Energy Technology Perspectives 2020: Special Report on Clean Technology Innovation.* Paris: International Energy Agency. www.iea.org/reports/clean-energy-innovation.

IEA (2020f). *Global EV Outlook 2020.* Paris: International Energy Agency. www.iea.org/reports/global-ev-outlook-2020.

IEA (2020g). *Energy Technology RD&D Statistics Overview.* Paris: International Energy Agency. www.iea.org/reports/energy-technology-rdd-budgets-2020.

IEA Bioenergy (2020). *Advanced Biofuels: Potential for Cost Reduction.* Paris: IEA Bioenergy. www.ieabioenergy.com/iea-publications.

IEA and UNEP (2013). *Modernising Building Energy Codes to Secure our Global Energy Future.* Paris: International Energy Agency.

Imperial College (2018). *Analysis of Alternative UK Heat Decarbonisation Pathways for the Committee on Climate Change.* London: Imperial College. www.theccc.org.uk/wp-content/uploads/2018/06/Imperial-College-2018-Analysis-of-Alternative-UK-Heat-Decarbonisation-Pathways.pdf.

IPCC (2013). Summary for policy makers. In T.F. Stoker et al. (Eds). *Climate Change 2013: The Physical Science Basis—Contribution of Working Group I to the Fifth Assessment Report of the Intergovernmental Panel on Climate Change* (pp. 3–29). Cambridge: Cambridge University Press.

IPCC (2014a). Summary for policy makers. In R.K. Pachauri et al. (Eds). *Climate Change 2014: Synthesis Report—Contribution of Working Groups I, II and III to the Fifth Assessment Report of the Intergovernmental Panel on Climate Change* (pp. 2–31). Cambridge: Cambridge University Press.

IPCC (2014b). Summary for policy makers. In O. Edenhofer et al. (Eds). *Climate Change 2014: Mitigation of Climate Change—Working Group III Contribution to the Fifth Assessment Report of the Intergovernmental Panel on Climate Change* (pp. 1–30). Cambridge: Cambridge University Press.

IPCC (2014c). Technology-specific cost and performance parameters. In O. Edenhofer et al. (Eds). *Climate Change 2014: Mitigation of Climate Change—Working Group III Contribution to the Fifth Assessment Report of the Intergovernmental Panel on Climate Change* (Annex III, pp. 1329–56). Cambridge: Cambridge University Press.

IPCC (2018). Summary for policy makers. In V. Masson-Delmotte et al. (Eds). *Global Warming of 1.5°C: An IPCC Special Report on the Impacts of Global Warming of 1.5°C above Pre-Industrial Levels and Related Global Greenhouse Gas Emission Pathways, in the Context of Strengthening the Global Response to the Threat of Climate Change, Sustainable Development, and Efforts to Eradicate Poverty* (pp. 3–24). www.ipcc.ch/site/assets/uploads/sites/2/2019/06/SR15_Full_Report_Low _Res.pdf.

IPCC (2019). Summary for policy makers. In H.-O. Pörtner et al. (Eds). *IPCC Special Report on the Ocean and Cryosphere in a Changing Climate* (pp. 3–35). www.ipcc .ch/srocc/.

IPCC (2021). Summary for policy makers. In V. Masson-Delmotte et al. (Eds). *Climate Change 2021: The Physical Science Basis. Contribution of Working Group I to the Sixth Assessment Report of the Intergovernmental Panel on Climate Change* (pp. SPM1–41). www.ipcc.ch/report/ar6/wg1/#FullReport.

IRENA (2018). *Corporate Sourcing of Renewables: Market and Industry Trends*. Abu Dhabi: International Renewable Energy Agency. www.irena.org/publications/2018/ May/Corporate-Sourcing-of-Renewable-Energy.

IRENA (2020). *Renewable Power Generation Costs in 2019*. Abu Dhabi: International Renewable Energy Agency. www.irena.org/publications/2020/Jun/Renewable -Power-Costs-in-2019.

Irwin, S., and Good, D. (2017). On the value of ethanol in the gasoline blend. *Farmdoc daily*, March 15, 2017, Department of Agricultural and Consumer Economics, University of Illinois at Urbana-Champaign. https://farmdocdaily.illinois.edu/wp -content/uploads/2017/04/fdd150317.pdf.

ITF (2015). Surface transport in the long run. In ITF, *Transport Outlook 2015* (pp. 47–61). Paris: OECD/International Transport Forum. https://doi.org/10.1787/ 9789282107782-6-en.

ITF (2019). *Transport Outlook 2019*. Paris: OECD/International Transport Forum. https://doi.org/10.1787/transp_outlook-en-2019-en.

Jacob, M. (2014). *The First Knowledge Economy: Human Capital and the European Economy, 1750–1850*. Cambridge: Cambridge University Press.

Jacobs, A.M. (2011). *Governing for the Long Term: Democracy and the Politics of Investment*. Cambridge: Cambridge University Press.

Jacobsen, G.D., and Kotchen, M.J. (2013). Are building codes effective at saving energy? Evidence from residential billing data in Florida. *Review of Economics and Statistics*, 95(1), 34–49. https://doi.org/10.1162/REST_a_00243.

Jacobsen, M.R. (2013). Evaluating US fuel economy standards in a model with pro- ducer and household heterogeneity. *American Economic Journal: Economic Policy*, 5(2), 148–87. https://doi.org/10.1257/pol.5.2.148.

Jaeger, C.C., and Jaeger, J. (2011). Three views of two degrees. *Regional Environmental Change*, 11(S1), S15–S26. https://doi.org/10.1007/s10113-010-0190-9.

Jaffe, A.B., Newell, R.G., and Stavins, R.N. (2005). A tale of two market failures: Technology and environmental policy. *Ecological Economics* 54(2–3), 164–74. https://doi.org/10.1016/j.ecolecon.2004.12.027.

Jamasb, T., Nepal, R., and Timilsina, G.R. (2017). A quarter century of effort yet to come of age: A survey of electricity sector reform in developing countries. *Energy Journal*, 38(3), 195–234. https://doi.org/10.5547/01956574.38.3.tjam.

Jamasb, T., and Pollitt, M.G. (2011). Electricity sector liberalisation and innovation: An analysis of the UK's patenting activities. *Research Policy*, 40(2), 309–24. https://doi .org/10.1016/j.respol.2010.10.010.

Jamasb, T., and Pollitt, M.G. (2015). Why and how to subsidise energy R+D: Lessons from the collapse and recovery of electricity innovation in the UK. *Energy Policy*, 83, 197–205. https://doi.org/10.1016/j.enpol.2015.01.041.

Jamasb. T., et al. (2005). Electricity sector reform in developing countries: A survey of empirical evidence on determinants and performance (World Bank Policy Research Working Paper 3549). Washington, DC: World Bank. https://doi.org/10.1596/1813 -9450-3549.

Jaworski, T., and Smyth, A. (2018). Shakeout of the early commercial airframe industry. *Economic History Review*, 71(2), 617–38. https://doi.org/10.1111/ehr.12430.

Jones, C.I., and Williams, J.C. (1998). Measuring the social return to R&D. *Quarterly Journal of Economics*, 113(4), 1119–35. https://doi.org/10.1162/003355398555856.

Joos, F., and Spahni, R. (2008). Rates of change in natural and anthropogenic radiative forcing over the past 20,000 years. *Proceedings of the National Academy of Sciences*, 105(5), 1425–30. https://doi.org/10.1073/pnas.0707386105.

Jorgenson, D.W., and Griliches, Z. (1967). The explanation of productivity change. *Review of Economic Studies*, 34(3), 249–83. https://doi.org/10.2307/2296675.

Joskow, P.L. (2008). Capacity payments in imperfect electricity markets: Need and design. *Utility Policy*, 16(3), 159–70. https://doi.org/10.1016/j.jup.2007.10.003.

Joskow, P.L. (2011). Comparing the costs of intermittent and dispatchable electricity generation technologies. *American Economic Review*, 101(3), 238–41. https://doi .org/10.1257/aer.101.3.238.

Joskow, P.L., and Tirole, J. (2007). Reliability and competitive electricity markets. *RAND Journal of Economics*, 38(1), 60–84. https://doi.org/10.1111/j.1756-2171 .2007.tb00044.x.

Just, M.G., Nichols, L.M., and Dunn, R.R. (2019). Human indoor climate preferences approximate specific geographies. *Royal Society Open Science*, 6(3), 180695. https://doi.org/10.1098/rsos.180695.

Kahneman, D. (2011). *Thinking Fast and Slow*. New York, NY: Farrar, Straus and Giroux.

Kander, A., Malanima, P., and Warde, P. (2013). *Power to the People: Energy in Europe over the Last Five Centuries*. Princeton, NJ: Princeton University Press.

Karlin, B., Zinger, J.F., and Ford, R. (2015). The effects of feedback on energy conservation: A meta-analysis. *Psychological Bulletin*, 141(6), 1205–27. https://doi.org/10 .1037/a0039650.

Kaufman, N., et al. (2020). A near-term to net zero alternative to the social cost of carbon for setting carbon prices. *Nature Climate Change*, 10, 1010–14. https://doi .org/10.1038/s41558-020-0880-3.

Kavlak, G., McNerney, J., and Trancik, J.E. (2018). Evaluating the causes of cost reduction in photovoltaic modules. *Energy Policy*, 123, 700–710. https://doi.org/10 .1016/j.enpol.2018.08.015.

Keller, W. (2004). International technology diffusion. *Journal of Economic Literature*, 42(3), 752–82. https://doi.org/10.1257/0022051042177685.

Kenny, G.P., et al. (2019). Towards establishing evidence-based guidelines on maximum indoor temperatures during hot weather in temperate continental climates. *Temperature*, 6(1), 11–36. https://doi.org/10.1080/23328940.2018.1456257.

Keohane, R.O. (1984). *After Hegemony*. Princeton, NJ: Princeton University Press.

Keohane, R.O., and Victor, D.G. (2016). Cooperation and discord in global climate policy. *Nature Climate Change*, 6(5), 570–75. https://doi.org/10.1038/nclimate2937.

King, M., Tarbusch, B., and Teytelboym, A. (2019). Targeted carbon tax reforms. *European Economic Review*, 119, 526–47. https://doi.org/10.1016/j.euroecorev .2019.08.001.

Kirschen, D.S., and Strbac, G. (2019). *Fundamentals of Power System Economics* (2nd Edition). New York, NY: John Wiley and Sons.

Kiso, T. (2019). Environmental policy and induced technological change: Evidence from automobile fuel economy regulations. *Environmental and Resource Economics*, 74(2), 785–810. https://doi.org/10.1007/s10640-019-00347-6.

Kittner, N., Lill, F., and Kammen, D.M. (2017). Energy storage deployment and innovation for the clean energy transition. *Nature Energy*, 2, 17125. https://doi.org/10.1038/nenergy.2017.125.

Klenert, D., et al. (2018a). Environmental taxation, inequality and Engel's law: The double dividend of redistribution. *Environmental and Resource Economics*, 71(3), 605–24. https://doi.org/10.1007/s10640-016-0070-y.

Klenert, D., et al. (2018b). Making carbon pricing work for citizens. *Nature Climate Change*, 8(7), 669–77. https://doi.org/10.1038/s41558-018-0201-2.

Knittel, C.R. (2011). Automobiles on steroids: product attribute trade-offs and technological progress in the automobile sector. *American Economic Review*, 101(7), 3368–99. https://doi.org/10.1257/aer.101.7.3368.

Kobayashi, H., et al. (2019). Science and technology of ammonia combustion. *Proceedings of the Combustion Institute*, 37, 109–33. https://doi.org/10.1016/j.proci.2018.09.029.

Kost, C., et al. (2018). *Levelized Costs of Electricity: Renewable Energy Technologies*. Freiburg: Fraunhofer Institute ISE.

Kotchen, M.J. (2017). Longer-run evidence on whether building energy codes reduce residential energy consumption. *Journal of the Association of Environmental and Resource Economists*, 4(1), 135–53. https://doi.org/10.1086/689703.

Krausmann, F., et al. (2017). Global socioeconomic material stocks rise 23-fold over the 20th century and require half of annual natural resource use. *Proceedings of the National Academy of Sciences*, 114(8), 1880–5. https://doi.org/10.1073/pnas.1613773114.

Kriegler, E., et al. (2014). The role of technology for achieving climate policy objectives: Overview of the EMF 27 study on global technology and climate policy strategies. *Climatic Change*, 123, 353–67. https://doi.org/10.1007/s10584-013-0953-7.

Krishnamurthy, C.K.B., and Kriström, B. (2015). How large is the owner-renter divide in energy efficient technology? Evidence from an OECD cross-section. *Energy Journal*, 36(4), 85–104. https://doi.org/10.5547/01956574.36.4.ckri.

Kroposki, B. (2017). Integrating high levels of variable renewable energy into electric power systems. *Journal of Modern Power Systems and Clean Energy*, 5(6), 831–7. https://doi.org/10.1007/s40565-017-0339-3.

Krugman, P. (1980). Scale economies, product differentiation, and the pattern of trade. *American Economic Review*, 70(5), 950–59. www.jstor.org/stable/1805774.

Krugman, P. (1991). *Geography and Trade*. Cambridge, MA: MIT Press.

Kumar, A., et al. (2011). Hydropower. In O. Edenhofer et al. (Eds). *IPCC Special Report: Renewable Energy Sources and Climate Change Mitigation* (pp. 437–96). Cambridge: Cambridge University Press.

Kydland, F.E., and Prescott, E.C. (1977). Rules rather than discretion: The inconsistency of optimal plans. *Journal of Political Economy*, 85(3), 473–91. https://doi.org/10.1086/260580.

Kyritsis, E., Andersson, J., and Serletis, S. (2017). Electricity prices, large-scale renewable integration and policy implications. *Energy Policy*, 101, 550–60. https://doi.org/10.1016/j.enpol.2016.11.014.

Lafond, F., et al. (2018). How well do experience curves predict technological progress? A method for making distributional forecasts. *Technological Forecasting and Social Change*, 128, 104–17. https://doi.org/10.1016/j.techfore.2017.11.001.

Laibson, D., and List, J.A. (2015). Principles of (behavioral) economics. *American Economic Review*, 105(5), 385–90. https://doi.org/10.1257/aer.p20151047.

Lall, S. (1992). Technological capabilities and industrialization. *World Development*, 20(2), 165–86. https://doi.org/10.1016/0305-750X(92)90097-F.

Lane, N. (2020). The new empirics of industrial policy. *Journal of Industry, Competition and Trade*, 20(2), 209–34. https://doi.org/10.1007/s10842-019-00323-2.

Lazard (2019a). *Lazard's Levelized Cost of Storage: Version 5.0*. www.lazard.com/perspective/lcoe2019.

Lazard (2019b). *Lazard's Levelized Cost of Energy Analysis: Version 13.0*. www.lazard.com/perspective/lcoe2019.

Lazkano, I., Nøstbakken, L., and Pelli, M. (2017). From fossil fuels to renewables: The role of electricity storage. *European Economic Review*, 99, 113–29. https://doi.org/10.1016/j.euroecorev.2017.03.013.

Léautier, T.-O. (2016). The visible hand: Ensuring optimal investment in electric power generation. *Energy Journal*, 37(2), 89–109. https://doi.org/10.5547/01956574.37.2.tlea.

Léautier, T.-O. (2019). *Imperfect Markets and Imperfect Regulation: An Introduction to the Microeconomics and Political Economy of Power Markets*. Cambridge, MA: MIT Press.

Lehne, J., and Preston, F. (2018). *Making Concrete Change: Innovation in Low-carbon Cement and Concrete*. London: Chatham House. www.chathamhouse.org/sites/default/files/publications/2018-06-13-making-concrete-change-cement-lehne-preston-final.pdf.

Le Quéré, C., et al. (2018). Global carbon budget 2018. *Earth System Science Data*, 10, 2141–94. https://doi.org/10.5194/essd-10-2141-2018.

Le Quéré, C., et al. (2019). Drivers of declining CO_2 emissions in 18 developed economies. *Nature Climate Change*, 9, 213–17. https://doi.org/10.1038/s41558-019-0419-7.

Lester, R.K., and Hart, D.M. (2012). *Unlocking Energy Innovation: How American Can Build a Low-Cost Low-Carbon Energy System*. Cambridge, MA: MIT Press.

Levi, P.G., and Cullen, J.M. (2018). Mapping global flows of chemicals: From fossil fuel feedstocks to chemical products. *Environmental Science and Technology*, 52, 1725–34. https://doi.org/10.1021/acs.est.7b04573.

Levinson, A. (2016). How much energy do building energy codes save? Evidence from California houses. *American Economic Review*, 106(10), 2867–94. https://doi.org/10.1257/aer.20150102.

Li, Z., et al. (2017). Air-breathing aqueous sulfur flow battery for ultralow-cost long-duration electoral storage. *Joule*, 1, 306–27. https://doi.org/10.1016/j.joule.2017.08.007.

Lichbach, M.I., and Zuckerman, A.S. (2009). Research traditions and theory in comparative politics: An introduction. In M.I. Lichbach and A.S. Zuckerman (Eds). *Comparative Politics: Rationality, Culture and Structure* (2nd edition, pp. 3–16). Cambridge: Cambridge University Press.

Lind, R.C. (1982). A primer on the major issues relating to the discount rate for evaluating national energy options. In Lind, R.C., et al. *Discounting for Time and Risk in Energy Policy* (pp. 21–114). Washington, DC: Resources for the Future.

Lockwood, M. (2018). Right-wing populism and the climate change agenda: Exploring the linkages. *Environmental Politics*, 27(4), 712–32. https://doi.org/10.1080/09644016.2018.1458411.

Lomas, K.J., et al. (2018). Do domestic heating controls save energy? A review of the evidence. *Renewable and Sustainable Energy Reviews*, 93, 52–75. https://doi.org/10.1016/j.rser.2018.05.002.

Lovins, A.B. (2018). How big is the energy efficiency resource? *Environmental Research Letters*, 13(9), 090401. https://doi.org/10.1088/1748-9326/aad965.

Lucon, O., et al. (2014). Buildings. In O. Edenhofer et al. (Eds). *Climate Change 2014: Mitigation of Climate Change—Contribution of Working Group III to the Fifth Assessment Report of the Intergovernmental Panel on Climate Change* (pp. 671–738). Cambridge: Cambridge University Press.

Lutsey, N., and Nicholas, M. (2019). Update on electric vehicle costs in the United States through 2030 (ICCT Working Paper 2019-06). Washington, DC: International Council on Clean Transportation. https://theicct.org/publications/update-US-2030-electric-vehicle-cost.

MacDougall, A.H., and P. Friedlingstein (2015). The origin and limits of the near proportionality between climate warming and cumulative CO_2 emissions. *Journal of Climate*, 28(10), 4217–30. https://doi.org/10.1175/JCLI-D-14-00036.1.

MacKay, D.J.C. (2008). *Sustainable Energy—Without the Hot Air*. Cambridge: UIT.

Magee, C.L., et al. (2016). Quantitative empirical trends in technical performance. *Technological Forecasting and Social Change*, 104, 237–46. https://doi.org/10.1016/j.techfore.2015.12.011.

Malanima, P. (2006). Energy crisis and growth 1650–1850: The European deviation in a comparative perspective. *Journal of Global History*, 1(1), 101–21. https://doi.org/10.1017/S1740022806000064.

Mancusi, M.L. (2008). International spillovers and absorptive capacity: a cross-country cross-sector analysis based on patents and citations. *Journal of International Economics*, 76(2), 155–65. https://doi.org/10.1016/j.jinteco.2008.06.007.

March, J.G. (1991). Exploration and exploitation in organizational learning. *Organization Science*, 2(1), 71–87. www.jstor.org/stable/2634940.

Markusen, J.R. (2002). *Multinational Firms and the Theory of International Trade*. Cambridge, MA: MIT Press.

Maskus, K.E. (2012). *Private Rights and Public Problems: The Global Economics of Intellectual Property in the 21st Century*. Washington, DC: Peterson Institute for International Economics.

Material Economics (2018). *The Circular Economy: A Powerful Force for Climate Mitigation*. Stockholm: Material Economics. https://materialeconomics.com.

Material Economics (2019). *Industrial Transformation 2050: Pathways to Net-Zero Emissions from EU Heavy Industry*. https://materialeconomics.com.

Matthews, H.D., and Caldeira, K. (2008). Stabilizing climate requires near-zero emissions. *Geophysical Research Letters*, 35(4), LO4705. https://doi.org/10.1029/2007GL032388.

Matthews, H.D., et al. (2009). The proportionality of global warming to cumulative carbon dioxide emissions. *Nature*, 459, 829–32. https://doi.org/10.1038/nature08047.

McCoy, D., and Kotsch, R. (2020). Quantifying the distributional impact of energy efficiency measures (Grantham Research Institute on Climate Change and the Environment Working Paper 306). London: London School of Economics. www.lse.ac.uk/granthaminstitute/publication/quantifying-the-distributional-impact-of-energy-efficiency-measures.

McGuckin, N., and Fucci, A. (2018). *Summary of Travel Trends: 2017 National Household Travel Survey* (Federal Highway Administration Technical Report).

Washington, DC: US Department of Transportation. https://nhts.ornl.gov/assets/ 2017_nhts_summary_travel_trends.pdf.

Mealy, P., Farmer, J.D., and Teytelboym, A. (2019). Interpreting economic complexity. *Science Advances*, 5(1), eaau1705. https://doi.org/10.1126/sciadv.aau1705.

Mealy, P., and Teytelboym, A. (2018). Economic complexity and the green economy (Institute for New Economic Thinking Working Paper 2018-03). Oxford: University of Oxford. www.inet.ox.ac.uk/files/green_complexity_draft_30_jan_2018.pdf.

Meckling, J., and Nahm, J. (2019). The politics of technology bans: Industrial policy competition for the auto industry. *Energy Policy*, 126, 470–79. https://doi.org/10 .1016/j.enpol.2018.11.031.

Meckling, J., et al. (2015). Winning coalitions for climate policy. *Science*, 349(6253), 1170–71. https://doi.org/10.1126/science.aab1336.

Meckling, J., Sterner, T., and Wagner, G. (2017). Policy sequencing toward decarbonization. *Nature Energy*, 2, 918–22. https://doi.org/10.1038/s41560-017-0025-8.

Mehta, J., Starmer, C., and Sugden, R. (1984a). Focal points in pure coordination games: An experimental investigation. *Theory and Decision*, 36(2), 163–85. https:// doi.org/10.1007/BF01079211.

Mehta, J., Starmer, C., and Sugden, R. (1984b). The nature of salience: An experimental investigation of pure coordination games. *American Economic Review*, 84(3), pp. 658–73. www.jstor.org/stable/2118074.

Mekonnen, M.M., et al. (2018). Water, energy, and carbon footprints of bioethanol from the U.S. and Brazil. *Environmental Science & Technology*, 52, 14508–18. https://doi.org/10.1021/acs.est.8b03359.

Mengel, M., et al. (2018). Committed sea-level rise under the Paris Agreement and the legacy of delayed mitigation action. *Nature Communications*, 9, 601. https://doi.org/ 10.1038/s41467-018-02985-8.

Metcalf, G.E. (2019). *Paying for Pollution: Why a Carbon Tax is Good for America*. Oxford: Oxford University Press.

Milford, R.L., et al. (2013). The roles of energy and material efficiency in meeting steel industry CO_2 targets. *Environmental Science & Technology*, 47(7), 3455–62. https:// doi.org/10.1021/es3031424.

Missirian, A., and Schlenke, W. (2017). Asylum applications respond to temperature fluctuations. *Science*, 358(6370), 1610–14. https://doi.org/10.1126/science.aao0432.

Mitrunen, M. (2019). War reparations, structural change and intergenerational mobility (Harris School of Public Policy mimeo). Chicago, IL: University of Chicago. https:// conference.nber.org/conf_papers/f131527.pdf.

Mokhtarian, P.L., and Chen, C. (2004). TTB or not TTB, that is the question: A review and analysis of the empirical literature on travel time (and money) budgets. *Transportation Research Part A: Policy and Practice*, 38 (9–10), 643–75. https://doi .org/10.1016/j.tra.2003.12.004.

Mokyr, J. (2000). The industrial revolution and the Netherlands: Why did it not happen? *De Economist*, 148, 503–20. https://doi.org/10.1023/A:1004134217178.

Mokyr, J. (2002). *The Gifts of Athena: Historical Origins of the Knowledge Economy*. Princeton, NJ: Princeton University Press.

Mokyr, J. (2009). *The Enlightened Economy: An Economic History of Britain 1700–1850*. New Haven, CT: Yale University Press.

Moore, F.C., et al. (2017). New science of climate change impacts on agriculture implies high social cost of carbon. *Nature Communications*, 8, 1607. https://doi.org/ 10.1038/s41467-017-01792-x.

Moore, G.E. (1965). Cramming more components onto integrated circuits. *Electronics*, 19 April 1965.

Moore, J., and Shabini, B. (2016). A critical study of stationary energy storage policies in Australia in an international context: The role of hydrogen and battery technologies. *Energies*, 9(9), 674. https://doi.org/10.3390/en9090674.

Morgenstern, P., et al. (2016). Benchmarking acute hospitals: Composite electricity targets based on departmental consumption intensities? *Energy and Buildings*, 118, 277–90. https://doi.org/10.1016/j.enbuild.2016.02.052.

Morris, A.C. (2018). Making border carbon adjustments work in law and practice. Washington, DC: Tax Policy Center, Urban Institute and Brookings Institution. www.brookings.edu/research/making-border-carbon-adjustments-work-in-law-and-practice.

Morris, I. (2010). *Why the West Rules—For Now*. New York, NY: Farrar, Straus and Giroux.

Morris, J., et al. (2020). Future energy: In search of a scenario reflecting current and future pressures and trends (MIT Joint Program Report 344). Cambridge, MA: Massachusetts Institute of Technology. https://globalchange.mit.edu/publications/joint-program.

Morris, S., and Shin, H.S. (2002). Social value of public information. *American Economic Review*, 92(5), 1521–34. https://doi.org/10.1257/000282802762024610.

Mourshed, M. (2016). Climatic parameters for building energy applications: A temporal-geospatial assessment of temperature indicator. *Renewable Energy*, 94, 55–71. https://doi.org/10.1016/j.renene.2016.03.021.

Mowery, D.C., Nelson, R.R., and Martin, B.R. (2010). Technology policy and global warming: Why new policy models are needed (or why putting new wine in old bottles won't work). *Research Policy*, 39(8), 1011–23. https://doi.org/10.1016/j.respol.2010.05.008.

Moynihan, M.C., and Allwood, J.M. (2014). Utilization of structural steel in buildings. *Proceedings of the Royal Society A*, 470, 20140170. https://doi.org/10.1098/rspa.2014.0170.

Muntean, M., et al. (2018). *Fossil CO_2 Emissions of All World Countries: 2018 Report*. Luxembourg: Publications Office of the European Union. https://doi.org/10.2760/30158.

Myers, E. (2018). Asymmetric information in residential rental markets: Implications for the energy efficiency gap (Energy Institute at Haas E2e Working Paper 021R). Berkeley, CA: University of California Berkeley. https://e2e.haas.berkeley.edu/pdf/workingpapers/WP021R.pdf.

Myers, E. (2019). Are home buyers inattentive? Evidence from capitalization of energy costs. *American Economic Journal: Economic Policy*, 11(2), 165–88. https://doi.org/10.1257/pol.20170481.

Myers, E., Puller, S.L., and West, J.D. (2019). Effects of mandatory energy efficiency disclosure in housing markets (NBER Working Paper No. 26436). Cambridge, MA: National Bureau of Economic Research. www.nber.org/papers/w26436.

Nagy, B., et al. (2013). Statistical basis for predicting technological progress. *PLoS One*, 8(2), e52669. https://doi.org/10.1371/journal.pone.0052669.

NASA (2020). *Global Climate Change: Vital Signs of the Planet* (Washington, DC: National Aeronautics and Space Administration). https://climate.nasa.gov.

National Academies of Sciences, Engineering and Medicine (2009). *Real Prospects for Energy Efficiency in the United States*. Washington, DC: National Academies Press.

www.nap.edu/catalog/12621/real-prospects-for-energy-efficiency-in-the-united -states.

National Academies of Sciences, Engineering and Medicine (2017). *Valuing Climate Damages: Updating Estimation of the Social Cost of Carbon Dioxide*. Washington, DC: National Academies Press.

Nemet, G.F. (2012a). Inter-technology spillovers for energy technologies. *Energy Economics*, 34(5), 1259–70. https://doi.org/10.1016/j.eneco.2012.06.002.

Nemet, G.F. (2012b). Subsidies for new technologies and knowledge spillovers from learning by doing. *Journal of Policy Analysis and Management*, 31(3), 601–22. https://doi.org/10.1002/pam.21643.

Nemet, G.F., and Kammen, D.M. (2007). US energy research and development: Declining investment, increasing need, and the feasibility of expansion. *Energy Policy*, 35(1),746–55. https://doi.org/10.1016/j.enpol.2005.12.012.

Nemet, G.F., et al. (2017). Addressing policy credibility problems for low-carbon investment. *Global Environmental Change*, 42, 47–57. https://doi.org/10.1016/j .gloenvcha.2016.12.004.

Newbery, D. (2016a). Towards a green energy economy? The EU Energy Union's transition to a low-carbon zero subsidy electricity system: Lessons from the UK's Electricity Market Reform. *Applied Energy*, 179, 1321–30. https://doi.org/10.1016/ j.apenergy.2016.01.046.

Newbery, D. (2016b). Missing money and missing markets: Reliability, capacity auctions and interconnectors. *Energy Policy*, 94, 401–10. https://doi.org/10.1016/j .enpol.2015.10.028.

Newbery, D. (2018). Evaluating the case for supporting renewable electricity. *Energy Policy*, 120, 684–96. https://doi.org/10.1016/j.enpol.2018.05.029.

Newbery, D., and Strbac, G. (2016). What is needed for battery electric vehicles to become socially cost competitive. *Economics of Transportation*, 5, 1–11. https://doi .org/10.1016/j.ecotra.2015.09.002.

Newbery, D., et al. (2018). Market design for a high-renewables European electricity system. *Renewable and Sustainable Energy Reviews*, 91, 695–707. https://doi.org/ 10.1016/j.rser.2018.01.125.

Newell, R.G., and Siikamäk, J. (2014). Nudging energy efficiency behavior: Role of information labels. *Journal of the Association of Environmental and Resource Economists*, 1(4), 555–98. https://doi.org/10.1086/679281.

Noailly, J. (2012). Improving the energy efficiency of buildings: The impact of environmental policy on technological innovation. *Energy Economics*, 34(3), 795–806. https://doi.org/10.1016/j.eneco.2011.07.015.

Noailly, J., and Smeets, R. (2015). Directing technological change from fossil-fuel to renewable energy innovation: An application using firm-level patent data. *Journal of Environmental Economics and Management*, 72, 15–37. https://doi.org/10.1016/ j.jeem.2015.03.004.

NOAO (2015). *Recommended Light Levels for Outdoor and Indoor Venues*. Tucson, AZ: National Optical and Atmospheric Observatory. www.noao.edu/education/ QLTkit/ACTIVITY_Documents/Safety/LightLevels_outdoor+indoor.pdf.

Nordhaus, W.D. (1994). *Managing the Global Commons: The Economics of Climate Change*. Cambridge, MA: MIT Press.

Nordhaus, W. (2008). *A Question of Balance: Weighing the Options on Global Warming Policies*. New Haven, CT: Yale University Press.

Nordhaus, W.D. (2014). The perils of the learning model for modeling endogenous technological change. *Energy Journal*, 35(1), 1–13. https://doi.org/10.5547/01956574.35.1.1.

Nordhaus, W. (2015). Climate clubs: Overcoming free-riding in international climate policy. *American Economic Review*, 105(4), 1339–70. https://doi.org/10.1257/aer.15000001.

Nordhaus, W.D., and Moffat, A. (2017). A survey of global impacts of climate change: Replication, survey methods and a statistical analysis (NBER Working Paper 23646). Cambridge, MA: National Bureau of Economic Research. www.nber.org/papers/w23646.

Nye, D.E. (1990). *Electrifying America: Social Meanings of a New Technology.* Cambridge, MA: MIT Press.

O'Connell, A., et al. (2019). *Sustainable Advanced Biofuels Technology Market Report 2018* (EUR 29929 EN). Luxemburg: European Commission Joint Research Centre. https://doi.org/10.2760/487802.

OECD (2018). *Effective Carbon Rates 2018: Pricing Carbon Emissions Through Taxes and Emissions Trading.* Paris: OECD. https://doi.org/10.1787/9789264305304-en.

OECD/NEA (2019). *The Costs of Decarbonisation: System Costs with High Shares of Nuclear and Renewables,* Paris: OECD/Nuclear Energy Agency. https://doi.org/10.1787/9789264312180-en.

Ogawa, T., Takeuchi, M., and Kajikawa, Y. (2018a). Comprehensive analysis of trends and emerging technologies in all types of fuel cells based on a computational method. *Sustainability*, 10(2), 458. https://doi.org/10.3390/su10020458.

Ogawa, T., Takeuchi, M., and Kajikawa, Y. (2018b). Analysis of trends and emerging technologies in water electrolysis research based on a computational method: A comparison with fuel cell research. *Sustainability*, 10(2), 478. https://doi.org/10.3390/su10020478.

Ohno, H., et al. (2017). Optimal recycling of steel scrap and alloying elements: Input-output based linear programming method with its application to end-of-life vehicles in Japan. *Environmental Science & Technology*, 51(22), 13086–94. https://doi.org/10.1021/acs.est.7b04477.

Olivier, J.G.J., and Peters, J.A.H.W. (2018). *Trends in Global CO_2 and Total Greenhouse Gas Emissions: 2018 Report.* The Hague: PBL Netherlands Environmental Assessment Agency. www.pbl.nl/en/publications/trends-in-global-co2-and-total-greenhouse-gas-emissions-2018-report.

Olson, M. (1971). *The Logic of Collective Action: Public Goods and the Theory of Groups* (revised edition). Cambridge, MA: Harvard University Press.

O'Neill, B.C., et al. (2017a). The roads ahead: Narratives for shared socioeconomic pathways describing world futures in the 21st century. *Global Environmental Change*, 42, 169–80. https://doi.org/10.1016/j.gloenvcha.2015.01.004.

O'Neill, B.C., et al. (2017b). IPCC reasons for concern regarding climate change risks. *Nature Climate Change*, 7(1), 28–37. https://doi.org/10.1038/nclimate3179.

Oppenheimer, M., et al. (2014). Emergent risks and key vulnerabilities. In C.B. Fields et al. (Eds). *Climate Change 2014: Impacts, Adaptation, and Vulnerability. Part A: Global and Sectoral Aspects—Contribution of Working Group II to the Fifth Assessment Report of the Intergovernmental Panel on Climate Change* (pp. 1039–99). Cambridge: Cambridge University Press.

Ostrom, E. (1990). *Governing the Commons: The Evolution of Institutions for Collective Action.* Cambridge: Cambridge University Press.

Ostrom, E. (2009). A polycentric approach for coping with climate change (Policy Research Working Paper 5095). Washington, DC: World Bank. https://documents .worldbank.org/curated/en/480171468315567893/pdf/WPS5095.pdf.

Oxford Economics (2019). *The Global Chemical Industry: Catalyzing Growth and Addressing Our World's Sustainability Challenges.* Washington, DC: International Council of Chemical Associations. www.icca-chem.org/EconomicAnalysis.

Parisi, C. (2018). Biorefineries distribution in the EU (Research Brief). Luxembourg: Publication Office of the European Union. https://doi.org/10.2760/119467.

Pauliuk, S., Sjöstrand, K., and Müller, D.B. (2013). Transforming the Norwegian dwelling stock to reach the 2 degrees Celsius climate target. Combining material flow analysis and life cycle assessment techniques. *Journal of Industrial Ecology*, 17(4), 542–54. https://doi.org/10.1111/j.1530-9290.2012.00571.x.

Pauliuk, S., Wang, T., and Müller, D.B. (2013). Steel all over the world: Estimating in-use stocks of iron for 200 countries. *Resources, Conservation and Recycling*, 71, 22–30. https://doi.org/10.1016/j.resconrec.2012.11.008.

Pearson, P.J.G., and Arapostathis, S. (2017). Two centuries of innovation, transformation and transition in the UK gas industry: Where next? *Proceedings of the Institution of Mechanical Engineers, Part A: Journal of Power and Energy*, 231(6), 478–97. https://doi.org/10.1177/0957650917693482.

Perez, C. (2002). *Technological Revolutions and Financial Capital: The Dynamics of Bubbles and Golden Ages.* Cheltenham, UK, and Northampton, MA, USA: Edward Elgar Publishing.

Pérez-Arriaga, I.J. (2011). Managing large-scale penetration of intermittent renewables. In *Managing Large-scale Penetration of Intermittent Renewables: An MIT Energy Initiative Symposium.* Cambridge, MA: Massachusetts Institute of Technology. https://energy.mit.edu/wp-content/uploads/2012/03/MITEI-RP-2011-001.pdf.

Peters, J.F., et al. (2017). The environmental impact of Li-ion batteries and the role of key parameters: A review. *Renewable and Sustainable Energy Reviews*, 67, 491–506. https://doi.org/10.1016/j.rser.2016.08.039.

Peters, M., et al. (2012). The impact of technology-push and demand-pull policies on technical change: Does the locus of policies matter? *Research Policy*, 41(8), 1296–308. https://doi.org/10.1016/j.respol.2012.02.004.

Pew Research Center (2019). *Climate Change Still Seen as the Top Global Threat, But Cyberattacks a Rising Concern* (Research Report). Washington, DC: Pew Research Center. www.pewresearch.org/global/2019/02/10/climate-change-still-seen-as-the -top-global-threat-but-cyberattacks-a-rising-concern.

Pfeiffer, A., et al. (2018). Committed emissions from existing and planned power plants and asset stranding required to meet the Paris Agreement. *Environmental Research Letters*, 13(5), 054019. https://doi.org/10.1088/1748-9326/aabc5f.

Pigou, A.C (1932). *The Economics of Welfare* (4th edition). London: Macmillan.

Pindyck, R.S. (2013). Climate change policy: What do the models tell us? *Journal of Economic Literature*, 51(3), 860–72. https://doi.org/10.1257/jel.51.3.860.

Pindyck, R.S. (2017). The use and misuse of models for climate policy. *Review of Environmental Economics and Policy*, 11(1), 100–14. https://doi.org/10.1093/reep/ rew012.

Pollitt, M.G, and Anaya, K.L. (2019). Competition in markets for ancillary services: The implications of rising distributed generation (Cambridge Working Papers in Economics 1973). Cambridge: University of Cambridge. www.econ.cam.ac.uk/ research-files/repec/cam/pdf/cwpe1973.pdf.

Popp, D. (2002). Induced innovation and energy prices. *American Economic Review*, 92(1), 160–80. https://doi.org/10.1257/000282802760015658.

Popp, D. (2004). ENTICE: Endogenous technological change in the DICE model of global warming. *Journal of Environmental Economics and Management*, 48(1), 742–68. https://doi.org/10.1016/j.jeem.2003.09.002.

Popp, D. (2017). From science to technology: The value of knowledge from different energy research institutions. *Research Policy*, 46(9), 1580–94. https://doi.org/10.1016/j.respol.2017.07.011.

Popp, D. (2019). Environmental policy and innovation: A decade of research (NBER Working Paper No. 25631). Cambridge, MA: National Bureau of Economic Research. www.nber.org/papers/w2563.

Popp, D., and Newell, R. (2012). Where does energy R&D come from? Examining crowding out from energy R&D. *Energy Economics*, 34(4), 980–91. https://doi.org/10.1016/j.eneco.2011.07.001.

Popp, D., Newell, R.G., and Jaffe, A.B. (2010). Energy, the environment and technological change. In B.H. Hall and N. Rosenberg (Eds). *Handbook of the Economics of Technological Innovation, Volume 2* (pp. 873–938). Amsterdam: North-Holland. https://doi.org/10.1016/S0169-7218(10)02005-8.

Popp, D., et al. (2013). Technology variation versus R&D uncertainty: What matters most for energy patent success. *Resources and Energy Economics*, 35(4), 505–33. https://doi.org/10.1016/j.reseneeco.2013.05.002.

Potomac Economics (2019). *2018 State of the Market Report for ERCOT Markets*. Fairfax, VA: Potomac Economics. www.potomaceconomics.com/wp-content/uploads/2019/06/2018-State-of-the-Market-Report.pdf.

Pozzi, A. (2013). The effect of internet distribution on brick-and-mortar sales. *RAND Journal of Economics*, 44(3), 569–83. https://doi.org/10.1111/1756-2171.12031.

Prüss-Üstün, A., et al. (2016). *Preventing Disease through Healthy Environments: A Global Assessment of the Burden of Disease from Environmental Risks*. Geneva: World Health Organization. https://apps.who.int/iris/handle/10665/204585.

Purdon, M. (2015). Advancing comparative climate change politics: Theory and method. *Global Environmental Politics*, 15(3), 1–26. https://doi.org/10.1162/GLEP_e_00309.

Putnam, R.D. (1988). Diplomacy and domestic politics: The logic of two-level games. *International Organization*, 42(3), 427–60. https://doi.org/10.1017/S0020818300027697.

Rainville. S., et al. (2005). World year of physics: A direct test of $E = mc^2$. *Nature*, 438(7071), 1096–7. https://doi.org/10.1038/4381096a.

Ramsey, F.P. (1927). A contribution to the theory of taxation. *Economic Journal*, 37(135), 47–61. https://doi.org/10.2307/2222721.

Ramsey, F.P. (1928). A mathematic theory of saving. *Economic Journal*, 38(152), 543–59. https://doi.org/10.2307/2224098.

Rausch, S., Metcalf, G.E., and Reilly, J.M. (2011). Distributional impacts of carbon pricing: A general equilibrium approach with micro-data for households. *Energy Economics*, 33(S1), S20–33. https://doi.org/10.1016/j.eneco.2011.07.023.

Rausch, S., and Mowers, M. (2014). Distributional and efficiency impacts of clean and renewable energy standards for electricity. *Resource and Energy Economics*, 36(2), 556–85. https://doi.org/10.1016/j.reseneeco.2013.09.001.

Rez, P. (2017). *The Simple Physics of Energy Use*. Oxford: Oxford University Press.

Riahi, K., et al. (2017). The Shared Socioeconomic Pathways and their energy, land use, and greenhouse gas emissions implications: An overview. *Global Environmental Change*, 42, 153–68. https://doi.org/10.1016/j.gloenvcha.2016.05.009.

Ricke, K.L., and Caldeira, K. (2014). Maximum warming occurs about one decade after a carbon dioxide emission. *Environmental Research Letters*, 9 (12), 124002. https://doi.org/10.1088/1748-9326/9/12/124002.

Robinson, J.P., and Gershuny, J. (2013). Visualizing multinational daily life via multidimensional scaling. *International Journal of Time Use Research*, 10(1), 76–90. https://doi.org/10.13085/eIJTUR.10.1.76-90.

Robiou du Pont, Y., and Meinshausen, M. (2018). Warming assessment of the bottom-up Paris Agreement emissions pledges. *Nature Communications*, 9, 4810. https://doi.org/10.1038/s41467-018-07223-9.

Rodrik, D. (2011). *The Globalization Paradox: Why Global Markets, States, and Democracy Can't Coexist*. Oxford: Oxford University Press.

Rogelj, J., et al. (2018a). Mitigation pathways compatible with 1.5°C in the context of sustainable development. In V. Masson-Delmotte et al. (Eds). *Global Warming of 1.5°C: An IPCC Special Report on the Impacts of Global Warming of 1.5°C above Pre-Industrial Levels and Related Global Greenhouse Gas Emission Pathways, in the Context of Strengthening the Global Response to the Threat of Climate Change, Sustainable Development, and Efforts to Eradicate Poverty* (pp. 93–174). www.ipcc.ch/site/assets/uploads/sites/2/2019/06/SR15_Full_Report_Low_Res.pdf.

Rogelj, J., et al. (2018b). Scenarios towards limiting global mean temperature increase below 1.5°C. *Nature Climate Change*, 8(3), 325–32. https://doi.org/10.1038/s41558-018-0091-3.

Rogge, K.S., and Schleich, J. (2018). Do policy mix characteristics matter for low-carbon Innovation? A survey-based exploration of renewable power generation technologies in Germany. *Research Policy*, 47(9), 1639–54. https://doi.org/10.1016/j.respol.2018.05.011.

Rudnick, H., and Velasquez, C. (2018). Taking stock of wholesale power markets in developing countries: A literature review (World Bank Policy Research Working Paper 8519). Washington, DC: World Bank. https://doi.org/10.1596/1813-9450-8519.

Saadi, F.H., Lewis, N.S., and McFarland, W.W. (2018). Relative costs of transporting electrical and chemical energy. *Energy and Environmental Science* 11(3), 469–75. https://doi.org/10.1039/C7EE01987D.

Salle, J.M. (2014). Rational inattention and energy efficiency. *Journal of Law and Economics*, 57(3), 781–820. https://doi.org/10.1086/676964.

Salle, J.M., West, S.E., and Fan, W. (2016). Do consumers recognize the value of fuel economy? Evidence from used car prices and gasoline price fluctuations. *Journal of Public Economics*, 135, 61–73. https://doi.org/10.1016/j.jpubeco.2016.01.003.

Sammed, K. A., et al. (2020). Reduced holey graphene oxide film and carbon nanotubes sandwich structure as a binder-free electrode material for supercapacitor. *Scientific Reports*, 10, 2315. https://doi.org/10.1038/s41598-020-58162-9.

Schäfer, A. (2000). Regularities in travel demand: An international perspective. *Journal of Transportation and Statistics*, 3(3), 1–31. https://doi.org/10.21949/1501657.

Schäfer, A., et al. (2009). *Transportation in a Climate Constrained World*. Cambridge, MA: MIT Press.

Schelling, T.C. (1960). *The Strategy of Conflict*. Cambridge, MA: Harvard University Press.

Schelling, T.C. (2002). What makes greenhouse sense? Time to rethink the Kyoto Protocol. *Foreign Affairs*, 81(3), 2–9. https://doi.org/10.2307/20033158.

Scherer, F.M., and Harhoff, D. (2000). Technology policy for a world of skew-distributed outcomes. *Research Policy*, 29(4–5), 559–66. https://doi.org/10.1016/S0048 -7333(99)00089-X.

Schmidt, O., et al. (2017). The future cost of electrical storage based on experience rates. *Nature Energy*, 2, 17100. https://doi.org/10.1038/nenergy.2017.110.

Schmidt, O., et al. (2018). The future cost of electrical energy storage based on experience rates: Dataset 2018 update. *Figshare Digital Repository*. https://doi.org/10 .6084/m9.figshare.7012202.

Schmidt, O., et al. (2019). Projecting the future levelized cost of electricity storage technologies. *Joule*, 3, 81–100. https://doi.org/10.1016/j.joule.2018.12.008.

Schumpeter, J.A. (1942). *Capitalism, Socialism and Democracy*. New York, NY: Harper & Bothers.

Schweppe, F.C., et al. (1988). *Spot Pricing of Electricity*. Boston, MA: Kluwer.

Scott, J.C. (2017). *Against the Grain: A Deep History of the Earliest States*. New Haven, CT: Yale University Press.

Seel, J., et al. (2018). *Impacts of High Variable Renewable Energy Futures on Wholesale Electricity Prices and on Electric-Sector Decision Making*. Berkeley, CA: Lawrence Berkeley National Laboratory. https://emp.lbl.gov/publications/ impacts-high-variable-renewable.

Seto, K.C., et al. (2016). Carbon lock-in: Types, causes, and policy implications. *Annual Review of Environment and Resources*, 41, 425–52. https://doi.org/10.1146/ annurev-environ-110615-085934.

Sheldon, T.L., and Dua, R. (2019). Measuring the cost-effectiveness of electric vehicle subsidies. *Energy Economics*, 84, 104545. https://doi.org/10.1016/j.eneco.2019 .104545.

Simonelli, F., et al. (2019). *Competitiveness of the Corporate Sourcing of Renewable Energy Sector: Part 2 of the Study on the Competitiveness of the Renewable Energy Sector* (Directorate General for Energy Report ENER/C2/2016-501). Brussels: European Commission. https://op.europa.eu/en/publication-detail/-/publication/ 618d5369-c48f-11e9-9d01-01aa75ed71a1/language-en.

Skea, J., Lechtenböhmer, S., and Asuka, J. (2013). Climate policies after Fukushima: Three views. *Climate Policy*, 13(S1), 36–54. https://doi.org/10.1080/14693062.2013 .756670.

Smil, V. (2017). *Energy and Civilization: A History*. Cambridge, MA: MIT Press.

Smith, P., et al. (2014). Agriculture, forestry and other land use. In O. Edenhofer et al. (Eds). *Climate Change 2014: Mitigation of Climate Change—Contribution of Working Group III to the Fifth Assessment Report of the Intergovernmental Panel on Climate Change* (pp. 811–922). Cambridge: Cambridge University Press.

Söderholm, P., and Tilton, J.E. (2012). Material efficiency: An economic perspective. *Resources, Conservation and Recycling*, 61, 75–82. https://doi.org/10.1016/j .resconrec.2012.01.003.

Solomon, S., et al. (2009). Irreversible climate change due to carbon dioxide emissions. *Proceedings of the Natural Academy of Sciences*, 106(6), 1704–9. https://doi.org/10 .1073/pnas.0812721106.

Sovacool, B.K. (2016). How long will it take? Conceptualizing the temporal dynamics of energy transitions. *Energy Research & Social Science*, 13, 202–13. https://doi .org/10.1016/j.erss.2015.12.020.

Spence, M. (1976). Product selection, fixed costs and monopolistic competition. *Review of Economic Studies*, 43(2), 217–35. https://doi.org/10.2307/2297319.

Spencer, T., et al. (2018). The 1.5°C target and coal sector transition: At the limits of societal feasibility. *Climate Policy*, 18(3), 335–51. https://doi.org/10.1080/14693062.2017.1386540.

Sperling, D. (2018). Electric vehicles: Approaching the tipping point. In D. Sperling (Ed.). *Three Revolutions: Steering Automated, Shared and Electric Vehicles to a Better Future*. Washington, DC: Island Press.

Staffell, I., et al. (2019). The role of hydrogen and fuel cells in the global energy system. *Energy & Environmental Science*, 12(2), 463–91. https://doi.org/10.1039/c8ee01157e.

Steger, M.B. (2017). *Globalization: A Very Short Introduction*. Oxford: Oxford University Press.

Stern, N. (2007). *The Economics of Climate Change: The Stern Review*. Cambridge: Cambridge University Press.

Stern, N. (2013). The structure of economic modeling of the potential impacts of climate change: Grafting gross underestimation of risk onto already narrow science models. *Journal of Economic Literature*, 51(3), 838–59. https://doi.org/10.1257/jel.51.3.838.

Stern, N. (2021). A time for action on climate change and a time for economics to change. Royal Economic Society Past President's Address, 12 April 2021. www.youtube.com/watch?v=zkFnP2RnE_I.

Stern, N., and Stiglitz, J.E. (2021). The social cost of carbon, risk, distribution, market failures: An alternative approach (NBER Working Paper 28472). Cambridge, MA: National Bureau of Economic Research. https://doi.org/10.3386/w28472.

Stiglitz, J.E. (2019). Addressing climate change through price and non-price interventions. *European Economic Review*, 119, 594–612. https://doi.org/10.1016/j.euroecorev.2019.05.007.

Strbac, G., and Aunedi, M. (2016). *Whole-System Cost of Variable Renewables in Future GB Electricity System*. London: Imperial College. https://doi.org/10.13140/RG.2.2.24965.55523.

Summers, C.M. (1971). The conversion of energy. *Scientific American*, 225(3), 148–63. www.jstor.org/stable/24923123.

Task Force on Climate-Related Financial Disclosures (2017). *Final Report: Recommendations of the Task Force on Climate-Related Financial Disclosures*. Basel: Financial Stability Board, Bank for International Settlements. www.fsb-tcfd.org/publications/final-recommendations-report.

Taylor, M., Spurlock, C.A., and Yang, H.-C. (2015). *Confronting Regulatory Cost and Quality Expectations: An Exploration of Technical Change in Minimum Efficiency Performance Standards* (Report LBNL-1000576). Berkeley, CA: Lawrence Berkeley National Laboratory. https://eta-publications.lbl.gov/sites/default/files/lbnl-1000576.pdf.

Thompson, P. (2012). The relationship between unit cost and cumulative quantity and the evidence for organizational learning-by-doing. *Journal of Economic Perspectives*, 26(3), 203–224. https://doi.org/10.1257/jep.26.3.203.

Tobin, P. (2017). Leaders and laggards: Climate policy ambition in developed states. *Global Environmental Politics*, 17(4), 28–47. https://doi.org/10.1162/GLEP_a_00433.

Tong, D., et al. (2019). Committed emissions from existing energy in jeopardize 1.5°C climate target. *Nature*, 572, 373–7. https://doi.org/10.1038/s41586-019-1364-3.

Tørstad, V., Sælen, H., and Bøyum, L.S. (2020). The domestic politics of international climate commitments: Which factors explain cross-country variation in NDC ambition? *Environmental Research Letters*, 15(2), 024021. https://doi.org/10.1088/1748 -9326/ab63e0.

Tsao, J., Lewis, N., and Crabtree, G. (2006). Solar FAQs (mimeo). Albuquerque, NM: Sandia National Laboratories, US Department of Energy. www.sandia.gov/~jytsao/ Solar%20FAQs.pdf.

Ueckedt, F., et al. (2013). System LCOE: What are the costs of variable renewables? *Energy*, 63, 61–75. https://doi.org/10.1016/j.energy.2013.10.072.

Ugur, M., et al. (2016). R&D and productivity in OECD firms and industries: A hierarchical meta-regression analysis. *Research Policy*, 45(10), 2069–86. https://doi.org/ 10.1016/j.respol.2016.08.001.

Ulph, A., and Ulph, D. (2013). Optimal climate policies when governments cannot commit. *Environmental and Resource Economics*, 56, 161–76. https://doi.org/10 .1007/s10640-013-9682-7.

UNCTAD (2019). *Review of Maritime Transport 2019*. Geneva: United Nations Conference on Trade and Development. https://unctad.org/en/PublicationsLibrary/ rmt2019_en.pdf.

UNEP (2017). *Towards a Zero-Emissions, Efficient and Resilient Buildings and Construction Sector: Global Status Report 2017*. New York, NY: United Nations Environment Program.

Ungemach, C., et al. (2018). Translated attributes as choice architecture: Aligning objectives and choices through decision signposts. *Management Science*, 64(5), 2445–59. https://doi.org/10.1287/mnsc.2016.2703.

United Nations (2019). *World Population Prospects: 2019*. New York, NY: United Nations. https://population.un.org/wpp.

Unruh, G.C. (2000). Understanding carbon lock-in. *Energy Policy*, 28(12), 817–30. https://doi.org/10.1016/S0301-4215(00)00070-7.

US EIA (2018). *Levelized Cost and Levelized Avoided Cost of New Generation Resources in the Annual Energy Outlook 2018*. Washington, DC: US Energy Information Agency. www.eia.gov/outlooks/aeo/electricity_generation.php.

US EIA (2020). *Levelized Cost and Levelized Avoided Cost of New Generation Resources in the Annual Energy Outlook 2020*. Washington, DC: US Energy Information Agency. www.eia.gov/outlooks/aeo/electricity_generation.php.

US EPA (2020). *Sustainable Management of Construction and Demolition Materials*. Washington, DC: Environmental Protection Agency. www.epa.gov/smm/sustainable -management-construction-and-demolition-materials.

USGS (2019). *Mineral Commodity Summaries* 2019. Washington, DC: US Geological Survey. www.usgs.gov/centers/nmic/mineral-commodity-summaries.

Van Buskirk, R.D., et al. (2014). A retrospective investigation of energy efficiency standards: Policies may have accelerated long-term declines in appliance costs. *Environmental Research Letters*, 9(11), 114010. https://doi.org/10.1088/1748-9326/ 9/11/114010.

Van der Ploeg, F. (2018). The safe carbon budget. *Climatic Change*, 147, 47–59. https://doi.org/10.1007/s10584-017-2132-8.

Victor, D.G. (2001). *Collapse of the Kyoto Protocol and the Struggle to Slow Global Warming*. Princeton, NJ: Princeton University Press.

Victor, D.G. (2011). *Global Warming Gridlock: Creating More Effective Strategies for Protecting the Planet*. Cambridge: Cambridge University Press.

Victor, D.G., et al. (2017). Prove Paris was more than paper promises. *Nature*, 548, 25–7. https://doi.org/10.1038/548025a.

Vogt-Schilb, A., Meunier, G., and Hallegatte, S. (2018). When starting with the most expensive option makes sense: Optimal timing, cost and sectoral allocation of abatement investment. *Journal of Environmental Economics and Management*, 88, 210–33. https://doi.org/10.1016/j.jeem.2017.12.001.

Vollset, S.E., et al. (2020). Fertility, mortality, migration, and population scenarios for 195 countries and territories from 2017 to 2100: A forecasting analysis for the Global Burden of Disease Study. *The Lancet*. Published online 14 July 2020. https://doi.org/10.1016/S0140-6736(20)30677-2.

Wagner, G., and Weitzman, M.L. (2015). *Climate Shock: The Economic Consequences of a Hotter Planet*. Princeton, NJ: Princeton University Press.

Walls, M. (2006). Extended producer responsibility and product design: Economic theory and selected case studies (RFF Discussion Paper No. 06-08). Washington, DC: Resources for the Future. https://media.rff.org/documents/RFF-DP-06-08-REV .pdf.

Walls, M., et al. (2017). Is energy efficiency capitalized into home prices? Evidence from three U.S. cities. *Journal of Environmental Economics and Management*, 82, 104–24. https://doi.org/10.1016/j.jeem.2016.11.006.

Wang, P., et al., (2019). Estimates of the social cost of carbon: A review based on meta-analysis. *Journal of Cleaner Production*, 209, 1494–1507. https://doi.org/10 .1016/j.jclepro.2018.11.058.

Warde, P. (2007). *Energy Consumption in England and Wales: 1540–2000*. Napoli: Istituto di Studi sulle Società del Mediterraneo, Consiglio Nazionale delle Ricerche.

Weaver, S., Lötjönen, S., and Ollikainen, M. (2019). *Overview of National Climate Change Advisory Councils*. Helsinki: Suomen Ilmastopaneeli. www.ilmastopaneeli .fi/wp-content/uploads/2019/05/Overview-of-national-CCCs.pdf.

Weidner, E., Ortiz Cebolla, R., and Davies, J. (2019). *Global Deployment of Large Capacity Stationary Fuel Cells: Drivers of, and Barriers to, Stationary Fuel Cell Deployment* (EUR 29693 EN). Luxembourg: Publications Office of European Union. https://doi.org/10.2760/372263.

Weitzman, M.L. (1974). Prices vs. quantities. *Review of Economic Studies*, 41(4), 477–91. www.jstor.org/stable/2296698.

Weitzman, M.L. (2009). On modeling the economic interpretation of catastrophic climate change. *Review of Economics and Statistics*, 91(1), 1–19. https://doi.org/10 .1162/rest.91.1.1.

Weyant, J.P. (2011). Accelerating the development and diffusion of new energy technologies: Beyond the valley of death. *Energy Economics*, 33(4), 674–82. https://doi .org/10.1016/j.eneco.2010.08.008.

Wiedenhofer, D., et al. (2019). Integrating material stock dynamics into economy-wide material flow accounting: Concepts, modelling, and global application for 1900–2050. *Ecological Economics*, 156, 121–33. https://doi.org/10.1016/j.ecolecon .2018.09.010.

Wilson, C., and Grübler, A. (2011). Lessons from the history of technological change for clean energy scenarios. *Natural Resources Forum*, 35, 165–84. https://doi.org/10 .1111/j.1477-8947.2011.01386.x.

Wiser, R., et al. (2011). Wind. In O. Edenhofer et al. (Eds). *IPCC Special Report: Renewable Energy Sources and Climate Change Mitigation* (pp. 535–608). Cambridge: Cambridge University Press.

Wolfson, R. (2016). *Energy, the Environment and Climate*. New York, NY: Norton.

World Steel Association (2019), *Steel Statistics Yearbook 2019*. Brussels: World Steel Association. www.worldsteel.org/en/dam/jcr:7aa2a95d-448d-4c56-b62b-b24 57f067cd9/SSY19%2520concise%2520version.pdf.

Worrell, E., Allwood, J., and Gutowski, T. (2016). The role of material efficiency in environmental stewardship. *Annual Review of Environment and Resources*, 41, 575–98. https://doi.org/10.1146/annurev-environ-110615-085737.

Wright, T.P. (1936). Factors affecting the cost of airplanes. *Journal of the Aeronautical Sciences*, 3(4), 122–8. https://doi.org/10.2514/8.155.

Xi, F., et al. (2016). Substantial global carbon uptake by cement carbonation. *Nature Geoscience*, 9, 880–83. https://doi.org/10.1038/ngeo2840.

Yu, Z., Li, S., and Tonga, L. (2016). Market dynamics and indirect network effects in electric vehicle diffusion. *Transportation Research Part D: Transport and Environment*, 47, 336–56, https://doi.org/10.1016/j.trd.2016.06.010.

Zachmann, G., and Kalcik, R. (2018). Export and patent specialization in low-carbon technologies. In S. Dutta, B. Lanvin and S. Wunsch-Vincent (Eds). *The Global Innovation Index 2018: Energizing the World with Innovation* (pp. 107–14). Ithaca, NY, Fontainebleau and Geneva: Cornell University, Institut Européen d'Administration des Affaires and World Intellectual Property Organization.

Zangheri, P., Serrenho, T., and Bertoldi, P. (2019). Energy savings from feedback systems: A meta-studies' review. *Energies*, 12, 3788. https://doi.org/10.3390/en12193788.

Zervas, G., Proserpio, D., and Byers, J.W. (2017). The rise of the sharing economy: Estimating the impact of Airbnb on the hotel industry. *Journal of Marketing Research*, 54(5), 687–705. https://doi.org/10.1509/jmr.15.0204.

Zickfeld, K., et al. (2009). Setting cumulative emissions targets to reduce the risk of dangerous climate change. *Proceedings of the National Academy of Sciences*, 106(38), 16129–34. https://doi.org/10.1073/pnas.0805800106.

Index